Medicinal Plants and Antimicrobial Therapies

Vinay Kumar • Varsha Shriram • Abhijit Dey
Editors

Medicinal Plants and Antimicrobial Therapies

Springer

Editors
Vinay Kumar
Department of Biotechnology
Modern College of Arts, Science and Commerce, Savitribai Phule Pune University
Pune, Maharashtra, India

Varsha Shriram
Department of Botany
Prof. Ramkrishna More College, Savitribai Phule Pune University
Pune, Maharashtra, India

Abhijit Dey
Presidency University
Kolkata, India

ISBN 978-981-99-7260-9 ISBN 978-981-99-7261-6 (eBook)
https://doi.org/10.1007/978-981-99-7261-6

This Springer imprint is published by the registered company Springer Nature Singapore Pte Ltd.
The registered company address is: 152 Beach Road, #21-01/04 Gateway East, Singapore 189721, Singapore

Paper in this product is recyclable.

Preface

Antimicrobial resistance (AMR), driven by the injudicious use of commonly available drugs, especially the lifesaving antibiotics, has become a global threat to human health and development. AMR has exploded in recent years and is rightly considered one of the greatest human health threats of the twenty-first century placed among the top 10 urgent threats by the World Health Organization. Initially considered as a menace primarily confined to the nosocomial settings, AMR is now spreading rapidly in the community and/or environment. Several drug-resistant bacterial strains, with multidrug-resistant (MDR), extensive drug-resistant (XDR), and even pan-drug-resistant (PDR) phenotypes, pose a major threat to human health and survival. This issue has been further aggravated by the drying pipeline of antibiotics, with few new members in sight. Global AMR trends are highly alarming with complex and dire threats but unfortunately with very few definite answers. The alarming spread of resistance to the commonly used antimicrobials warrants the exploration of alternative strategies. Complementary and alternative therapies are being hailed as novel and effective, long-term strategies for addressing the AMR problem.

This book may serve as an excellent and timely reference resource on this hot topic for wide-ranging readers giving them comprehensive material covering the current understandings and updates on AMR. A panel of authors with credentials and impactful work has contributed to the excellently written chapters.

Chapters cover important issues related to AMR and address them with medicinal plant resources, including an overview of the current AMR problem, AMR and One Health, and medicinal plant resources including essential oils, endophytes, and other metabolites as effective agents against drug-resistant strains. Emphasis has been given on presenting the plant resources that target effectively the major determinants of AMR such as drug efflux, biofilm, and quorum sensing. This book presents the current understanding and updates on medicinal plant resources and their effective use in combating AMR and pathogens.

We express our sincerest thanks and appreciation to our eminent authors for their contributions. We gratefully acknowledge the reviewers for their valuable comments that helped in the improvement of the scientific content and quality of the chapters.

We also thank the Springer publishing team comprising the Publisher, Editorial Project Manager, and the entire Springer production team for their consistent hard work in the publication of this book.

Pune, Maharashtra, India — Vinay Kumar
Pune, Maharashtra, India — Varsha Shriram
Kolkata, India — Abhijit Dey

Contents

Chapter 1
One Health Perspectives for Addressing Antimicrobial Resistance

Kawaljeet Kaur, Pramod Barathe, Sagar Reddy, Vartika Mathur, and Vinay Kumar

Abstract Injudicious and irrelevant use of antimicrobials for human health, hygiene, and in animal husbandry and allied fields has induced microbial resistance to wide-spectrum antimicrobials or antibiotics a condition referred to as antimicrobial resistance (AMR). It is challenging for scientists, researchers, and governments to tackle these situations via novel and effective approaches. The increase in the usage of antimicrobials in sectors of animal, aquatic, human, and environment has increased the cases of multi-drug resistant (MDR) pathogens. Major drivers of AMR in these sectors are found to be mobile genetic elements (MGEs) and antibiotic-resistant genes (ARGs) that transfer horizontally from one health sector to another via horizontal gene transfer (HGT) affecting the whole food chain or food web. Considering the current situation of AMR, its emergence, and its prevalence, one health approach has been characterized as a collaborative effort by multiple sectors to develop effective solutions for humans, animals, and environmental health. According to the "One Health Initiative Task Force," the one health strategy advocates for the collaboration of many disciplines working locally, regionally, and worldwide to achieve optimal health for humans, animals, and the environment. This chapter highlights the AMR as a global concern and the effects of excess use of antimicrobial drugs in each one health sectors with major resistance drivers. Furthermore, we discuss the initiated and effective one health strategies for combating AMR in human, animal, and environmental health. Finally, glimpses of the research gap in one health and antimicrobial resistance such as sector-specific financing, research and development investments, and AMR surveillance have been addressed.

K. Kaur · P. Barathe · S. Reddy · V. Kumar (✉)
Department of Biotechnology, Modern College of Arts, Science and Commerce, Savitribai Phule Pune University, Pune, Maharashtra, India
e-mail: vinay.kumar@moderncollegegk.org

V. Mathur
Animal Plant Interactions Lab, Department of Zoology, Sri Venkateswara College, Benito Juarez Marg, New Delhi, India

V. Kumar et al. (eds.), *Medicinal Plants and Antimicrobial Therapies*,
https://doi.org/10.1007/978-981-99-7261-6_1

Keywords One health · Antimicrobial resistance · Antibiotic-resistant genes · AMR surveillance · Strategies

1.1 Introduction

Credible research in recent years has proved that the uncontrollable and excessive use of antimicrobials in animals, humans, and the environment has led to the development of several strains of microbes resistant to antibiotics. In order to treat these antimicrobial-resistant pathogens, concentrations of antimicrobials are often increased which has further hindered the global micro-biota leading to the emergence of multi-drug resistant (MDR), extensively drug-resistant (XDR), and even the pan-drug resistant (PDR) microorganisms. Studies have shown that many of the antimicrobials used commonly in animals, humans, and environments belonging to the same antimicrobial classes are leading to the prevalence of the same MDR microorganisms in each sector making the current situation worse and diversifying the problem of "Antimicrobial Resistance" (AMR) globally (Hernando-Amado et al. 2020; Pokharel et al. 2020; Ma et al. 2021). Recently, many researchers have proven the common usage of antimicrobials in each sector to revolve around the food chain and food web (Bandyopadhyay and Samanta 2020; Samtiya et al. 2022). The transfer of these antimicrobials from the lower trophic level to the higher trophic level has led to the bioaccumulation of these antimicrobials in each feeding level leading the adverse effects and consequences to human, animal, and environmental health (Hu et al. 2023). Studies have reported antibiotic-resistant genes (ARGs) and mobile genetic elements (MGEs) to be the major drivers of antimicrobial resistance in animals, humans, and environmental health that get transferred via horizontal gene transfer (HGT) among different bacterial species (Kaur et al. 2022; Reddy et al. 2022). Apart from ARGs and MGEs, heavy metal-resistant genes (HMGs) co-occurring with ARGs and MGEs are also been found to elevate AMR and co-selection in the environment (Li et al. 2017). Concerning the current situation, various one health strategies have been initiated or developed such as global awareness campaigns, improvement of hygiene measures and prevention of infection measures to reduce infection, surveillance and research for AMR clarification, promotion of the use of vaccines and alternatives, and training of skilled professionals in the field of AMR to sublime the AMR in one health (McEwen and Collignon 2018). AMR is a key concern in India, and numerous attempts have been made to control it, including AMR policies such as the National Action Plan on Antimicrobial Resistance (NAP-AMR), Chennai Declaration, Jaipur Declaration on Antimicrobial Resistance," "Redline" campaign, and many others (Mutua et al. 2020). However, with such amended government policies, various research gaps in one health and AMR limit future action plans. Some of the research gaps such as poor AMR surveillance systems, sector-specific financing, treatment optimization, rapid and accurate diagnostics, and investment strategies in research and development must be handled on a priority basis.

1.2 Antimicrobial Resistance: A Global Concern

The AMR or bacterial pathogens' ability to resist the actions of antimicrobials to which they were sensitive before developing resistance, has emerged as a serious global health threat (Ahmad and Khan 2019) causing devastating effects such as increased mortality, long-term stay, treatment failures, and increased health care cost (Dadgostar 2019). A recent report (Antimicrobial Resistance Collaborators 2022) highlighted 1.27 million deaths directly attributing to AMR in 2019. Globally, AMR contributes to 700,000 deaths annually, and if current trends of AMR continue, 10 million deaths are projected to occur annually by 2050 (Manesh and Varghese 2021). The economy is also significantly impacted by this increase in AMR. By 2050, global economic production or gross domestic product (GDP) is expected to shrink by 1.1% in the case of a low AMR scenario and 3.8% in the case of a high AMR scenario annually (Ahmad and Khan 2019). Furthermore, additional healthcare expenditure is anticipated to reach $0.33 trillion in low-AMR scenarios and $1.2 trillion in high-income situations by 2050. Livestock production plays an important role in nutrition, and export in low- and high-income countries is expected to decline from 2.6% to 7.5% (Jonas et al. 2017).

The negative impacts of AMR on economic growth result in a significant increase in extreme poverty. In India, the situation is even more worrisome and a cause for concern due to the high consumption of antibiotics and the subpar surveillance system that hinders the tracking and management of AMR (Farooqui et al. 2018; Manesh and Varghese 2021). As a result, the World Health Organization (WHO) identified AMR as one of the top ten threats (World Health Organization 2014; EClinicalMedicine 2021). Antibiotics are being prescribed more frequently to treat secondary infections in clinical settings during the COVID-19 pandemic, which increases the global burden of antimicrobial resistance (Pulingam et al. 2022). AMR is not just a problem in clinical settings, but it is also a problem in the environment, including urban rivers, sediments, groundwater, and agricultural land. Antibiotic resistance genes and pathogens are emerging, becoming more prevalent, and spreading throughout these environments (Reddy et al. 2022; Wu et al. 2023), for instance, the presence of fluoroquinolones and ARGs (*qnrS*, *ermB*, *sul1*, and *Tetw*) at the highest concentrations in river water receiving hospital and urban wastewater effluents (Rodriguez-Mozaz et al. 2015). Similar to this, eight ARGs (*ermB, ermC, ermE, ermF, qepA, qnrA, qnrB*, and *qnrS*) were discovered in farmland soil fertilized with chicken manure (Wang et al. 2019), raising concerns over the dissemination of ARGs and ARBs in the food chain and clinical pathogens via HGT (Wright 2010). This is particularly alarming as it could lead to the development of antibiotic-resistant bacteria, making it difficult to treat infections in humans. Hence, it is crucial for policymakers and stakeholders to take action to prevent the spread of ARGs and ARBs in the food chain and clinical settings.

1.3 Antimicrobial Usage and Its Impact on One Health

To date, a number of antibiotics have been developed to treat various bacterial infections in humans, animals, and the environment. The injudicious use of these antibiotics has increased the prevalence of antibiotic-resistant bacteria in one health. Some antimicrobials have been employed to one health for decades before the emergence of resistance, while others developed resistance much more quickly (Velazquez-Meza et al. 2022). Therefore, the increasing AMR has become a major concern and priority for the research sector to reduce their adverse impact on one health.

1.3.1 Humans

Third-generation cephalosporins including cefixime, ceftibuten, cefdinir, cefpodoxime, and cefditoren are widely used in treating Gram-negative meningitis, Lyme disease, *Pseudomonas* pneumonia, Gram-negative sepsis, Streptococcal endocarditis, melioidosis, penicillinase-producing *Neisseria gonorrhea*, chancroid, and Gram-negative osteomyelitis in humans (Arumugham et al. 2022). Third-generation cephalosporin antibiotics containing beta-lactams are used for treating drug-resistant infections and hence are designated as "significant" to public health (Das et al. 2019). Combinations of meropenem with β-lactam-metallo-β-lactamase inhibitors are used as antimicrobials to combat carbapenem resistance in *Klebsiella pneumoniae* (Reddy et al. 2023). Other includes last resort beta-lactam antibiotics (such as ceftazidime-avibactam and aztreonam) for treating metallo-beta-lactamase-producing enterobacterales (Larcher et al. 2022); colistin for treating *Acinetobacter baumannii–calcoaceticus* infections (Kaye et al. 2023); and polymyxin B for carbapenem-resistant *Acinetobacter baumannii* and *Klebsiella pneumoniae* (Zha et al. 2023). However, the increased and continuous dosage of some of the antimicrobials for a long period of time can have adverse or opposite effects on the environment. Some of the examples of antimicrobials used in human health with their adverse effects are listed in Table 1.1.

1.3.2 Animals

Despite the fact that antimicrobials are used in comparable doses in animals and humans, the likelihood of mutations in animals is greater than in humans due to their bigger animal biomasses (Van Boeckel et al. 2017). The volume of antimicrobials used in animals is larger than that in humans, and most of them in animal livestock are used as therapeutic, prophylactic, and developmental boosters (Velazquez-Meza et al. 2022). However, frequent exposure and overuse of some of the medicinal

Table 1.1 Different examples of antimicrobials used in human health with their adverse effects

Antimicrobial used	Class of antimicrobial	Maximum dosage of antimicrobial	Mode of action	Drug-resistant bacteria	Resistance mechanism	Antibiotic adverse effects on human health	References
Amoxicillin	β-Lactam	750–1750 mg/day	Inhibits transpeptidation, binding to beta-lactamase enzyme	*Helicobacter pylori, Staphylococcus aureus*	Inactivation of amoxicillin by the enzyme beta-lactamases, diminishing efflux and permeability, and PBP alterations, beta-lactamase production	Mucocutaneous candidiasis, *Clostridium difficile* associated diarrhea, Crystalluria, interstitial nephritis, hypersensitivity vasculitis, neutropenia	Sodhi et al. (2021), Akhavan et al. (2022)
Azithromycin	Macrolides	600 mg	Binds to the 23S portion of the 50S bacterial ribosomal subunit and inhibits bacterial protein synthesis by preventing the transit of aminoacyl-tRNA	*Mycoplasma pneumoniae; Escherichia coli*	Mutation in the 23S rRNA gene; efflux pump (AcrAB-TolC) activation	Cardiac arrhythmias like Torsades de pointes, ventricular tachycardia, ventricular fibrillation, toxic epidermal necrolysis	Liu et al. (2014), Patel and Hashmi (2023), Al-Marzooq et al. (2023)
Gentamicin	Aminoglycoside	3–5 mg/kg/day	Binds to the 16 s rRNA at the 30 s ribosomal subunit, disturbs mRNA translation, and leads to the formation of non-functional proteins	*Listeria monocytogenes; Klebsiella pneumoniae*	Transposition of Tn3706, presence of gentamicin resistance gene (aac6'-aph2) and efflux pump ErmC; biofilm production, efflux pump activation	Neuromuscular blockade, Myasthenia gravis, Vecuronium, urticarial	Baquero et al. (2020), Chaves and Tadi (2023), Karami-Zarandi et al. (2023)

(continued)

Table 1.1 (continued)

Antimicrobial used	Class of antimicrobial	Maximum dosage of antimicrobial	Mode of action	Drug-resistant bacteria	Resistance mechanism	Antibiotic adverse effects on human health	References
Ampicillin	β-Lactam	250–500 mg q6hr	Binds to membrane-bound penicillin-binding proteins (PBPs), inhibition of cell wall synthesis	*Listeria monocytogenes*	Mutations in susceptible penicillin-binding proteins (PBPs)	Enterocolitis, black hairy tongue, thrombocytopenic purpura, thrombocytopenia, eosinophilia, nephrotoxicity, thrombophlebitis	Baquero et al. (2020), Peechakara et al. (2022)
Cephalosporins	Third-generation cephalosporins	125 or 250 mg	Inhibit the synthesis of the bacterial cell wall, structural binding of cephalosporin antibiotics to the active site of PBPs, inhibition of their enzymatic activity	Enterobacteriaceae	Production of serine-carbapenemases or metallo-carbapenemases, expression of extended-spectrum beta-lactamases (ESBLs) and/or AmpC, upregulation of efflux pumps	Drug-induced immune hemolytic anemia, disulfiram-like reaction, vitamin K deficiency, pseudomembranous colitis, hypersensitivity reaction	Bui and Preuss (2023), Kaye et al. (2023)
Carbapenems	β-Lactam	15–25 mg/kg/dose IV every 6 h	Inhibits the peptidase domain of PBPs, inhibit peptidoglycan synthesis	*Klebsiella pneumoniae*	β-Lactamases or aminoglycoside modifying enzymes, decreased cell permeability through loss of Omps, overexpression of efflux pumps (KpnGH)	Nosocomial infections-pneumonia, intra-abdominal infections, urinary tract infections, meningitis	Pandey and Cascella (2022), Karampatakis et al. (2023)

antimicrobials such as tetracycline, streptomycin, ampicillin, fluroroquinolones, and sulfonamides have been linked to antimicrobial resistance (Kasimanickam et al. 2021). MDR bacterial strains such as *Salmonella enterica* serotype Enteritidis, methicillin-resistant *Staphylococcus aureus*, and *Escherichia coli* have been found in various animal farms and their vicinity in Poland and Ukraine further concerning animal husbandry, animal origin food supply chain institutions, and slaughterhouses (Jeżak and Kozajda 2022). Other studies have confirmed the spread of resistance from animal farms to the environment via animal manure and water resources increasing their potential exposure risk to the food chain (Bai et al. 2022; Min et al. 2023). Some of the drug-resistance bacteria found in animals, livestock, and food supply are listed in Table 1.2.

1.3.3 Environment

The environment acts as a substantial reservoir for the transmission of many bacterial species from land to water or vice versa. Many bactericides and pesticides with antimicrobial properties have been developed that play an important rolc in plant disease management (Bai et al. 2022). Triclosan (2,4,4′-trichloro-2′--hydroxydiphenyl ether), which is an antimicrobial agent approved by The United States Environmental Protection Agency (EPA), is used in the United States as pesticides. Studies on the toxicity level of triclosan pesticides reported their safe usage in microalgae aquatic environmental protection (Taştan et al. 2017). Antimicrobials such as oxytetracycline, doxycycline, chloramphenicol, and streptomycin are frequently used in aquaculture for the production of catfishes, lobsters, shrimp farming, Chilean salmon, and mussels (Leal et al. 2019) for increased biomass production and livestock (Table 1.3). Recent studies have reported the dissemination of microplastics in the environment and their associated drug-resistant pathogens driving AMR not only in aquatic ecosystems but also in wastewater, groundwater, marine, freshwater, and urban river ecosystems (Kaur et al. 2022).

1.4 Antimicrobial-Resistance Drivers in One Health

ARGs are widely identified in rivers, soil, livestock farms, drinking water, glacial environments, and even in the Antarctic. ARGs can be transferred from bacterial pathogens to their original hosts through local confluences between bacteria colonizing various hosts (including humans and animals) and their shared surroundings (Fig. 1.1). Clinically significant resistance genes connected to MGEs are capable of crossing habitat boundaries, despite resisters across habitats being linked to the phylogeny of microbial populations along ecological gradients. Additionally, non-clinical ecosystem-associated microorganisms are frequently the original hosts

Table 1.2 Various drug-resistance bacteria found in animals

Drug-resistant bacteria	Antibiotic resistance	Animal	Resistance mechanism	References
Pleuromutilin-resistant *Enterococcus*	Tetracycline (92.4%), Streptomy cin (92.4%), and erythromycin (91.4%)	Breeding chicken, Chick, young chicken, and commercial laying hen	Transposons (Tn554, Tn558, Tn6261, and Tn6674); ARGs (*erm (A), ant(9)-la, fex(A),* and *optrA*)	Lin et al. (2024)
Campylobacter jejuni	Quinolone resistance	Cattle, dogs, and chickens	The-86-Ile mutation in the gyrA gene	Aydin et al. (2023)
Salmonella, Shigella, and *Escherichia coli*	Ampicillin (91.7%), gentamicin (25.6%), cefotaxime (32.1%), erythromycin (40.3%), neomycin (33.9%), streptomy-cin (34.8%), and sul-famethoxazole (52.2%)	Poultry chicken meat	Presence of mobile genetic elements and antibiotic-resistance genes	Tagar and Qambrani (2023)
Proteobacteria	Aminoglycoside, tet racycline, MLS (macrolide-lincosamides-streptogramin), and beta-lactam	Chickens	*Mcr-1* and *tet*(X3)	Wang et al. (2021)
Linezolid (LZD)-resistant *Enterococcus faecalis*	Macrolides, tetracy clines, and aminoglycosides	Chicken meat	*optrA*-carrying plas-mid, transposase genes (*tnpA, tnpB,* and *tnpC*)	Yoon et al. (2020)
β-Lactamase-producing *Escherichia coli*	Third-generation cephalosporins	Layer hens	Plasmid-mediated AmpC (pAmpC), ESBL/pAmpC genes $bla_{CTX\text{-}M\text{-}1}$, $bla_{CTX\text{-}M\text{-}14}$, $bla_{CTX\text{-}M\text{-}15}$, and $bla_{CMY\text{-}2}$	Shim et al. (2019)
Staphylococcus aureus	Penicillins (51.2%), tetracycline (38.8%), and ciprofloxacin (CIP; 33.9%)	Broiler	Double mutations of S84L/S80F in *gyr* A/*par* C	Kim et al. (2018)

of clinically significant ARGs that have been transmitted from environmental micro-organisms to human pathogens.

Soil being a natural large resistome plays an important role in one health perspective which receives ARB and ARGs from both human and animal wastes (Thompson et al. 2017). It contains a diverse population of microorganisms and human activities such as fecal application on land and agricultural practices, which are the key activities of antibiotic concern (Wang et al. 2018). There have been

Table 1.3 Different examples of antimicrobials used in the environment and their impact on environmental health

Antimicrobial used	Antimicrobial concentration	Sample	Drug-resistant bacteria	Consequences in environmental health	References
Florfenicol	469 mg/kg	Chilean salmon aquaculture	*Piscirickettsia salmoni*	Poor smolt quality, unhealthy high densities of fish in pens, dysbiosis triggered by unrelenting heavy use of antimicrobials	Hossain et al. 2022; Cabello et al. (2023)
Oxytetracycline dehydrate	100 mg/kg	Nile tilapia (*Oreochromis niloticus*)	*Aeromonas liquifaciens* and *Pseusomonas* spp., *Plesiomonas*	Dysbiosis of gut-microbiome	Payne et al. (2021)
Ormetoprim	50 mg/kg	Tilapia (*Oreochromis niloticus*)	*Edwardsiella ictuluri*	Elevation of toxic and biochemical lesions	Ferri et al. (2022), Akshaya et al. (2023)
Oxytetracycline	125 μg/L	Mussels	*Vibrio parahaemolyticus*, *Enterococcus faecium*	Activation of major vault protein (MVP) gene expression, inflammatory responses in tissues such as gills, digestive system, and mantle (gonads) underlining the worsening of bivalves' general health.	Hallmann et al. (2023)
Oxytetracycline and streptomycin	–	Cherry radish	Streptomycin-resistant bacteria	Higher antibiotic-accumulated radish tissues, hormesis of antibiotic leaves, fruits, and roots of radishes	Yin et al. (2023)
Oxytetracycline and streptomycin	200 μg/mL	Citrus plants	*Candidatus Liberibacter asiaticus*	Deposits in phloem, xylem, leaves, and root tissues	Al-Rimawi et al. (2019)
Endosulfan	800 μg/mL	Wastewater-irrigated soil	*Pseudomonas*	Deposits on okra fruits; fractions of endosulfan components ingested by humans after crop harvest	Shafiani and Malik (2003), Nath et al. (2022)
Lindane, carbaryl, and methyl parathion	0.1%	Agricultural fields	*Rhizobium* sp., *Pseudomonas aeruginosa*, *Bacillus subtilis*, *Escherichia coli*	Influences plant phenotype, growth, yield, and quality by contributing to plant resistance toward diseases	Sangiorgio et al. (2022), Shanthi et al. (2022)
Tetracyclines	102.7 μg/mL	Irrigation water	ESBL-producing *Escherichia coli*	Antibiotics detected in spinach and radish at concentrations of 6.3–330 μg/kg tissue	Gekenidis et al. (2018), Jalloul et al. (2021)

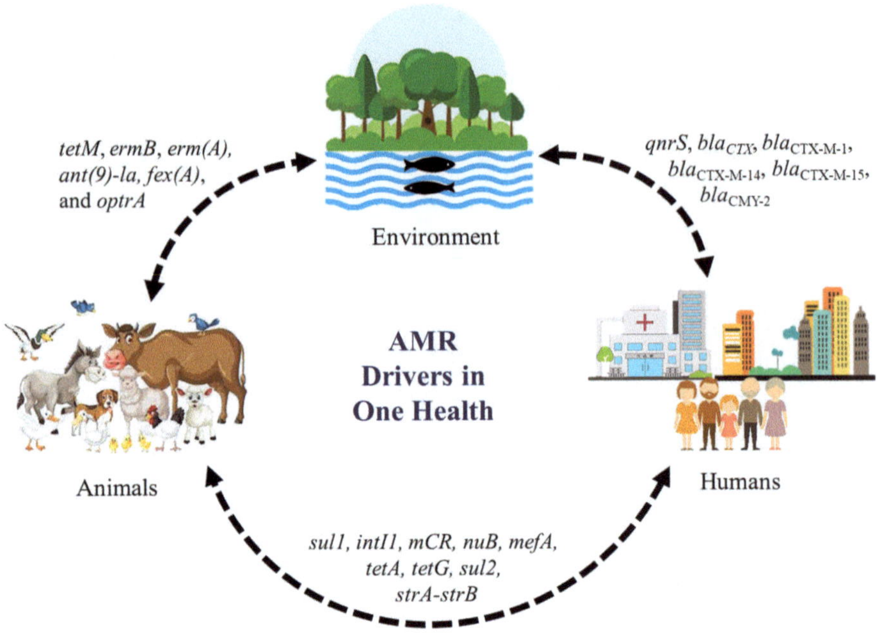

Fig. 1.1 Antibiotic resistant genes (ARGs) and mobile genetic elements (MGEs) as antimicrobial resistance drivers in one health

reports of numerous antibiotics and resistance genes in the soil near pig farms and even rice fields, including the bla_{CTX-M} gene (which confers resistance against cefotaxime) (Xiao et al. 2016) (Table 1.2). Soil irrigation practice is also a major concern that potentiates antibiotic resistance in soil. Manure application from the animals that consume feed supplement containing metals can co-select and confer resistance to both metals and antibiotics (Poole 2017). This manure contamination of soil exposes them to vegetables and reclaimed water making humans vulnerable to antibiotic resistance.

The excessive use of vaccines and hygienic standards, which were partially the consequence of ongoing scientific research in this area, are the key driving forces for AMR development. As the use of antibiotics in aquaculture varies greatly, even within the same fish species, there are definitely challenges beyond those that are scientific and technological. Although the exact cause of the difference is still unknown, it is likely a combination of factors including not only a lack of vaccination but also high fish density, poor fishing techniques, including poor hygiene, and feeding with unidentified ingredients that may contain antibiotics or other agents that put pressure on the evolution of antibiotic resistance. Additionally, aquaculture systems that use livestock production wastes are efficient at cycling nutrients but may face issues with antibiotic resistance. In terms of the ARGs, Caputo et al. (2023) reviewed studies in which they found that the sulfonamide-resistant gene *sul1*, the tetracycline-resistant gene *tetM*, the beta-lactamase-resistant gene *bla*XXXXXXXXX, the fluoroquinolone-resistant gene *qnrS*, and the erythromycin-resistant gene *ermB* were all found in 6%, 4%, 3%, and 2% of the studies, respectively. Furthermore, their attention was the last resort antibiotic colistin for the treatment of cystic fibrosis and other conditions in humans. The presence of colistin residues and/or resistance genes (i.e., mcr 1–4) persists despite the fact that colistin is illegal in the majority of aquaculture-producing nations.

Inhalation of urban airborne fine particulate matter (PM2.5) by humans accounts for a comparable portion of daily consumption of several prevalent environmental ARGs same as drinking water and meals (Li et al. 2020). Recent studies on snowfall samples from several countries revealed the precipitation of ARGs in fresh snow by air pollution (Zhou et al. 2021). Other seasonal and environmental factors are also known to affect the total ARG abundance in the urban and rural regions, with urban regions seeing more ARG abundance than rural areas in the summer and vice versa in the winter (Kormos et al. 2022). Further studies have enlisted several genes, including *sul1, intI1*, –lactam ARGs, and tetracycline ARGs that are often examined and generally found abundant (Table 1.4).

1.5 One Health Strategies for Combating Antimicrobial Resistance

1.5.1 Global Awareness Campaign

At the most basic level, everyone should know the principle behind maintaining hygiene and need to follow the antimicrobial prescriber's instructions for treatment in order to prevent AMR spread. This can be useful to all the dimensions of one health which includes, animal owners, farmers, veterinarians, and those involved in the food industry. Henceforth, understanding of these groups varies; for example, animal owners should understand the infection and transmission risk of disease given by veterinary, and farmers should understand how to cultivate crops and ease the animals without or with minimal use of antimicrobials and only be used to treat clinically ill animals. Veterinarians prescribing antimicrobials and advising farmers on treatment should also consider one health dimensions of AMR. They also need to understand how to reduce the use of antimicrobials by improving the overall level of animal husbandry and minimizing the unsanitary and stressful conditions that promote disease spread to animals and plants (WHO 2000, 2004). After the understanding of the one health perspective of AMR, it will protect the health and welfare of the wider community (OIE 2016; Food and Agriculture Organization 2021). Opportunities to better understand one health aspects of AMR include farmer outreach activities, veterinary consultations, professional development programs for physicians and veterinarians, and more importantly programs offered by public health and animal health organizations (McEwen and Collignon 2018). Also, educating people about harm caused by the overuse and misuse of antimicrobials by

Table 1.4 Various drug-resistant bacteria found in the environment

Antimicrobial	Sample	Location	Drug-resistant bacteria	Resistance mechanism	References
Third generation cephalosporins	River sample	Yamuna River, India	ESBL-producing *Escherichia coli*, *Klebsiella pneumonia*, *Aeromona* sp., *Klebsiella oxytoca*, *Klebsiella georgiana*, and *Acinetobacter junii*	$bla_{CTX-M-71}$ (5%), $bla_{CTX-M-3}$ (7.5%), $bla_{CTX-M-32}$ (2.5%), $bla_{CTX-M-152}$ (7.5%), $bla_{CTX-M-55}$ (2.5%), and mercury tolerance determinants (merP, merT, and merB)	Azam et al. (2016)
Penicillin	Wastewater streams	Pakistan	ESBL producers (*Aeromonas* spp. and *Escherichia* sp.), MBL producers (*Stenotrophomonas* sp. and *Citrobacter* sp.), and AmpC producers (*Pseudomonas* spp. and *Morganella* sp.)	*blaTEM* and *qnrS*	Saima et al. (2020)
Tetracycline, trimethoprim, sulfamethizole, amoxicillin, streptomycin	Chilean salmon aquaculture	Calbuco archipelago, Los Lagos region in Southern Chile	*Vibrio* sp., *Serratia* sp., *Marinobacter litoralis*	*tetA*, *tetG*, *dfrA1*, *dfrA5*, *dfrA12*, *sul1*, *sul2*, *strA-strB*, bla_{TEM}, Integron integrase (intl1), *aad9* gene cassettes	Shah et al. (2014)
Chloramphenicol, florfenicol, and fosfomycin	Fish multitrophic farming	Portugal	*Enterobacter ludwigii* INSAq77	β-Lactams (blaACT-88), chloramphenicol (catA4-type), fosfomycin (fosA2-type) and colistin (mcr-9.1), efflux pumps (oqxAB-type and mar operon)	Manageiro et al. (2022)
Fluoroquinolones	Livestock farming	Northern Germany	Fluoroquinolone-resistant *Escherichia coli*	Fluoroquinolone-resistant genes	Schulz et al. (2019)

conducting a global awareness campaign will reduce the unnecessary prescription of antimicrobials.

1.5.2 *Improvement of Hygiene Measures and Preventing Infection Measures to Reduce Infection*

It is well-recognizing and important means to limit the spread of antimicrobial resistance in humans. Poultry and swine sectors in farm and food animal industries are important by means of biosecurity and disease control (McEwen and Fedorka-Cray 2002; Aarestrup et al. 2008). It is crucial to put policies in place to increase the safety of food and drinking water, especially in developing nations, and to reduce environmental pollution from the pharmaceutical industry to lessen human exposure to the spread of AMR from environmental sources and pathways (Gaze et al. 2013; Collignon 2015).

1.5.3 *Antimicrobial Resistance Surveillance and Research*

AMR surveillance is essential, as it is important for people to know more about AMR and how to prevent it (World Health Organization 2014, 2016). Surveillance is carried out to know the burden pattern and extent of AMR at the global as well as national and regional levels (Collignon and Voss 2015; World Health Organization 2017). Surveillance must be able to detect emerging trends, inform antimicrobial policies and antimicrobial stewardship programs, and measure the effectiveness of measures that are taken to control AMR. One health surveillance should include the sampling of bacteria from each and every aspect of the environment, humans, animals, and plants (Collignon and McEwen 2019). With this, it should provide the rate of antimicrobial consumption and use in agriculture, humans, and animals at national and regional levels to compare with the other countries. The promotion of new and rapid disease diagnosis tools allows clinicians to administer exact antimicrobials to patients, which could stop unnecessary prescriptions.

1.5.4 *Promotion of Use of Vaccines and Alternatives, and Training of Skilled Professionals*

Vaccines and alternatives of antimicrobials against the bacteria that cause serious infections will decrease the number of patients who require antimicrobials. Additionally, investments and additional attention must be given to alternatives such as phage therapy, probiotics, antibodies, and so on. Dealing with AMR requires skilled professionals, such as microbiologists, pharmacists, infectious disease specialists, nurses, infection control specialists, veterinarians, and epidemiologists. To do so, countries must invest in the training of human resources. Building a global coalition for making significant progress in the fight against AMR and addressing it using the one health approach is important for effective change (McEwen and Collignon 2018).

1.6 Research Gap in Antimicrobial Resistance and One Health

The primary research gap that needs considerable attention for the successful implementation of one health strategy is the absence of adequate AMR surveillance. AMR surveillance is crucial to demonstrate trends and monitor the emergence and prevalence of drug-resistant pathogens, antimicrobial use, resistant determinants, and resistance mechanisms (McCubbin et al. 2021). Weak surveillance systems particularly in low- and middle-income countries (LMICs) due to limited resources and a lack of health information systems result in a lack of comprehensive data on AMR limiting the use of surveillance data in policy making, allocating funds, and financial resources (Iskandar et al. 2021; Bulteel et al. 2021).

Furthermore, sector-specific financing of AMR control activities primarily prioritizes human health. Activities to combat AMR in non-human health sectors are either underfunded or undeveloped. In addition, antimicrobial management in the non-human sector is frequently unregulated (Frost et al. 2021). Henceforth, a need for an integrated surveillance system that would track antimicrobial use in both human and non-human health, as well as environmental sectors is must. Without a proper surveillance system, there will be no benchmarks to evaluate established mitigation strategies or identify important areas of policy development concerning AMR (McCubbin et al. 2021). In addition, the development of clear AMR data-sharing protocols is required for sharing the surveillance data across regions, countries, and so on for effective comparison of AMR status (Matee et al. 2023). Effective risk assessment is a component that is essential for predicting the threat to humans and animals due to the pathogen's resistance to antimicrobial therapy. Lack of consideration of the negative effects of AMR on vulnerable populations, the possibility of treatment failure, and the necessity of including the phenomena of cross-resistance, co-resistance, and inherent resistance in assessments of the risk of AMR list the significant research gaps that exist at the level of risk assessment (Caffrey et al. 2019). Economic impact due to AMR is another category that affects the successful implementation of one health strategy. The critical gap identified in this category includes a lack of sufficient investment in research and development that focuses on the development of novel alternative strategies and antimicrobials that outperform existing drugs. Also, there is a lack of investment in rewarding and training specialists who work with infectious diseases (Akhtar et al. 2022). The unnecessary use of antimicrobials is the major cause of the spread of AMR and subsequent treatment failure; therefore, the development of rapid and accurate diagnostics highlights another area of research in which gaps exist (Samreen et al. 2021). The rapid and accurate diagnostic will be helpful to target diseases with suitable antimicrobials which will reduce the unnecessary use of antimicrobials in the human and animal health sector.

Environment reservoirs have a significant influence on the emergence, spread, and prevalence of AMR (Reddy et al. 2022). There is a gap between the pathway by which antimicrobial drugs move within the environmental reservoirs and the understanding of the role of anthropogenic inputs in the evolution of resistance (Bulteel et al. 2021). Wastewater from hospitals, aquaculture, agriculture, wastewater treatment plants, and so on represent a major source of the spread of ARGs and drug-resistant pathogens into the environment reservoirs. This demands comprehensive research for the development of efficient treatment strategies that effectively remove ARGs, anthropogenic waste, and drug-resistant pathogens (Reddy et al. 2022). McCubbin et al. (2021) reviewed crucial gaps in three important categories, i.e., treatment optimization, surveillance of antimicrobial use, AMR, and prevention of transmission of AMR. They also highlighted urgent gaps in the AMR prevalence estimation between livestock species, wildlife, companion animals, and risk associated with the close proximity of companion animals to humans. Finally, to implement successful one health strategy gaps concerning weak surveillance systems, study design, statistical measures, producing reliable data, and public awareness regarding antimicrobial use and resistance are needed to address.

1.7 Conclusion

For decades, extensive use of antimicrobials in one health has led to the production of drug-resistant microorganisms resistant to multiple antimicrobial classes. Majority of the antimicrobial classes are used commonly in humans, animals, and the environment making it challenging for the scientists and government to the optimal use of antimicrobials in each sector that needs urgent attention. Henceforth, to combat AMR, one health approach compromising collaborative solutions for humans, animals, and the environment is used for accelerating global development to find long-term solutions and further enhance overall governance. One health strategies such as global awareness programs, AMR surveillance, improvement in hygienic measures, and prevention of infectious diseases by using vaccines and eco-friendly alternatives are strategized for the same. With the increased intervention of AMR, existing scientific research has substantial constraints regarding techniques, policies, and scope for tackling AMR in one health. Therefore, research gaps such as sector-specific financing, research and development investments, and poor AMR surveillance systems particularly in low- and middle-income countries (LMICs) need to be addressed on an urgent basis.

Acknowledgments The authors acknowledge the funding support under DBT-BUILDER (BT/INF/22/SP45363/2022) from the Department of Biotechnology (DBT), Government of India, and the DST-FIST (SR/FST/COLLEGE-/19/568) from the Department of Science and Technology (DST), Government of India implemented at Modern College, Ganeshkhind, Pune, India.

References

Aarestrup FM, Wegener HC, Collignon P (2008) Resistance in bacteria of the food chain: epidemiology and control strategies. Expert Rev Anti-Infect Ther 6:733–750. https://doi.org/10.1586/14787210.6.5.733

Ahmad M, Khan AU (2019) Global economic impact of antibiotic resistance: a review. J Glob Antimicrob Resist 19:313–316. https://doi.org/10.1016/j.jgar.2019.05.024

Akhavan BJ, Khanna NR, Vijhani P (2022) Amoxicillin. StatPearls, Treasure Island, FL

Akhtar N, Singh KS, Prerna GD (eds) (2022) Emerging modalities in mitigation of antimicrobial resistance. Springer, Cham

Akshaya L, Sarathchandra G, Shanmugam SA et al (2023) Adverse effects of Oxytetracycline hydrochloride, Florfenicol, Sulphadimethoxine and Ormetoprim in tilapia (*Oreochromis niloticus*) at therapeutic dose and maximum residual limit (MRL) level. Indian J Anim Res 57(5):656–661. https://doi.org/10.18805/IJAR.B-5073

Al-Marzooq F, Ghazawi A, Daoud L, Tariq S (2023) Boosting the antibacterial activity of Azithromycin on multidrug-resistant Escherichia coli by efflux pump inhibition coupled with outer membrane Permeabilization induced by phenylalanine-arginine β-Naphthylamide. Int J Mol Sci 24:8662. https://doi.org/10.3390/ijms24108662

Al-Rimawi F, Hijaz F, Nehela Y et al (2019) Uptake, translocation, and stability of Oxytetracycline and Streptomycin in citrus plants. Antibiotics 8:196. https://doi.org/10.3390/antibiotics8040196

Antimicrobial Resistance Collaborators (2022) Global burden of bacterial antimicrobial resistance in 2019: a systematic analysis. Lancet 399(10325):629–655. https://doi.org/10.1016/S0140-6736(21)02724-0

Arumugham VB, Gujarathi R, Cascella M (2022) Third-generation Cephalosporins. StatPearls [Internet], Treasure Island, FL

Aydin F, Kayman T, Abay S et al (2023) MLST genotypes and quinolone resistance profiles of Campy lobacter jejuni isolates from various sources in Turkey. Int J Food Microbiol 391–393: 110137. https://doi.org/10.1016/j.ijfoodmicro.2023.110137

Azam M, Jan AT, Haq QMR (2016) blaCTX-M-152, a novel variant of CTX-M-group-25, identified in a study performed on the prevalence of multidrug resistance among natural inhabitants of river Yamuna, India. Front Microbiol 7:176. https://doi.org/10.3389/fmicb.2016.00176

Bai H, He L-Y, Wu D-L et al (2022) Spread of airborne antibiotic resistance from animal farms to the environment: dispersal pattern and exposure risk. Environ Int 158:106927. https://doi.org/10.1016/j.envint.2021.106927

Bandyopadhyay S, Samanta I (2020) Antimicrobial resistance in Agri-food chain and companion animals as a re-emerging menace in post-COVID epoch: low-and middle-income countries perspective and mitigation strategies. Front Vet Sci 7:620. https://doi.org/10.3389/fvets.2020.00620

Baquero F, Lanza VF, Duval M, Coque TM (2020) Ecogenetics of antibiotic resistance in listeria monocytogenes. Mol Microbiol 113:570–579. https://doi.org/10.1111/mmi.14454

Bui T, Preuss CV (2023) Cephalosporins. StatPearls [Internet], Treasure Island, FL

Bulteel AJB, Larson EL, Getahun H (2021) Identifying global research gaps to mitigate antimicrobial resistance: a scoping review. Am J Infect Control 49:818–824. https://doi.org/10.1016/j.ajic.2020.11.024

Cabello FC, Millanao AR, Lozano-Muñoz I, Godfrey HP (2023) Misunderstandings and misinterpretations: antimicrobial use and resistance in salmon aquaculture. Environ Microbiol Rep 15:245. https://doi.org/10.1111/1758-2229.13147

Caffrey N, Invik J, Waldner CL et al (2019) Risk assessments evaluating foodborne antimicrobial resistance in humans: a scoping review. Microb Risk Anal 11:31–46. https://doi.org/10.1016/j.mran.2018.08.002

Caputo A, Bondad-Reantaso MG, Karunasagar I et al (2023) Antimicrobial resistance in aquaculture: a global analysis of literature and national action plans. Rev Aquac 15:568–578. https://doi.org/10.1111/raq.12741

Chaves BJ, Tadi P (2023) Gentamicin. StatPearls [Internet], Treasure Island, FL

Collignon P (2015) Antibiotic resistance: are we all doomed? Intern Med J 45:1109–1115. https://doi.org/10.1111/imj.12902

Collignon P, McEwen S (2019) One health—its importance in helping to better control antimicrobial resistance. Trop Med Infect Dis 4:22. https://doi.org/10.3390/tropicalmed4010022

Collignon P, Voss A (2015) China, what antibiotics and what volumes are used in food production animals? Antimicrob Resist Infect Control 4:16. https://doi.org/10.1186/s13756-015-0056-5

Das N, Madhavan J, Selvi A, Das D (2019) An overview of cephalosporin antibiotics as emerging contaminants: a serious environmental concern. 3 Biotech 9:231. https://doi.org/10.1007/s13205-019-1766-9

Dadgostar P (2019) Antimicrobial resistance: implications and costs. Infect Drug Resist 12:3903–3910. https://doi.org/10.2147/IDR.S234610

EClinicalMedicine (2021) Antimicrobial resistance: a top ten global public health threat. eClinicalMedicine 41:101221. https://doi.org/10.1016/j.eclinm.2021.101221

Farooqui HH, Selvaraj S, Mehta A, Heymann DL (2018) Community level antibiotic utilization in India and its comparison Vis-à-Vis European countries: evidence from pharmaceutical sales data. PLoS One 13:e0204805. https://doi.org/10.1371/journal.pone.0204805

Ferri G, Lauteri C, Vergara A (2022) Antibiotic resistance in the finfish aquaculture industry: a review. Antibiotics 11:1574. https://doi.org/10.3390/antibiotics11111574

Food and Agriculture Organization (2021) The FAO action plan on antimicrobial resistance 2021–2025. Food and Agriculture Organization of the United Nations FAO, Rome. ISBN: 978-92-5-134673-0

Frost I, Kapoor G, Craig J et al (2021) Status, challenges and gaps in antimicrobial resistance surveillance around the world. J Glob Antimicrob Resist 25:222–226. https://doi.org/10.1016/j.jgar.2021.03.016

Gaze WH, Krone SM, Larsson DGJ et al (2013) Influence of humans on evolution and mobilization of environmental antibiotic Resistome. Emerg Infect Dis 19:e120871. https://doi.org/10.3201/eid1907.120871

Gekenidis M-T, Qi W, Hummerjohann J et al (2018) Antibiotic-resistant indicator bacteria in irrigation water: high prevalence of extended-spectrum beta-lactamase (ESBL)-producing Escherichia coli. PLoS One 13:e0207857. https://doi.org/10.1371/journal.pone.0207857

Hallmann A, Leszczyńska D, Czumaj A et al (2023) Oxytetracycline-induced inflammatory process without oxidative stress in blue mussels Mytilus trossulus. Environ Sci Pollut Res 30:80462. https://doi.org/10.1007/s11356-023-28057-z

Hernando-Amado S, Coque TM, Baquero F, Martínez JL (2020) Antibiotic resistance: moving from individual health norms to social norms in one health and Global Health. Front Microbiol 11:1914. https://doi.org/10.3389/fmicb.2020.01914

Hossain A, Habibullah-Al-Mamun M, Nagano I et al (2022) Antibiotics, antibiotic-resistant bacteria, and resistance genes in aquaculture: risks, current concern, and future thinking. Environ Sci Pollut Res 29:11054–11075. https://doi.org/10.1007/s11356-021-17825-4

Hu T, Zhang J, Xu X et al (2023) Bioaccumulation and trophic transfer of antibiotics in the aquatic and terrestrial food webs of the Yellow River Delta. Chemosphere 323:138211. https://doi.org/10.1016/j.chemosphere.2023.138211

Iskandar K, Molinier L, Hallit S et al (2021) Surveillance of antimicrobial resistance in low- and middle-income countries: a scattered picture. Antimicrob Resist Infect Control 10:63. https://doi.org/10.1186/s13756-021-00931-w

Jalloul G, Keniar I, Tehrani A, Boyadjian C (2021) Antibiotics contaminated irrigation water: an overview on its impact on edible crops and visible light active Titania as potential photocatalysts for irrigation water treatment. Front Environ Sci 9:767963. https://doi.org/10.3389/fenvs.2021.767963

Jeżak K, Kozajda A (2022) Occurrence and spread of antibiotic-resistant bacteria on animal farms and in their vicinity in Poland and Ukraine—review. Environ Sci Pollut Res 29:9533–9559. https://doi.org/10.1007/s11356-021-17773-z

Jonas OB, Irwin A, Berthe FCJ, et al (2017) Drug-resistant infections: a threat to our economic future (Vol. 2): finalreport (English). HNP/Agriculture Global Antimicrobial Resistance Initiative Washington, D.C.:World Bank Group

Karami-Zarandi M, Rahdar HA, Esmaeili H, Ranjbar R (2023) Klebsiella pneumoniae : an update on antibiotic resistance mechanisms. Future Microbiol 18:65–81. https://doi.org/10.2217/fmb-2022-0097

Karampatakis T, Tsergouli K, Behzadi P (2023) Carbapenem-resistant Klebsiella pneumoniae: virulence factors, molecular epidemiology and latest updates in treatment options. Antibiotics 12:234. https://doi.org/10.3390/antibiotics12020234

Kasimanickam V, Kasimanickam M, Kasimanickam R (2021) Antibiotics use in food animal production: escalation of antimicrobial resistance: where are we now in combating AMR? Med Sci 9:14. https://doi.org/10.3390/medsci9010014

Kaur K, Reddy S, Barathe P et al (2022) Microplastic-associated pathogens and antimicrobial resistance in environment. Chemosphere 291:133005. https://doi.org/10.1016/j.chemosphere.2021.133005

Kaye KS, Naas T, Pogue JM, Rossolini GM (2023) Cefiderocol, a siderophore cephalosporin, as a treatment option for infections caused by carbapenem-resistant Enterobacterales. Infect Dis Ther 12:777–806. https://doi.org/10.1007/s40121-023-00773-6

Kaye KS, Shorr AF, Wunderink RG et al (2023) Efficacy and safety of sulbactam–durlobactam versus colistin for the treatment of patients with serious infections caused by Acinetobacter baumannii–calcoaceticus complex: a multicentre, randomised, active-controlled, phase 3, non-inferiority clinical tri. Lancet Infect Dis 23:1072. https://doi.org/10.1016/S1473-3099(23)00184-6

Kim YB, Seo KW, Jeon HY et al (2018) Characteristics of the antimicrobial resistance of Staphylococcus aureus isolated from chicken meat produced by different integrated broiler operations in Korea. Poult Sci 97:962–969. https://doi.org/10.3382/ps/pex357

Kormos D, Lin K, Pruden A, Marr LC (2022) Critical review of antibiotic resistance genes in the atmosphere. Environ Sci Process Impacts 24:870–883. https://doi.org/10.1039/D2EM00091A

Larcher R, Laffont-Lozes P, Roger C et al (2022) Last resort beta-lactam antibiotics for treatment of new-Delhi Metallo-Beta-lactamase producing Enterobacterales and other difficult-to-treat resistance in gram-negative bacteria: a real-life study. Front Cell Infect Microbiol 12:1048633. https://doi.org/10.3389/fcimb.2022.1048633

Leal JF, Santos EBH, Esteves VI (2019) Oxytetracycline in intensive aquaculture: water quality during and after its administration, environmental fate, toxicity and bacterial resistance. Rev Aquac 11:1176–1194. https://doi.org/10.1111/raq.12286

Li L, Wang Q, Bi W et al (2020) Municipal solid waste treatment system increases ambient airborne bacteria and antibiotic resistance genes. Environ Sci Technol 54:3900–3908. https://doi.org/10.1021/acs.est.9b07641

Li L-G, Xia Y, Zhang T (2017) Co-occurrence of antibiotic and metal resistance genes revealed in complete genome collection. ISME J 11:651–662. https://doi.org/10.1038/ismej.2016.155

Lin C, Feng Y, Xie X et al (2024) Antimicrobial resistance characteristics and phylogenetic relationships of pleuromutilin-resistant Enterococcus isolates from different environmental samples along a laying hen production chain. J Environ Sci 137:195–205. https://doi.org/10.1016/j.jes.2023.01.012

Liu X, Jiang Y, Chen X et al (2014) Drug resistance mechanisms of Mycoplasma pneumoniae to macrolide antibiotics. Biomed Res Int 2014:1–7. https://doi.org/10.1155/2014/320801

Ma F, Xu S, Tang Z et al (2021) Use of antimicrobials in food animals and impact of transmission of antimicrobial resistance on humans. Biosaf Heal 3:32–38. https://doi.org/10.1016/j.bsheal.2020.09.004

Manageiro V, Salgueiro V, Rosado T et al (2022) Genomic analysis of a mcr-9.1-Harbouring Inc HI2-ST1 plasmid from Enterobacter ludwigii isolated in fish farming. Antibiotics 11:1232. https://doi.org/10.3390/antibiotics11091232

Manesh A, Varghese GM (2021) Rising antimicrobial resistance: an evolving epidemic in a pandemic. Lancet Microbe 2:e419–e420. https://doi.org/10.1016/S2666-5247(21)00173-7

Matee M, Mshana SE, Mtebe M et al (2023) Mapping and gap analysis on antimicrobial resistance surveillance systems in Kenya, Tanzania, Uganda and Zambia. Bull Natl Res Cent 47:12. https://doi.org/10.1186/s42269-023-00986-2

McCubbin KD, Anholt RM, de Jong E et al (2021) Knowledge gaps in the understanding of antimicrobial resistance in Canada. Front Public Health 9:726484. https://doi.org/10.3389/fpubh.2021.726484

McEwen SA, Collignon PJ (2018) Antimicrobial resistance: a one health perspective. Microbiol Spectr 111:255. https://doi.org/10.1128/microbiolspec.ARBA-0009-2017

McEwen SA, Fedorka-Cray PJ (2002) Antimicrobial use and resistance in animals. Clin Infect Dis 34:S93–S106. https://doi.org/10.1086/340246

Min J, Kim P, Yun S et al (2023) Zoo animal manure as an overlooked reservoir of antibiotic resistance genes and multidrug-resistant bacteria. Environ Sci Pollut Res 30:710–726. https://doi.org/10.1007/s11356-022-22279-3

Mutua F, Sharma G, Grace D et al (2020) A review of animal health and drug use practices in India, and their possible link to antimicrobial resistance. Antimicrob Resist Infect Control 9:103. https://doi.org/10.1186/s13756-020-00760-3

Nath R, Komala G, Fantke P, Mukherjee S (2022) Dissipation kinetics, residue modeling and human intake of endosulfan applied to okra (Abelmoschus esculentus). Sci Total Environ 835:155591. https://doi.org/10.1016/j.scitotenv.2022.155591

OIE (2016) The OIE strategy on antimicrobial resistance and the prudent use of antimicrobials

Pandey N, Cascella M (2022) Beta-lactam Antibiotics. StatPearls [Internet], Treasure Island, FL

Patel PH, Hashmi MF (2023) Macrolides. StatPearls [Internet], Treasure Island, FL

Payne CJ, Turnbull JF, MacKenzie S, Crumlish M (2021) Investigating the effect of an Oxytetracycline treatment on the gut microbiome and antimicrobial resistance gene dynamics in Nile tilapia (*Oreochromis niloticus*). Antibiotics 10:1213. https://doi.org/10.3390/antibiotics10101213

Peechakara BV, Hajira B, Gupta M (2022) Ampicillin. StatPearls [Internet], Treasure Island, FL

Pokharel S, Shrestha P, Adhikari B (2020) Antimicrobial use in food animals and human health: time to implement 'one health' approach. Antimicrob Resist Infect Control 9:181. https://doi.org/10.1186/s13756-020-00847-x

Poole K (2017) At the nexus of antibiotics and metals: the impact of cu and Zn on antibiotic activity and resistance. Trends Microbiol 25:820–832. https://doi.org/10.1016/j.tim.2017.04.010

Pulingam T, Parumasivam T, Gazzali AM et al (2022) Antimicrobial resistance: prevalence, economic burden, mechanisms of resistance and strategies to overcome. Eur J Pharm Sci 170:106103. https://doi.org/10.1016/j.ejps.2021.106103

Reddy N, Girdhari L, Shungube M et al (2023) Neutralizing Carbapenem resistance by co-administering Meropenem with novel β-lactam-Metallo-β-lactamase inhibitors. Antibiotics 12:633. https://doi.org/10.3390/antibiotics12040633

Reddy S, Kaur K, Barathe P et al (2022) Antimicrobial resistance in urban river ecosystems. Microbiol Res 263:127135. https://doi.org/10.1016/j.micres.2022.127135

Rodriguez-Mozaz S, Chamorro S, Marti E et al (2015) Occurrence of antibiotics and antibiotic resistance genes in hospital and urban wastewaters and their impact on the receiving river. Water Res 69:234–242. https://doi.org/10.1016/j.watres.2014.11.021

Saima S, Fiaz M, Manzoor M et al (2020) Molecular investigation of antibiotic resistant bacterial strains isolated from wastewater streams in Pakistan. 3 Biotech 10:378. https://doi.org/10.1007/s13205-020-02366-3

Samreen AI, Malak HA, Abulreesh HH (2021) Environmental antimicrobial resistance and its drivers: a potential threat to public health. J Glob Antimicrob Resist 27:101–111. https://doi.org/10.1016/j.jgar.2021.08.001

Samtiya M, Matthews KR, Dhewa T, Puniya AK (2022) Antimicrobial resistance in the food chain: trends, mechanisms, pathways, and possible regulation strategies. Foods 11:2966. https://doi.org/10.3390/foods11192966

Sangiorgio D, Spinelli F, Vandelle E (2022) The unseen effect of pesticides: the impact on phytobiota structure and functions. Front Agron 4:936032. https://doi.org/10.3389/fagro.2022.936032

Schulz J, Kemper N, Hartung J et al (2019) Analysis of fluoroquinolones in dusts from intensive livestock farming and the co-occurrence of fluoroquinolone-resistant Escherichia coli. Sci Rep 9:5117. https://doi.org/10.1038/s41598-019-41528-z

Shafiani S, Malik A (2003) Tolerance of pesticides and antibiotic resistance in bacteria isolated from wastewater-irrigated soil. World J Microbiol Biotechnol 19:897–901. https://doi.org/10.1023/B:WIBI.0000007290.94694.4f

Shah SQA, Cabello FC, L'Abée-Lund TM et al (2014) Antimicrobial resistance and antimicrobial resistance genes in marine bacteria from salmon aquaculture and non-aquaculture sites. Environ Microbiol 16:1310–1320. https://doi.org/10.1111/1462-2920.12421

Shanthi TR, Hatha M, Satyakeerthy TR (2022) A study on the diversity of pesticide-resistant bacterial population from different agricultural fields of Manjoor. Nat Environ Pollut Technol 21:1209–1216. https://doi.org/10.46488/NEPT.2022.v21i03.026

Shim JB, Seo KW, Bin KY et al (2019) Molecular characteristics of extended-spectrum and plasmid-mediated Amp C β-lactamase-producing Escherichia coli isolated from commercial layer in Korea. Poult Sci 98:949–956. https://doi.org/10.3382/ps/pey417

Sodhi KK, Kumar M, Singh DK (2021) Insight into the amoxicillin resistance, ecotoxicity, and remediation strategies. J Water Process Eng 39:101858. https://doi.org/10.1016/j.jwpe.2020.101858

Tagar S, Qambrani NA (2023) Bacteriological quality assessment of poultry chicken meat and meat contact surfaces for the presence of targeted bacteria and determination of antibiotic resistance of Salmonella spp. in Pakistan. Food Control 151:109786. https://doi.org/10.1016/j.foodcont.2023.109786

Taştan BE, Tekinay T, Çelik HS et al (2017) Toxicity assessment of pesticide triclosan by aquatic organisms and degradation studies. Regul Toxicol Pharmacol 91:208–215. https://doi.org/10.1016/j.yrtph.2017.10.030

Thompson LR, Sanders JG, McDonald D et al (2017) A communal catalogue reveals Earth's multiscale microbial diversity. Nature 551:457–463. https://doi.org/10.1038/nature24621

Van Boeckel TP, Glennon EE, Chen D et al (2017) Reducing antimicrobial use in food animals. Science 80(357):1350–1352. https://doi.org/10.1126/science.aao1495

Velazquez-Meza ME, Galarde-López M, Carrillo-Quiróz B, Alpuche-Aranda CM (2022) Antimicrobial resistance: one health approach. Vet World 15:743–749. https://doi.org/10.14202/vetworld.2022.743-749

Wang Y, Lyu N, Liu F et al (2021) More diversified antibiotic resistance genes in chickens and workers of the live poultry markets. Environ Int 153:106534. https://doi.org/10.1016/j.envint.2021.106534

Wang F, Xu M, Stedtfeld RD et al (2018) Long-term effect of different fertilization and cropping systems on the soil antibiotic resistome. Environ Sci Technol 52:13037–13046. https://doi.org/10.1021/acs.est.8b04330

Wang L, Zhao X, Wang J et al (2019) Macrolide- and quinolone-resistant bacteria and resistance genes as indicators of antibiotic resistance gene contamination in farmland soil with manure application. Ecol Indic 106:105456. https://doi.org/10.1016/j.ecolind.2019.105456

World Health Organization (2000) WHO global principles for the containment of antimicrobial resistance in animals intended for food: report of a WHO consultation with the participation of the food and agriculture Organization of the United Nations and the office international des Epizoo. World Health Organization, Geneva. https://apps.who.int/iris/handle/10665/68931

World Health Organization (2004) Second joint FAO/OIE/WHO expert workshop on non human antimicrobial usage and antimicrobial resistance: management options. World Health Organization, Geneva. https://apps.who.int/iris/handle/10665/68701

World Health Organization (2014) Antimicrobial resistance: global report on surveillance. World Health Organization, Geneva. https://apps.who.int/iris/handle/10665/112642

World Health Organization (2016) Global action plan on antimicrobial resistance. World Health Organization, Geneva. https://www.who.int/publications/i/item/9789241509763

World Health Organization (2017) Integrated surveillance of antimicrobial resistance in foodborne bacteria: application of a one health approach: guidance from the WHO advisory group on integrated surveillanec of antimicrobial resistance (AGISAR). World Health Organization, Geneva. https://apps.who.int/iris/handle/10665/255747

Wright GD (2010) Antibiotic resistance in the environment: a link to the clinic? Curr Opin Microbiol 13:589–594. https://doi.org/10.1016/j.mib.2010.08.005

Wu J, Wang J, Li Z et al (2023) Antibiotics and antibiotic resistance genes in agricultural soils: a systematic analysis. Crit Rev Environ Sci Technol 53:847–864. https://doi.org/10.1080/10643389.2022.2094693

Xiao K-Q, Li B, Ma L et al (2016) Metagenomic profiles of antibiotic resistance genes in paddy soils from South China. FEMS Microbiol Ecol 92:fiw023. https://doi.org/10.1093/femsec/fiw023

Yin L, Wang X, Li Y et al (2023) Uptake of the plant agriculture-used antibiotics Oxytetracycline and streptomycin by cherry radish—effect on plant microbiome and the potential health risk. J Agric Food Chem 71:4561–4570. https://doi.org/10.1021/acs.jafc.3c01052

Yoon S, Son SH, Bin KY et al (2020) Molecular characteristics of optrA-carrying Enterococcus faecalis from chicken meat in South Korea. Poult Sci 99:6990–6996. https://doi.org/10.1016/j.psj.2020.08.062

Zha L, Zhang X, Cheng Y et al (2023) Intravenous polymyxin B as adjunctive therapy to high-dose tigecycline for the treatment of nosocomial pneumonia due to Carbapenem-resistant Acinetobacter baumannii and Klebsiella pneumoniae: a propensity score-matched cohort study. Antibiotics 12:273. https://doi.org/10.3390/antibiotics12020273

Zhou Z-C, Shuai X-Y, Lin Z-J et al (2021) Prevalence of multi-resistant plasmids in hospital inhalable particulate matter (PM) and its impact on horizontal gene transfer. Environ Pollut 270: 116296. https://doi.org/10.1016/j.envpol.2020.116296

Chapter 2
Plant Essential Oils as Potent Antimicrobials

Sagar Reddy, Kawaljeet Kaur, Pramod Barathe, Varsha Shriram, Atish T. Paul, and Vinay Kumar

Abstract Antimicrobial resistance (AMR) is spreading at an alarming rate, reducing the effectiveness of antibiotics and producing undesirable results such as increased mortality and significant economic loss. Public health is seriously threatened by this global problem, which restricts our capacity to treat common infections and increases risk for vulnerable groups. To address this expanding issue, a novel and effective antimicrobial agent or treatment strategy is needed. Essential oils represent an important source of a diverse range of bioactive constituents with potent antimicrobial activity. Wider acceptance due to its traditional use, lower toxicity, and ability to target multiple determinants of resistance makes essential oils a potent candidate for effectively tackling AMR and eradicating drug-resistant pathogens. Essential oil-loaded nanomaterials have also shown improved efficacy in treating antimicrobial resistance due to increased bioavailability, stability, and solubility and reduced degradation of the active principles of essential oils. Furthermore, combining essential oils with antibiotics has a synergistic impact, helping to revitalize an otherwise depleted antibiotic arsenal. This chapter gives a comprehensive summary of the antibacterial properties of essential oils and their active principles. The chapter also highlights the major bacterial AMR-determinants targeted by plant essential oils besides discussing the successful experiments on the combination of essential oils with antibiotics and nanomaterials for combating drug-resistant microbes.

S. Reddy · K. Kaur · P. Barathe · V. Kumar (✉)
Department of Biotechnology, Modern College of Arts, Science and Commerce, Savitribai Phule Pune University, Pune, Maharashtra, India
e-mail: vinay.kumar@moderncollegegk.org

V. Shriram
Department of Botany, Prof. Ramkrishna More College, Savitribai Phule Pune University, Pune, Maharashtra, India

A. T. Paul
Laboratory of Natural Product Chemistry, Department of Pharmacy, Birla Institute of Technology and Science Pilani, Pilani Campus, Pilani, Rajasthan, India

V. Kumar et al. (eds.), *Medicinal Plants and Antimicrobial Therapies*,
https://doi.org/10.1007/978-981-99-7261-6_2

Keywords Antimicrobials · Antimicrobial resistance · Antibiotic potentiation · Biofilm · Essential oils · Multi-drug resistance

2.1 Introduction

Antimicrobial resistance (AMR) is defined as microorganisms' ability to survive and thrive in the presence of wide-ranging antimicrobial agents. It has emerged as one of the greatest threats to human health and the environment, and the situation of AMR is unfortunately worsening by the day. Microbes acquire and demonstrate resistance to routinely used and often even against last-resort antibiotics resulting in increasing treatment failure, mortality, and morbidity as well as huge economic losses (Shriram et al. 2018). Annual global economic production, or gross domestic product (GDP), is predicted to fall by 1.1% in the case of a low AMR scenario and 3.8% in the case of a high AMR scenario (Ahmad and Khan 2019; The Global Risks Report 2018). Furthermore, additional healthcare expenditure in low-AMR scenarios is expected to reach \$0.33 trillion by 2050 and \$1.2 trillion in high-AMR scenarios.

Globally, AMR causes 700,000 deaths per year, and if current AMR rates continue, 10 million deaths per year are expected by 2050 (Reddy et al. 2022). As a result, the World Health Organization (WHO) correctly identified AMR as one of the top ten urgent concerns (EClinicalMedicine (2021)). AMR has been observed at three escalating levels: multi-drug resistance (MDR), extensive drug resistance (XDR), and pan drug resistance (PDR), all of which make infectious illness treatment challenging (Shriram et al. 2018). Overuse or misuse of antimicrobials in health care, livestock as a growth promoter, or other sectors is the leading source of resistance emergence, prevalence, and spread (Yu et al. 2020). The drying pipeline of new and potent antibiotics further aggravated this issue, leaving only a few treatment options available to treat infectious diseases. Therefore, it is necessary to identify or develop novel, natural, effective, and less toxic antimicrobials. Additionally, it has been demonstrated that conventional antibiotics with a single target are vulnerable to mutational resistance (Shriram et al. 2018). Scientists can access a vast array of compounds that have evolved over millions of years to fight different pathogens by looking into natural products. These compounds frequently target several different pathways, which makes it more challenging for pathogens to evolve resistance against these natural metabolites. Additionally, natural products can serve as a source of inspiration for the creation of new antimicrobials with distinct modes of action, enhancing our toolbox for fighting drug-resistant bacteria.

Since antiquity, plant-based chemicals and essential oils (EOs) have been known to have biological activity, notably antibacterial action. EOs are aromatic, volatile liquids obtained from plant material through steam distillation or hydrodistillation and have antiviral, antimycotic, anti-toxigenic, and insecticidal properties (Perricone et al. 2015). EOs distinguish themselves from other antimicrobial drugs due to their biological constituents, which include anti-pathogenic phytomolecules such as monoterpenes, sesquiterpenes, oxygenated derivatives, phenylpropanoids, and fatty acids, as well as their esters (Swamy et al. 2016).

This chapter focuses on EOs from plant families Asteraceae, Lamiaceae, and Myrataceae and their potentiation effect. Major EOs for their antimicrobial property and how they have targeted various determinants such as efflux pumps, biofilm formation, and quorum sensing have been discussed in this chapter. The use of EOs with nanomaterials and antibiotics as a novel and effective approach to potentiate ineffective antibiotics has also been discussed.

2.2 Antimicrobial Potential of Plants of Three Major Essential Oil-Yielding Family

2.2.1 *Lamiaceae*

Lamiaceae also known as Labiatae or Mint family are a family of flowering plants that consist of more than 230 genera and over 7000 species of flowering plants (Zhao et al. 2021). It is also recognized as the sixth-largest angiosperm family and the largest family in the order Lamiales (Zhao et al. 2021). The largest genera from this family are *Salvia, Thymus, Vitex, Hyptis, Scutellaria, Stachys, Plectranthus, and Teucrium* (Kuete 2017). Lamiaceae are known for their characteristic features such as a four-sided stem, opposite leaves, a verticillaster or thyme type of inflorescence, bilabiate flowers with deeply four-lobed superior ovary, and a gynobasic style (Simpson 2019). *Ocimum basilicum* L., *Origanum vulgare* L., *Thymus vulgaris* L., *Mentha* × *piperita* L., *Lavandula angustifolia* Mill., and *Rosmarinus officinalis* are some well-known essential oil-yielding plants of the family Lamiaceae. The members typically have the characteristic components of striking therapeutic importance such as carvacrol, *1,8*-cineole, α-terpineol α-pinene, camphor, β-geraniol, thymol, eugenol, and Patcholene alcohol (Hussain et al. 2011; Moumni et al. 2020). For centuries, Lamiaceae family plants have been used to treat bacterial infections, for example, complete inhibition of *Staphylococcus aureus* (including methicillin-resistant *S. aureus* (MRSA)), and *Escherichia coli* at the 4.5% and 2.25% of Tulsi (*Ocimum tenuiflorum)* essential oil, and similarly, the strong bactericidal and inhibitory activity of *Thymus vulgaris* against MDR *S. aureus*. *Thymus vulgaris* showed minimum inhibitory concentration (MIC) and minimum bactericidal concentration (MBC) values of (0.09–078 mg/mL) and total killing of bacterial cells at the concentration of 2 MIC after 6 h (Kot et al. 2018). Likewise, in a study conducted by Mohammed et al. (2020), *R. officinalis* essential oil showed antibacterial activity against all the tasted uropathogenic isolates with inhibition zone and MIC in the range of 7.00 ± 0.00 to 9.6 ± 0.32 mm and 0.06 ± 0.00 to 0.16 ± 0.07 mg/mL, respectively. The highest activity was recorded against *Klebsiella pneumoniae, Proteus vulgaris,* and *S. aureus*. Furthermore, Jovanovi and Sokovi (2014) assessed the antimicrobial activities of essential oils of five Lamiaceae species (*Mentha pulegium, Mentha piperita, Satureja montana, Salvia lavandulifolia, and Lavandula angustifolia*) against seven clinical bacterial isolates. All essential oils showed excellent antibacterial activity against all the tested clinical isolates, with *S. montana* being the most potent one.

Targeting bacterial biofilm is a powerful resistance reversal strategy because it represents one of the key mechanisms of resistance used by bacteria to deal with the stress of antibiotics. EOs are being explored for their potent antibiofilm activities against drug-resistant pathogens. Snoussi et al. (2016) demonstrated strong antibacterial activity of parsley and basil essential oils against vibrio spp., with zones of inhibition ranging from 8.67 to 23.33 mm and MIC and MBC ranging from 0.023 to 0.047 mg/mL and >3 to >24 mg/mL, respectively. Additionally, both EOs were successful in inhibiting *Vibro* spp. biofilm at very low concentrations. *Ocimum gratissimum* was tested against four gastrointestinal pathogens by Chimnoi et al. (2018) for possible killing and antibacterial effects. They found that all of the tested isolates displayed outstanding antibacterial activity at MICs of 1–2 mg/mL. Furthermore, a rapid lethal effect was seen to occur within 5 s. Further research pointed to membrane rupture as a potentially viable mode of action for EOs. Furthermore, not only single EO but also their blends or combinations have shown striking antibacterial activity, for instance, the strong synergistic antibacterial activity of an essential oil, consisting of *Blepharis cuspidate* and *Thymus schimper* against both Gram-positive and Gram-negative bacterial isolates (Gadisa et al. 2019).

2.2.2 *Asteraceae*

Asteraceae plant family also known as the Compositae family or sunflower family is an exceedingly large family of flowering plants having more than 23,600 species, i.e., 1620 genera and 13 subfamilies (Kuete 2017). Due to the widespread occurrence of species in different eco-geographic zones such as temperate, cold temperate, and subtropical, and thus, this family is also known as the cosmopolitan family (Michel et al. 2020). Distinctive features of this family include a head or capitulum type of inflorescence surrounded by the involucre of phyllaries, flowers either bilabiate, ray, or disk with pappus calyx, synoecious androecium, and fruit a multiple of achenes (Simpson 2019). *Artemisia annua, Achillea cretica, Tanacetum parthenium, Matricaria recutita, Helichrysum cymosum* subsp. *Cymosum, Helichrysum italicum, Tanacetum trifurcatum,* and *Artemisia arborescens* are some essential oil-yielding plants of the family that show significant antibacterial activity (Bedoya and Bermejo 2013). Camphor, caryophylladienol, 1,8-cineole, eudesmol, geraniol, limonene, methyl eugenol, a-pinene, thujone, and vulgarone B are the major components of EOs in this member that are attributed to potent antibacterial activity (Bedoya and Bermejo 2013). By using the essential oil from three different Artemisia species, namely, *Artemisia herba* Alba, *Artemisia absinthium*, and *Artemisia campestris*, Mathlouthi et al. (2021) showed a considerable reduction of bacterial biofilm formation without impacting the planktonic growth. Additionally, chamazulene, b-pinene, and a-thujone were reported as the main ingredients in the EOs. Similar results were seen when antibiotic-resistant *Enterococcus faecalis* was exposed to *Baccharis psiadioides* (Less.) Joch. Muller essential oil, significantly reduced its viability and microbial adherence. Because bacterial adhesion is a crucial step in the development of bacterial biofilms, inhibiting bacterial adhesion may be a useful therapeutic approach (de Negreiros et al. 2016). Likewise, Mohammadi et al. (2021) observed that the MIC/2 concentration of *Artemisia dracunculus* essential oil resulted in the downregulation of genes associated with quorum sensing (*luxS* and *pfs* for *Salmonella typhimurium* and *hld* for *S. aureus*). Additionally, it has been observed that essential oils exhibit non-toxicity at sub-MIC and MBC concentrations and significantly inhibit and disrupt the formation of biofilms.

2.2.3 *Myrtaceae*

The Myrtaceae family is the world's most prominent commercial fruit tree family, with roughly 121 genera and over 5800 species dispersed throughout tropical and subtropical locations (de Paulo Farias et al. 2020). This family is distinctive in being shrubs and trees with glandular-punctate or pellucid leaves, and epigynous flowers with numerous stamens (Simpson 2019). Some common and economically important species in this family include *Eugenia uniflora, Eugenia sulcate, Syzygium aromaticum, Syzygium cumini, Myrtus communis, Melaleuca alternifolia,* and others. *Syzygium aromaticum*, sometimes known as clove, is one of the most potent members of the family with outstanding antibacterial activity. It is widely used as a food preservative, in spices, cosmetics, and personal care products, and, most critically, in medical therapy because of its powerful anti-inflammatory, antioxidant, and antibacterial properties (Zhang et al. 2017). In one of their investigations, Zhang et al. (2017) investigated the antibacterial activity and mode of action of clove essential oil and its primary component, eugenol, against the periodontopathogen *Porphyromonas gingivalis*. Clove EOs were found to target a variety of resistance mechanisms, including the permeabilization of cell membranes, the degradation of plasma membrane integrity, the inhibition of biofilm formation, and the reduction of preformed biofilm. Furthermore, clove EOs significantly reduced the expression of virulence factor genes (*fimA, hagB, rgpB, hagA, rgpA,* and *kgp*) involved in biofilm formation. In a similar investigation, De Oliveira et al. (2020) evaluated 13 EOs for their antibacterial and antibiofilm properties against the microorganisms linked to oral disorders. Clove, cinnamon, thyme, and oregano strongly inhibited all bacterial strains (De Oliveira et al. 2020). Additionally, the *Streptococcus mutans* biofilm was eradicated by toothpaste that had clove, thyme, oregano, and cinnamon added to the formulation. *Myrtus communis* L. is another potent antibacterial plant species of the Myrtaceae family that is similar to the clove essential oil, which can target multiple determinants of resistance. For example, complete reduction of MDR *Acinetobacter baumannii* bacterial count by the alone as well as synergistic combination of myrtle essential oil and polymyxin B with FIC index under or equal to 0.5 (Aleksic et al. 2014). These results confirmed the Lamiaceae, Asteraceae, and

Myrtaceae families as potent sources of a diverse range of bioactive metabolites that can be looked up for screening of potent antimicrobials as well as for the formulation of natural therapeutics. Table 2.1 enlists the examples of EOs-producing plants, as well as their main constituents and antibacterial properties.

2.3 Major Determinants of Antimicrobial Resistance Targeted by the Essential Oils

Bacteria have evolved multiple determinants of resistance such as biofilm, efflux pumps, quorum sensing, membrane permeabilization, and enzyme-degradation that give bacteria the ability to survive under the stress of antimicrobials, thus developing resistance to particular antimicrobials. As a result, in order for antimicrobials to be effective, they must be able to target and affect these determinants of resistance. This section deals with the major determinants of resistance targeted by the EOs (Fig. 2.1).

2.3.1 Efflux Pumps

Bacterial Efflux pumps are the proteins that are localized and embedded in the plasma membrane of bacteria and whose function is to extrude the antimicrobial agents before they reach their intended target (Amaral et al. 2014). Efflux pumps are found in both Gram-positive and Gram-negative bacteria. It has six main families, including ABC ATP-binding cassette, MATE (multidrug and toxic compound extrusion), RND (resistance-nodulation-division), PACE proteobacterial antimicrobial compound efflux, and MFS (major facilitator superfamily) (Sun et al. 2014). Efflux pumps are known to carry out physiological tasks such as bile salt export, cell homeostasis participation, intracellular signaling, and stress adaptation and are involved in bacterial virulence and pathogenicity (Sun et al. 2014). Several studies demonstrated the key role of the efflux pump in biofilm formation, quorum sensing, and beta-lactam resistance and highlighted reduction in bacterial biofilm and quorum sensing via the application of efflux pump inhibitors (He et al. 2015; Pages et al. 2009; Wang et al. 2019), suggesting the use of efflux pump inhibitor as a possible strategy for biofilm as well as quorum sensing inhibition and subsequently the AMR. Studies show that substances found in EOs such as carvacrol, eugenol, thymol, trans-cinnamaldehyde, α-pinene, and limonene act as potent efflux pump inhibitors and significantly reduce AMR brought on by excessively expressed efflux pumps (de Araújo et al. 2021; Miladi et al. 2017a; Wenqian Yuan 2019). Similarly, other studies highlighted the strong efflux pump inhibitory potential of EOs, for instance, the effective inhibition of *Staphylococcus aureus* SA1199B NorA efflux pump by very low concentration (1/32 MIC) of *Rhus albida* EOs (Elhidar et al. 2021).

Table 2.1 Antibacterial properties of essential oils from Lamiaceae, Myrtaceae, and Asteraceae

Plant family	Essential oil-yielding plants	Identified chemical constituents	Targeted bacteria	MIC	MBC	Main findings	References
Lamiaceae	*Origanum compactum* Benth.	Carvacrol, *p*-cymene, thymol, and γ-terpinene	*Escherichia coli* K12, *listeria innocua* 4030, *Staphylococcus aureus* 25,923, *Bacillus subtilis* 6633	0.06 to 0.25% (v/v)	0.12 to 0.5% (v/v).	Strong antibacterial activity against all four tested bacteria *Origanum compactum* Benth. Essential oil most active against *listeria innocua* with a zone of inhibition of 49 ± 1.00 mm	Laghmouchi et al. (2018)
	Thymus daenensis, Nepeta sessilifolia, Hymenocrater incanus, and *Stachys infata*	Thymol, oleic acid, (−)-caryophyllene, 1,8-cineole, palmitic acid, germacrene	*Staphylococcus epidermidis, S. Aureus, Bacillus subtilis, Klebsiella pneumonia, Pseudomonas aeruginosa, Proteus vulgaris*	16–2000 μg/mL	16–2000 μg/mL	Essential oils of *Hymenocrater incanus* and *Stachys infata* were most active against *Pseudomonas aeruginosa* with a MIC of 16 and 63 μg/mL	Ghavam et al. (2022)
Lamiaceae	*Origanum compactum* L. and *Mentha piperita* L.	Carvacrol, thymol, p-cymene, α-terpinene, menthofuran, menthol, methyl acetate, 1,8 cineole, and β-pinene	*Escherichia coli* ATCC 25,922	0.0625% (v/v) and 1%		A mixture of both *Origanum compactum* L. and *Mentha piperita* L. essential oils provided better antimicrobial effects against *E. coli* than the essential oil alone	El Amrani et al. (2022)
Lamiaceae	*Thymbra capitata*	Thymol	*P. aeruginosa* 10,145	1.11%		Inhibition of bacterial biofilm, swarming motility, aggregation ability, and hydrophobicity of *P. aeruginosa* at sub-MIC concentration Reduced production of three QS-regulated factors such as pyocyanin, rhamnolipid, and LasA protease	Qaralleh (2019)

(continued)

Table 2.1 (continued)

Plant family	Essential oil-yielding plants	Identified chemical constituents	Targeted bacteria	MIC	MBC	Main findings	References
Lamiaceae and Myrtaceae	*Salvia sclarea* and *Melaleuca alternifolia*		*Chromobacterium violaceum* CV026 *and P. Aeruginosa*	5 μL/mL and 20 μL/mL		Significant reduction in QS-controlled factors such as violacein, pyocyanin, alginate, and protease production, along with the swarming motility at MIC/4 and MIC/8 concentration of both the essential oils Inhibition of bacterial biofilm and reduced expression of QS regulatory genes *las R, lasI, rhlI,* and *rhlR*	Srivastava et al. (2023)
Myrtaceae, Lamiaceae, Asteraceae	*Thymus mastichina* L., *Mentha pulegium* L., *Calamintha nepeta* L., *Rosmarinus officinalis* L., *Dittrichia viscosa, Eucalyptus globulus* LABILL. ssp. *globulus*	Isopulegol, isopulegone, 1.8-cineole, pulegone, β-myrcene, camphor, α-pinene	*S. aureus* ATCC 25923, *B. subtilis* ATCC 6633, *E. coli* ATCC 25922, *P. aeruginosa* ATCC 27853	6–25 mg/mL	12–70 mg/mL	With *M. pulegium* and *E. globulus* being the most effective, all essential oils showed antibacterial activity against the tested isolates. It was discovered that the combinations of *D. viscosa/C. nepeta* and *E. globulus/T. mastichina* was synergistic against *S. aureus*. The most effective essential oil for preventing biofilm formation was determined to be *C. nepeta*'s	Vieira (2017)
Myrtaceae	*Myrciaria pilosa*	Sesquiterpenes guaiol and (E)-b-caryophyllene	*S. aureus*	5 μg/mL	10–20 μg/mL	92.0% and 47.2% reduction in staphyloxanthin and hemolytic action of *S. aureus* treated with *M. Pilosa* essential oil	Costa et al. (2020)

	Eugenia stipitata	Guaiol, trans-caryophyllene, b-eudesmol, and	*S. aureus*	128–512 µg/mL	256–1024 µg/mL	Essential oil of *E. stipitate* showed synergistic action with gentamycin and ciprofloxacin antibiotics. Significant reduction in hemolysis (78.18–92.48%) and staphyloxanthin (67.27–91.89%). An increase in efflux of k + ion, protein, and nucleic acid is detected indicating cell membrane rapture. The essential oil of *E. stipitate* demonstrated no cytotoxicity in human fibroblast and macrophage cells while protecting *T. mollitor* larvae from the negative effects of *S. aureus*	Costa et al. (2022)
	Eugenia brejoensis L.		*S. aureus* ATCC 29312	8–516 µg/mL		Decrease in hemolytic activity, staphyloxantin, and ability to survive in human blood by application of sub-MIC concentration of *E. brejoensis* essential oil. Essential oil application increased the life span of *Caenorhabditis elegans* and *Galleria mellonella* infected with *S. aureus* by reducing the bacterial load	Macêdo et al. (2020)

(continued)

Table 2.1 (continued)

Plant family	Essential oil-yielding plants	Identified chemical constituents	Targeted bacteria	MIC	MBC	Main findings	References
Asteraceae	*Artemisia dracunculus* var. qinghaiensis Y. R.	Sabinene, β-terpinene, terpinen-4-ol, α-pinene	*Staphylococcus aureus, Micrococcus luteus, Bacillus subtilis, Klebsiella pneumoniae, Pseudomonas aeruginosa, Salmonella paratyphi*, and *Saccharomyces cerevisiae*	1.25–5.0 μL/mL		A high proportion of monoterpenoids (42.12%) were identified in the essential oil. Moderate inhibitory activity was detected against the seven bacterial isolates	Liu et al. (2018)
	Helichrysum italicum	Neryl acetate, α-curcumene	*Haemophilus influenzae, H. parainfluenzae, Streptococcus pneumoniae*, and *Pseudomonas aeruginosa*	0.312 mg/mL against *Haemophilus* spp., and 0.625 mg/mL against *Pseudomonas aeruginosa*		The antibiofilm activity of *Helichrysum italicum* essential oil was much higher against *H. influenzae* (75.33%) and *H. parainfluenzae* (78.67%) than the *Pseudomonas aeruginosa* and *Streptococcus pneumoniae*. The treatment above the MIC and above the MIC value caused membrane damage, whereas full cell lysis was recorded at the MIC × 4 value	Balázs et al. (2022)
Asteraceae and Lamiaceae	*Thymus daenensis* and *Satureja hortensis*	Carvacrol, terpinene and thymol	*Staphylococcus aureus*	0.0625 μL/mL (for *Thymus daenensis*) and 0.125 μL/mL (for *S. hortensis*)	0.125 μL/mL	Significant *hld* gene downregulation after *S. hortensis* EO treatment at MIC/2 concentration. Inhibition of bacterial biofilm at sub-MIC concentration	Sharifi et al. (2018)

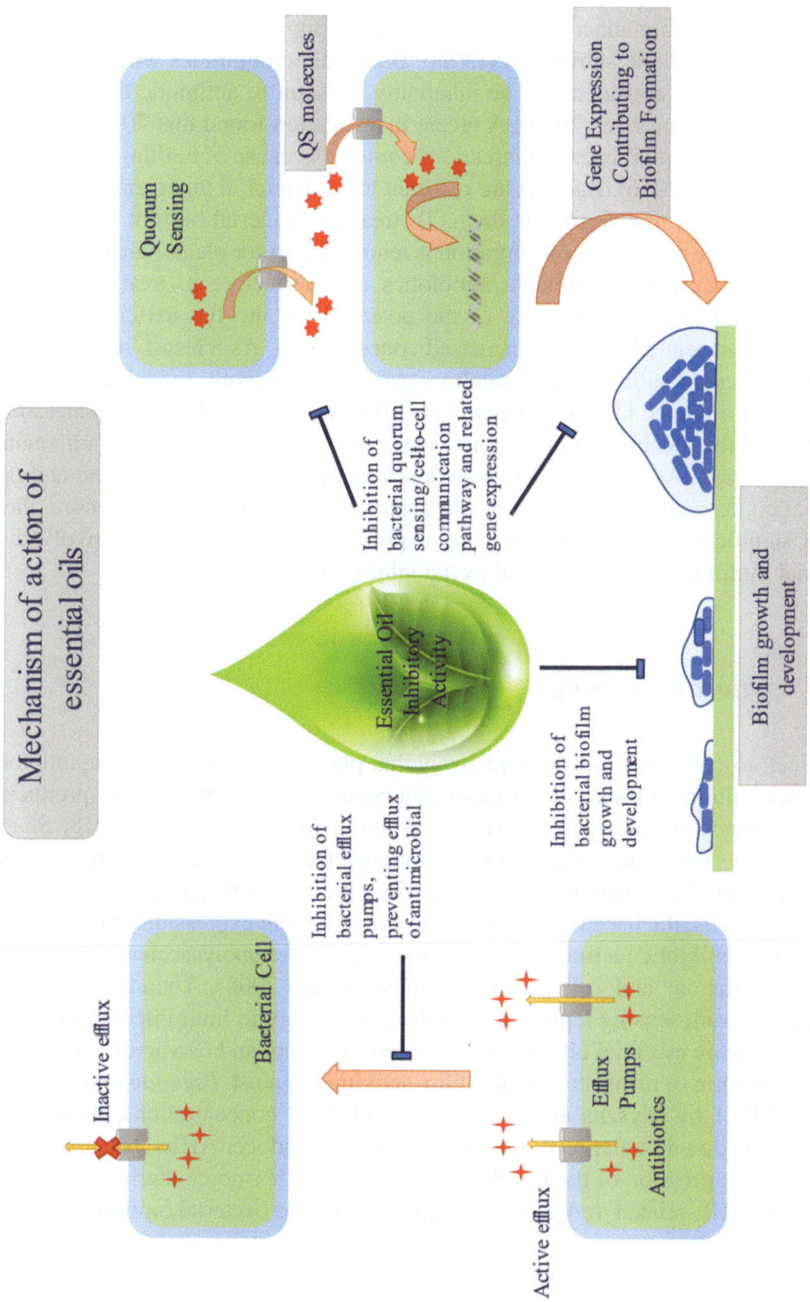

Fig. 2.1 Mechanism of antimicrobial action of essential oils

2.3.2 *Bacterial Biofilms*

Bacterial biofilms are adherent, complex bacterial populations that are enclosed in extracellular polymeric substances (EPSs). Bacterial communities within the extracellular matrix are considered to be inherently resistant to antibiotic treatment and host defenses (Wang et al. 2019). A recent investigation found that 70% of chronic diseases and 65% of all bacterial infections could be because of biofilm involvement (Fastenberg et al. 2016). Bacteria are resistant to antibacterial treatments because of the biofilm matrix that surrounds them. To treat the bacterial biofilm successfully, 1000 times more antibiotic concentration is required than for planktonic cells (Wang et al. 2019). Currently, available antibiotics are ineffective in treating bacterial biofilm-related infections because of the potential for in vivo toxicity and other side effects associated with this increased concentration. As a result, it is critical to develop or screen antibiofilm molecules capable of effectively reducing and eliminating biofilm-related infections (Roy et al. 2018). In this regard, Sharifi et al. (2021) investigated the mechanism of action of *Cuminum cyminum* L. essential oil against MDR *S. aureus*. At ½ MIC, there was deformation of the cell membrane and cell destruction, as well as a decrease in *hld* expression (3.13-fold). Furthermore, there was a significant decrease in *ica* gene expression levels, which are involved in bacterial biofilm formation, as well as the inhibition of efflux pumps.

2.3.3 *Quorum Sensing*

It is a cell-to-cell communication pathway that plays an important role in antibiotic resistance, biofilm formation, bacterial pathogenesis, and virulence by producing and secreting molecules that are known as autoinducers (Poli et al. 2018; Sharifi et al. 2018). The concentration of autoinducer that they secrete increases with population density. When the autoinducer hits a crucial level signifying a "quorum" of bacterial cells, the bacterium responds by changing the expression of many genes involved in antibiotic resistance, exoenzyme, capsular exopolysaccharide synthesis, biofilm formation, and cell growth (Pumbwe et al. 2008). Therefore, targeting bacterial quorum sensing represents an effective strategy to limit infectious disease. Thymus vulgare essential oil and its constituents (thymol and carvacrol) were tested for anti-quorum sensing and antibiofilm activities against *Pseudomonas fluorescence* KM121 by Myszka et al. (2016). At sub-MIC concentrations, there was a considerable downregulation of quorum-sensing autoinducer AHLs, bacterial motility, and the expression of the flagella gene (*flgA*). Furthermore, altering the expression of the AHL-related *flgA* gene significantly inhibited bacterial biofilm formation.

2.4 Antibiotic Potentiation Effect of Essential Oils and Their Major Constituents

Increasing AMR among a wide variety of bacteria from clinical to environmental setups has emerged as one of the greatest global health threats of the twenty-first century and a major hurdle for the global drug discovery program (Reddy et al. 2022). It necessitates the need to find out novel and alternative approaches to tackle this looming threat. One of the strategies that help revive an otherwise faded antibiotic arsenal is the synergistic combination of antibiotics with other compounds including EOs. Several articles have reported essential oil's antibiotic potentiation effect by targeting multiple resistance mechanisms employed by bacteria (Aumeeruddy-Elalfi et al. 2016). For instance, the potentiation of norfloxacin by *Aloysia gratissima* essential oil and its major component β-caryophyllene against the *Pseudomonas aeruginosa, Staphylococcus aureus*, and *E. coli* (Santos et al. 2021). In this case, *Aloysia gratissima* essential oil significantly lowers norfloxacin's minimum inhibitory concentration (MIC) by (six-fold for *S. aureus* and eightfold for *P. aeruginosa*). Similarly, *Eucalyptus camaldulensis* Dehn EOs was shown to have antibacterial (MIC = 1000 μg/mL) and β-lactams optimizer potential activities against *S. aureus*, methicillin-resistant *S. aureus* (MRSA), and *E. coli*, respectively; the synergistic effect was primarily shown for amoxicillin, ampicillin, cephalexin, and cefuroxime (Chaves et al. 2018). Limaverde et al. (2017) found no direct antibacterial activity of essential oil from *Chenopodium ambrosioides* leaves against *S. aureus* IS-58 when used alone. However, when combined in subinhibitory concentration (¼ MIC) with tetracycline essential oil of *Chenopodium ambrosioides* significantly reduced the MIC of tetracycline by twofold. In addition, they also reported the efflux pump inhibitory potential of essential oil as the probable cause of such MIC reduction. Similar to these results, Miladi et al. (2017b) reported the synergistic effect of essential oil components (eugenol, carvacrol, thymol, γ-terpinene, and p-cymene) and tetracycline, with two- to eightfold reduction in tetracycline MIC. Additionally, antibiofilm activity as well as inhibition of efflux EtBr was also reported. It highlights the antibiotic potentiating potentials of natural EOs, which can be attributed to their abilities of targeting multiple determinants of resistance such as efflux pump and bacterial biofilms. Particularly, helping in reviving the efficacy of antibiotics and also lowering the side effect of antibiotics by reducing the minimum inhibitory concentration. Examples of essential oils' antibiotic potentiation effect are shown in Table 2.2.

Table 2.2 Antibiotic potentiation effect of essential oils and their major constituents

Essential oil-yielding plants	Targeted bacteria	MIC of essential oil	Antibiotics show synergy with essential oils	The main finding of checkerboard analysis	References
Periploca laevigata	*Bacillus cereus* ATCC 14579, *Bacillus subtilis* ATCC 9524, *Micrococcus luteus* ATCC 381, *Escherichia coli* ATCC 8739, *Klebsiella pneumonia, Pseudomonas aeruginosa* DSM 50090, and *Staphylococcus aureus* 209 PCIP 53156	3.75–7.5 mg/mL and 15–30 mg/mL against Gram-positive and Gram-negative isolates	Ciprofloxacin and gentamicin	Gram-positive strains were more sensitive than Gram-negative isolates to the essential oils. The combination of gentamicin and essential oil exhibited a synergistic effect against both Gram-positive and Gram-negative isolates (FICI = 0.28–0.50), followed by the essential oil and ciprofloxacin combination (FICI= 0.31–0.38)	Ait Dra et al. (2017)
Satureja montana L.	*S. aureus* ATCC 25923, *S. aureus* ATCC 6538P, *E. coli* ATCC 25922, *E. coli* strains (ECP19 and ECP32), *Listeria monocytogenes* ATCC 7644, and *Listeria monocytogenes* (LM2 and LM9)	0.3–0.78 mg/mL (for *S. aureus*), 0.78–1.56 mg/mL (for *L. monocytogenes*, 1.5–3.12 mg/mL (for *E. coli*)	Gentamicin	Depending on the strain, the presence of *Satureja montana* L. essential oil at a 1/4 MIC value allowed gentamicin to be reduced from its MIC value to 1/16 MIC with an FIC index of less than 0.5. When 1/4 and 1/8 MIC of oil were synergistically combined with 1/8 MIC of gentamicin (FICI 0.38), the combination inhibited the LM2 strain of *L. monocytogenes*	Vitanza et al. (2018)
Origanum vulgare L.	*E. coli*	0.5 μL/mL	Amoxicillin, fluoroquinolones, doxycycline, polymyxin, lincomycin, florfenicol	FIC indices in the range of 0.625–0.750 demonstrated an additive effect of OEO's antibacterial properties when paired with amoxicillin, polymyxin, and lincomycin against *E. coli*. With FIC indices ranging from 0.375 to 0.500, the antibacterial activities of OEO in conjunction with fluoroquinolones, doxycycline, lincomycin, and florfenicol demonstrated synergism against *E. coli*	Trifan et al. (2018)

Oregano essential oil purchased	*Acinetobacter baumannii*	1.75–3.50 mg/mL	Polymyxin B	Using a checkerboard experiment, a synergistic interaction between OEO and polymyxin B (FICI: 0.18–0.37) was discovered. When combined, OEO showed a 16-fold reduction in the MIC for polymyxin B	Amaral et al. (2020)
Myrtus communis L.	*Acinetobacter baumannii*	0.25 to 4 μL/mL	Polymyxin B and ciprofloxacin	The significant synergy between *M. communis* L. essential oils and both antibiotics was observed to be effective against MDR *A. baumannii* wound isolates with an FIC value of less than or equal to 0.50. The time-kill curve method confirmed the efficiency of the combination of myrtle essential oil and polymyxin B, with a total reduction of bacterial count after 6 h	Aleksic et al. (2014)

2.5 Efficacy of Essential Oil-Loaded Nanomaterials Against Drug-Resistant Pathogens

EOs being used as a traditional natural and medicinal compound due to their antibacterial, antifungal, and antiviral properties have been applied to the medicinal field for centuries (Ramsey et al. 2020). However, in recent years, researchers have tried to increase EOs' efficacy and therapeutic properties by encapsulating them into nanocarriers and delivering them to the target (Guidotti-Takeuchi et al. 2022). The World Health Organization (WHO) and the Department of Biotechnology (DBT), Government of India (Department of Biotechnology 2021), have identified major drug-resistant priority pathogens have the potential to needing urgent attention and curing including *Pseudomonas aeruginosa*, *Escherichia coli*, *Klebsiella pneumoniae,* and *Staphylococci aureus* (Asokan et al. 2019). Recent publications on the nanodelivery of EOs as an effective tool to restrict AMR have boomed the medical field for combating infectious diseases caused by MDR pathogens (Dupuis et al. 2022). Studies on the antibacterial activity of chitosan-loaded essential oil from the medicinal plant of guava leaves for combating MDR *Klebsiella pneumoniae* resulted in 13%, 52%, and 96% inhibition rates at 10 μg/mL, 60 μg/mL, and 100 μg/mL concentrations of nanoparticles, respectively (Zhang et al. 2020). Similarly, Rajivgandhi et al. (2023)studied chitosan-loaded essential oils (CsEOs) extracted from the medicinal plant of *Solanum nigrum* against MDR *Pseudomonas aeruginosa*, *Escherichia coli*, *Klebsiella pneumoniae*, and *Staphylococci aureus* resulting in 92%, 89%, 88%, and 94% inhibition, respectively, at 500 μg/mL concentration of CsEOs. Furthermore, in vitro studies on human A549 lung carcinoma cells revealed their both antimicrobial and anticancer properties with their potential usage in biomedical applications. *Cinnamomum verum* is widely used as a food spice in India, and its essential oil has shown promising antibacterial effects against resistant strains. Investigations on the formulations of *Cinnamomum verum* oil-loaded solid lipid nanoparticles (CO-SLN) as nanocarriers (average size of 337.6 nm, and zeta potential of −26.6 mV) against drug-resistant *E. coli* resulted in the noteworthy bacterial inhibition at 60–75 μg/mL MIC values (Nemattalab et al. 2022). To increase the efficacy and stability of various essential oils in water, nanoemulsions have been synthesized using crosslinkers; such as Tea tree oil (TTO)-in-water host–guest crosslinked nanoemulsions (HGCTNs) showed a MIC of 12.5 v/v% (0.13 μL/mL TTO) with slow release effect of TTO revealing their sustainable antibacterial ability against MDR bacteria (Jiang et al. 2023). Some of the other examples of essential oil-loaded nanomaterials against MDR strains are listed in Table 2.3.

Table 2.3 List of different essential oil-loaded nanomaterials against drug-resistant pathogens

Sr. no.	Nanomaterial	Essential oil parameters			Nanoparticle parameters				Target microorganism	Minimum inhibitory concentration (MIC)	Minimum bactericidal concentration (MBC)	Bacterial inhibition (%)	References
		Plant source	Extraction method	Components	Preparation method	Size	Polydispersity index (PDI)	Zeta potential (Z-PT)					
1.	Eucalyptus essential oil nanoemulsions (EEO NEs)	Eucalyptus	–	–	Low-energy emulsification	15.34 ± 3.77 nm	0.173 ± 0.058	–	*Staphylococcus aureus*	15 mg/mL	25 mg/mL	77.530 ± 7.292%	Cai et al. (2023)
2.	*Ferulago angulate* essential oil chitosan nanoparticles (EO CSNPs)	Fennel	Steam-distillation	α-Pinene (21.85%), β-ocimene (19.37%), bornyl acetate (10.50%), and thymol (6.80%)	Ionic-gelation	175 to 350 nm	0.184–0.32	+43.5 to +19.2 mV	*Staphylococcus aureus* and *Escherichia coli*	0.45 and 2.12 mg/mL	–	–	Kaboudi et al. (2023)
3.	Alkaline tea tree oil nanoemulsion	Tea tree	–	α-Terpineol (2.96%), 1,3-cineole (4.125%), α-terpinene (11.91%), γ-terpinene (23.58%), terpinen-4-ol (35.93%)	Spontaneous-emulsification	293.2 nm	–	−15.0 mV	*Acinetobacter baumannii*	–	6.50 mg/mL	–	Zhang et al. 2023)
4.	Fennel essential oil encapsulated polylactic acid and glycolic acid (PLGA) nanoparticles (FEO–PLGANPs)	*Alhagi maurorum*	–	–	Solvent emulsion diffusion	60.21 ± 3.0105 nm	0.024 ± 0.003	23 ± 1.15 mV	*Staphylococcus aureus*	3.00 μg/mL	–	Log reduction for three times MIC value	Alam et al. (2022)
5.	Mexican oregano encapsulated polyhydroxybutyrate (PHB) and poly-3-hydroxybutyrate-co-hydroxyhexanoate (PHB-HHx) nanoparticles (EO loaded PHB-HHx-NPs)	*Lippia graveolens* (Mexican oregano)	Hydro distillation	Thymol (>50%), Carvacrol (<0.1%)	Microemulsification	150–210 nm	0.29 ± 0.04	−40.1 ± 0.9 mV	*Micrococcus luteus*	0.25 mg/mL	0.4 mg/mL	–	Corrado et al. (2022)
6.	Rosemary essential oil-loaded multiple lipid nanoparticles	Cinnamon	–	1,8-Cineole (23.8%), α-pinene (16.2%), camphor (15.3%), and camphene (9.8%)	Ultrasonication	~110 nm	0.843 ± 0.002	−58.1 ± 0.7 mV	*Pseudomonas aeruginosa*	0.6 mg/mL	–	–	Ben-Khalifa et al. (2021)
7.	*Thymus capitatus* essential oil-loaded chitosan nanoparticles (Th-CNPs) and *Origanum vulgare* essential oil-loaded chitosan nanoparticles (Or-CNPs)	*Thymus capitatus* and *Origanum vulgare*	Hydrodistillation	Thymol (more than 43%), p-cymene (14.4%), and γ-terpinene (15.4%) in Or-EO and Carvacrol (73.0%), thymol (0.3%), p-cymene (6.9%) γ-terpinene (2.5%)	Ionic gelation	86 nm and 449 nm for Th-NCPs, 62 nm and 407 nm for or-NCPs	0.42 for Th-NCPs and 0.53 for or-NCPs	> 40 mV	*Staphylococcus aureus*, *Listeria monocytogenes*, and *Escherichia coli*	0.06, 0.03, and 0.12 mg/mL for Th-CNPs, 0.03, 0.03, and 0.06 mg/mL for Or-CNPs	Same as MIC	–	Granata et al. (2021)

(continued)

Table 2.3 (continued)

Sr. no.	Nanomaterial	Essential oil parameters: Plant source	Extraction method	Components	Nanoparticle parameters: Preparation method	Size	Polydispersity index (PDI)	Zeta potential (Z-PT)	Target microorganism	Minimum inhibitory concentration (MIC)	Minimum bactericidal concentration (MBC)	Bacterial inhibition (%)	References
8.	*Alhagi maurorum* essential oil-loaded chitosan nanoemulsion	Rosemary	Dry steam distillation	2-Nonadecanone, octadecane, propanamide, trichloroacetic acid, tetramethyl-2-hexadecen-1-ol, squalene, and octacosane	Ionotropic gelation	172 ± 4 nm	–	+28.6 mv	*Bacillus cereus*, *Pseudomonas aeruginosa*, *Acinetobacter baumannii*, *Escherichia coli*, *Klebsiella pneumonia*, *Staphylococcus aureus*	1.75, 3.25, 12.5, 1.75, 6.25, 6.25 mg/mL respectively	6.25, 12.5, 25, 3.25, 25, 25 mg/mL respectively	82%	Hassanshahian et al. 2020)
9.	Zein-pectin composite nanoparticles stabilized cinnamon essential oil nanoemulsion (ZCCPEs)	*Origanum vulgare*	–	Camphene (4.426 min), benzaldehyde (4.578 min), 4-isopropyltoluene (5.476 min), limonene (5.550 min), cineole (5.606 min), salicylaldehyde (5.791 min), nonanal (6.639 min), borneol (7.737 min), phenethyl acetate (8.875 min), and cinnamaldehyde (9.214 min)	Ultrasound anti-solvent precipitation	660.8 ± 8.1 nm	0.271 ± 0.004	31.23 ± 0.70 mV	*Alternaria alternata* and *Botrytis cinerea*	0.14 lL/mL and 0.08 lL/mL, respectively	–	Nearly 100%	Jiang et al. (2020)

2.6 Conclusion

Plant-based EOs have received comparatively less attention in the context of antimicrobial medication development. However, they have recently picked the interest of drug development scientists because of their pharmacological uniqueness, abundance, variety, and bioactivity. EOs from plant families such as Asteraceae, Lamiaceae, and Myrataceae are known to have antimicrobial activity with their potential use in medicine and combating antimicrobial resistance. Some of the commercially available EOs extracted from *Allium sativum*, *Artemisia dracunculus*, *Cananga odorata*, and *Guaiacum officinale* are applied for treating acne, inflammations, and skin infections. These EOs possessing potent antimicrobial activity are found to target various drug-resistant determinants of pathogens such as biofilm formation, efflux pumps, cell membrane permeability, and quorum sensing. However, increasing AMR conditions and drying pipeline of novel antibiotics necessitate advanced approaches and technologies. Henceforth, EOs-loaded nanodelivery with increased efficacy and potentiation effect has boomed the medical community efforts for combating antimicrobial resistance in hospital and community settings. Even though scientific data show several EOs with beneficial antimicrobial potency and efficacy for various infection treatments, more study is still required to effectively explore their medicinal properties.

Acknowledgments The authors acknowledge the funding support under DBT-BUILDER (BT/INF/22/SP45363/2022) from the Department of Biotechnology (DBT), Government of India, and the DST-FIST (SR/FST/COLLEGE-/19/568) from the Department of Science and Technology (DST), Government of India implemented at Modern College, Ganeshkhind, Pune, India.

References

Ahmad M, Khan AU (2019) Global economic impact of antibiotic resistance: a review. J Glob Antimicrob Resis 19:313–316. https://doi.org/10.1016/j.jgar.2019.05.024

Ait Dra L, Ait Sidi Brahim M, Boualy B, Aghraz A, Barakate M, Oubaassine S, Markouk M, Larhsini M (2017) Chemical composition, antioxidant and evidence antimicrobial synergistic effects of Periploca laevigata essential oil with conventional antibiotics. Ind Crop Prod 109 (September):746–752. https://doi.org/10.1016/j.indcrop.2017.09.028

Alam A, Foudah AI, Salkini MA, Raish M, Sawale J (2022) Herbal fennel essential oil nanogel: formulation, characterization and antibacterial activity against Staphylococcus aureus. Gels 8(11):736. https://doi.org/10.3390/gels8110736

Aleksic V, Mimica-Dukic N, Simin N, Nedeljkovic NS, Knezevic P (2014) Synergistic effect of Myrtus communis L. essential oils and conventional antibiotics against multi-drug resistant Acinetobacter baumannii wound isolates. Phytomedicine 21(12):1666–1674. https://doi.org/10.1016/j.phymed.2014.08.013

Amaral L, Martins A, Spengler G, Molnar J (2014) Efflux pumps of Gram-negative bacteria : what they do, how they do it, with what and how to deal with them. Front Pharmocol 4 (January):1–11. https://doi.org/10.3389/fphar.2013.00168

Amaral SC, Pruski BB, de Freitas SB, Allend SO, Ferreira MRA, Moreira C, Pereira DIB, Junior ASV, Hartwig DD (2020) Origanum vulgare essential oil: antibacterial activities and synergistic effect with polymyxin B against multidrug-resistant Acinetobacter baumannii. Mol Biol Rep 0123456789:9615. https://doi.org/10.1007/s11033-020-05989-0

Asokan G, Ramadhan T, Ahmed E, Sanad H (2019) WHO global priority pathogens list: a bibliometric analysis of medline-pubmed for knowledge mobilization to infection prevention and control practices in Bahrain. Oman Med J 34(3):184–193. https://doi.org/10.5001/omj.2019.37

Aumeeruddy-Elalfi Z, Gurib-Fakim A, Mahomoodally MF (2016) Chemical composition, antimicrobial and antibiotic potentiating activity of essential oils from 10 tropical medicinal plants from Mauritius. J Herb Med 6(2):88–95. https://doi.org/10.1016/j.hermed.2016.02.002

Balázs VL, Filep R, Répás F, Kerekes E, Szabó P, Kocsis B, Böszörményi A, Krisch J, Horváth G (2022) Immortelle (Helichrysum italicum (Roth) G. Don) essential oil showed antibacterial and biofilm inhibitory activity against respiratory tract pathogens. Molecules 27(17):5518. https://doi.org/10.3390/molecules27175518

Bedoya LM, Bermejo P (2013) Essential oils from the Asteraceae family active against multidrug-resistant bacteria. Academic, Cambridge, MA, pp 205–221. https://doi.org/10.1016/B978-0-12-398539-2.00014-8

Ben-Khalifa R, Gaspar FB, Pereira C, Chekir-Ghedira L, Rodríguez-Rojo S (2021) Essential oil and hydrophilic antibiotic co-encapsulation in multiple lipid nanoparticles: proof of concept and in vitro activity against Pseudomonas aeruginosa. Antibiotics 10(11):1300. https://doi.org/10.3390/antibiotics10111300

Cai K, Liu Y, Yue Y, Liu Y, Guo F (2023) Essential oil Nanoemulsion hydrogel with anti-biofilm activity for the treatment of infected wounds. Polymers 15(6):1376. https://doi.org/10.3390/polym15061376

Chaves TP, Pinheiro REE, Melo ES, Soares MJ dos S, Souza JSN, Andrade TB de, Lemos TLG de, Coutinho HDM (2018) Essential oil of Eucalyptus camaldulensis Dehn potentiates β-lactam activity against Staphylococcus aureus and Escherichia coli resistant strains. Ind Crop Prod 112:70–74. https://doi.org/10.1016/j.indcrop.2017.10.048

Chimnoi N, Reuk-ngam N, Chuysinuan P, Khlaychan P, Khunnawutmanotham N, Chokchaichamnankit D, Thamniyom W, Klayraung S, Mahidol C, Techasakul S (2018) Characterization of essential oil from Ocimum gratissimum leaves: antibacterial and mode of action against selected gastroenteritis pathogens. Microb Pathog 118:290. https://doi.org/10.1016/j.micpath.2018.03.041

Corrado I, Di Girolamo R, Regalado-González C, Pezzella C (2022) Polyhydroxyalkanoates-based nanoparticles as essential oil carriers. Polymers 14(1):166. https://doi.org/10.3390/polym14010166

Costa K, Mac A, Oliveira D, Bianca I, Karine E, Alves DO, Paula A, Anna S, Vera L, Lima DM, Tereza M (2022) Antibacterial mechanism of Eugenia stipitata McVaugh essential oil and synergistic effect against Staphylococcus aureus. S Afr J Bot 147:724. https://doi.org/10.1016/j.sajb.2022.03.012

Costa WK, Ohana N, Souza B, Macêdo C, Filho B, De Oliveira AM, Cláudia G, De Veras BO, Granja F, César J, De Oliveira R, De Aguiar F, Maria D, Navarro AF, Tereza M, Vanusa M (2020) First report on the chemical composition of leaf essential oil of Myrciaria pilosa Sobral & Couto and its antimicrobial and antivirulence activities against Staphylococcus aureus antimicrobial and antivirulence activities against. Nat Prod Res 36:1–5. https://doi.org/10.1080/14786419.2020.1837805

de Araújo ACJ, Freitas PR, dos Santos Barbosa CR, Muniz DF, de Almeida RS, Alencar de Menezes IR, Ribeiro-Filho J, Tintino SR, Coutinho HDM (2021) In vitro and in silico inhibition of Staphylococcus aureus efflux pump NorA by α-Pinene and limonene. Curr Microbiol 78(9):3388–3393. https://doi.org/10.1007/s00284-021-02611-9

de Negreiros MO, Pawlowski Â, Zini CA, Soares GLG, de Motta AS, Frazzon APG (2016) Antimicrobial and antibiofilm activity of Baccharis psiadioides essential oil against antibiotic-resistant Enterococcus faecalis strains. Pharm Biol 54(12):3272–3279. https://doi.org/10.1080/13880209.2016.1223700

De Oliveira I, Aparecida G, Soares M, Ramos V, Ribeiro R, Diaz-muñoz G, Alves M, Diaz N (2020) Archives of Oral biology in vitro anticariogenic and antibio fi lm activities of toothpastes formulated with essential oils. Arch Oral biol 117:104834. https://doi.org/10.1016/j.archoralbio.2020.104834

de Paulo Farias D, Neri-Numa IA, de Araújo FF, Pastore GM (2020) A critical review of some fruit trees from the Myrtaceae family as promising sources for food applications with functional claims. Food Chem 306:125630. https://doi.org/10.1016/j.foodchem.2019.125630

Department of Biotechnology (2021) Indian Priority Pathogen List, pp 1–22. https://dbtindia.gov.in/sites/default/files/IPPL_final.pdf

Dupuis V, Cerbu C, Witkowski L, Potarniche A-V, Timar MC, Żychska M, Sabliov CM (2022) Nanodelivery of essential oils as efficient tools against antimicrobial resistance: a review of the type and physical-chemical properties of the delivery systems and applications. Drug Deliv 29(1):1007–1024. https://doi.org/10.1080/10717544.2022.2056663

EClinicalMedicine (2021) Antimicrobial resistance : a top ten global public health threat. EClinicalMedicine 41:101221. https://doi.org/10.1016/j.eclinm.2021.101221

El Amrani S, Sanae L, Ezzoubi Y, Evrendilek GA, Mouhcine F, Hicham K, Rabia B, Abdelhakim EOL (2022) Combined antibacterial effect of Origanum compactum and Mentha piperita (Lamiaceae) essential oils against ATCC Escherichia coli and Staphylococcus aureus. Vegetos 35(1):74–82. https://doi.org/10.1007/s42535-021-00276-0

Elhidar N, Soulaimani B, Abbad A, Hassani L, Mezrioui N (2021) chemical composition, antibacterial activity and effect of Rhus albida Schousb essential oil on the inhibition of NorA ef fl ux pump in Staphylococcus aureus. S Afr J Bot 142:19. https://doi.org/10.1016/j.sajb.2021.05.025

Fastenberg JH, Hsueh WD, Mustafa A, Akbar NA, Abuzeid WM (2016) Biofilms in chronic rhinosinusitis: pathophysiology and therapeutic strategies. World J Otorhinolaryngol Head Neck Surg 2(4):219–229. https://doi.org/10.1016/j.wjorl.2016.03.002

Gadisa E, Weldearegay G, Desta K, Tsegaye G, Hailu S, Jote K, Takele A (2019) Combined antibacterial effect of essential oils from three most commonly used Ethiopian traditional medicinal plants on multidrug resistant bacteria. BMC Complement Altern Med 19(1):1–9. https://doi.org/10.1186/s12906-019-2429-4

Ghavam M, Bacchetta G, Castangia I, Manca ML (2022) Evaluation of the composition and antimicrobial activities of essential oils from four species of Lamiaceae Martinov native to Iran. Sci Rep 12(1):1–12. https://doi.org/10.1038/s41598-022-21509-5

Granata G, Stracquadanio S, Leonardi M, Napoli E, Malandrino G, Cafiso V, Stefani S, Geraci C (2021) Oregano and thyme essential oils encapsulated in chitosan nanoparticles as effective antimicrobial agents against foodborne pathogens. Molecules 26(13):4055. https://doi.org/10.3390/molecules26134055

Guidotti-Takeuchi M, de Morais Ribeiro LN, Dos Santos FAL, Rossi DA, Lucia FD, de Melo RT (2022) Essential oil-based nanoparticles as antimicrobial agents in the food industry. Microorganisms 10(8):1504. https://doi.org/10.3390/microorganisms10081504

Hassanshahian M, Saadatfar A, Masoumipour F (2020) Formulation and characterization of nanoemulsion from Alhagi maurorum essential oil and study of its antimicrobial, antibiofilm, and plasmid curing activity against antibiotic-resistant pathogenic bacteria. J Environ Health Sci Eng 18(2):1015–1027. https://doi.org/10.1007/s40201-020-00523-7

He X, Lu F, Yuan F, Jiang D, Zhao P, Zhu J, Cheng H, Cao J, Lu G (2015) Biofilm formation caused by clinical acinetobacter baumannii isolates is associated with overexpression of the AdeFGH efflux pump. Antimic Agents Chemother 59(8):4817–4825. https://doi.org/10.1128/AAC.00877-15

Hussain AI, Anwar F, Nigam PS, Sarker SD, Moore JE, Rao JR, Mazumdar A (2011) Antibacterial activity of some Lamiaceae essential oils using resazurin as an indicator of cell growth. Lwt 44(4):1199–1206. https://doi.org/10.1016/j.lwt.2010.10.005

Jiang JX, Li QY, Zhu DY, Xiao W, Chen ZP, Lan MH, Gan C, Duan X, Yang Y (2023) Host–guest assembly-crosslinked nanoemulsions with dual-antimicrobial mechanism for treating multidrug-resistant bacterial biofilms. ACS Appl Mate Interfaces 15(22):27046–27055. https://doi.org/10.1021/acsami.3c01760

Jiang Y, Wang D, Li F, Li D, Huang Q (2020) Cinnamon essential oil Pickering emulsion stabilized by zein-pectin composite nanoparticles: characterization, antimicrobial effect and advantages in storage application. Int J Biol Macromol 148:1280–1289. https://doi.org/10.1016/j.ijbiomac.2019.10.103

Jovanovi KK, Sokovi M (2014) Chemical composition, antimicrobial, and cytotoxic properties of five Lamiaceae essential oils. Indust Crop Product 61:225–232. https://doi.org/10.1016/j.indcrop.2014.07.011

Kaboudi Z, Peighambardoust SH, Nourbakhsh H, Soltanzadeh M (2023) Nanoencapsulation of Chavir (Ferulago angulata) essential oil in chitosan carrier: investigating physicochemical, morphological, thermal, antimicrobial and release profile of obtained nanoparticles. Int J Biol Macromol 237:123963. https://doi.org/10.1016/j.ijbiomac.2023.123963

Kot B, Wierzchowska K, Piechota M, Czerniewicz P, Chrzanowski G (2018) Antimicrobial activity of five essential oils from lamiaceae against multidrug-resistant Staphylococcus aureus. Nat Prod Res 6419:1–5. https://doi.org/10.1080/14786419.2018.1486314

Kuete V (2017) Medicinal spices and vegetables from Africa: therapeutic potential against metabolic, inflammatory, infectious and systemic diseases. In: Medicinal spices and vegetables from Africa: therapeutic potential against metabolic, inflammatory, infectious and systemic diseases. Elsevier, Amsterdam. https://doi.org/10.1016/B978-0-12-809286-6/00008-X

Laghmouchi Y, Belmehdi O, Senhaji NS, Abrini J (2018) Chemical composition and antibacterial activity of Origanum compactum Benth. Essential oils from different areas at northern Morocco. S Afr J Bot 115:120–125. https://doi.org/10.1016/j.sajb.2018.02.002

Limaverde PW, Campina FF, da Cunha FAB, Crispim FD, Figueredo FG, Lima LF, de M. Oliveira-Tintino CD, de Matos YMLS, Morais-Braga MFB, Menezes IRA, Balbino VQ, Coutinho HDM, Siqueira-Júnior JP, Almeida JRGS, Tintino SR (2017) Inhibition of the TetK efflux-pump by the essential oil of Chenopodium ambrosioides L. and α-terpinene against Staphylococcus aureus IS-58. Food Chem Toxicol 109:957–961. https://doi.org/10.1016/j.fct.2017.02.031

Liu T, Lin P, Bao T, Ding Y, Lha Q, Nan P, Huang Y, Gu Z, Zhong Y (2018) Essential oil composition and antimicrobial activity of Artemisia dracunculus L. var. qinghaiensis Y. R. Ling (Asteraceae) from Qinghai-Tibet plateau. Indust Crop Prod 125:1–4. https://doi.org/10.1016/j.indcrop.2018.08.085

Macêdo C, Filho B, Cláudio L, Vanusa M, Løbner-olesen A, Struve C, Krogfelt KA, Tereza M, Luiza M, Oliva V (2020) Antimicrobial and Antivirulence action of Eugenia brejoensis essential oil in vitro and in vivo invertebrate models. Front Microbial 11:1–11. https://doi.org/10.3389/fmicb.2020.00424

Mathlouthi A, Saadaoui N, Pennacchietti E, De Biase D, Ben-Attia M (2021) Essential oils from Artemisia species inhibit biofilm formation and the virulence of Escherichia coli EPEC 2348/69. Biofouling 37(2):174–183. https://doi.org/10.1080/08927014.2021.1886278

Michel J, Zahirah N, Rani A, Husain K (2020) A review on the potential use of medicinal plants from Asteraceae and Lamiaceae plant family in cardiovascular diseases. Front Pharmacol 11:1–26. https://doi.org/10.3389/fphar.2020.00852

Miladi H, Zmantar T, Kouidhi B, Chaabouni Y, Mahdouani K, Bakhrouf A, Chaieb K (2017a) Use of carvacrol, thymol, and eugenol for biofilm eradication and resistance modifying susceptibility of Salmonella enterica serovar typhimurium strains to nalidixic acid. Microb Pathog 104:56–63. https://doi.org/10.1016/j.micpath.2017.01.012

Miladi H, Zmantar T, Kouidhi B, Mohammed Y, Qurashi A, Al Bakhrouf A, Chaabouni Y, Mahdouani K, Chaieb K (2017b) Synergistic effect of eugenol, carvacrol, thymol, p-cymene and γ-terpinene on inhibition of drug resistance and biofilm formation of oral bacteria. Microb Pathog 112:156. https://doi.org/10.1016/j.micpath.2017.09.057

Mohammadi S, Leila P, Zarehshuran K, Babaeekhou L, Ghane M (2021) Antibacterial, anti - biofilm and anti - quorum sensing activities of Artemisia dracunculus essential oil (EO): a study against Salmonella enterica serovar typhimurium and Staphylococcus aureus. Arch Microbiol 203(4):1529–1537. https://doi.org/10.1007/s00203-020-02138-w

Mohammed JJ, Al Zuhairi J, Jookar F, Rahimi-moghaddam A (2020) Antioxidant, cytotoxic and antibacterial activity of Rosmarinus officinalis L. essential oil against bacteria isolated from urinary tract infection. Eur J Integr Med 38(August):101192. https://doi.org/10.1016/j.eujim.2020.101192

Moumni S, Elaissi A, Trabelsi A, Merghni A, Chraief I, Jelassi B, Chemli R, Ferchichi S (2020) Correlation between chemical composition and antibacterial activity of some lamiaceae species essential oils from Tunisia. BMC Complement Med Ther 20(1):1–15. https://doi.org/10.1186/s12906-020-02888-6

Myszka K, Schmidt MT, Majcher M, Juzwa W, Olkowicz M, Czaczyk K (2016) Inhibition of quorum sensing-related biofilm of Pseudomonas fluorescens KM121 by Thymus vulgare essential oil and its major bioactive compounds. Int Biodeterior Biodegradation 114:252–259. https://doi.org/10.1016/j.ibiod.2016.07.006

Nemattalab M, Rohani M, Evazalipour M, Hesari Z (2022) Formulation of cinnamon (Cinnamomum verum) oil loaded solid lipid nanoparticles and evaluation of its antibacterial activity against multi-drug resistant Escherichia coli. BMC Complement Med Ther 22(1):289. https://doi.org/10.1186/s12906-022-03775-y

Pages JM, Lavigne JP, Leflon-Guibout V, Marcon E, Bert F, Noussair L, Nicolas-Chanoine MH (2009) Efflux pump, the masked side of ß-lactam resistance in Klebsiella pneumoniae clinical isolates. PLoS One *4*(3):1. https://doi.org/10.1371/journal.pone.0004817

Perricone M, Arace E, Corbo MR, Sinigaglia M, Bevilacqua A (2015) Bioactivity of essential oils: a review on their interaction with food components. Front Microbiol 6:1–7. https://doi.org/10.3389/fmicb.2015.00076

Poli J-P, Guinoiseau E, de Rocca Serra D, Sutour S, Paoli M, Tomi F, Quilichini Y, Berti L, Lorenzi V (2018) Anti-quorum sensing activity of 12 essential oils on chromobacterium violaceum and specific action of cis-cis-p-Menthenolide from Corsican Mentha suaveolens ssp Insularis. Molecules 23(9):2125. https://doi.org/10.3390/molecules23092125

Pumbwe L, Skilbeck CA, Wexler HM (2008) Presence of quorum-sensing systems associated with multidrug resistance and biofilm formation in Bacteroides fragilis. Microbiol Ecol 56:412–419. https://doi.org/10.1007/s00248-007-9358-3

Qaralleh H (2019) Thymol rich Thymbra capitata essential oil inhibits quorum sensing, virulence and biofilm formation of beta lactamase producing Pseudomonas aeruginosa. Nat Product Sci 25(2):172–180

Rajivgandhi G, Kadaikunnan S, Ramachandran G, Chackaravarthi G, Kanisha Chelliah C, Maruthupandy M, Natesan M, Quero F, Li W-J (2023) Chitosan loaded plant essential oils efficiently eradicate the multi-drug resistant bacterial infection and lung cancer cells. J King Saud Univ Science 35(5):102662. https://doi.org/10.1016/j.jksus.2023.102662

Ramsey JT, Shropshire BC, Nagy TR, Chambers KD, Li Y, Korach KS (2020) Essential oils and health. Yale J Biol Med 93(2):291–305

Reddy S, Kaur K, Barathe P, Shriram V, Govarthanan M, Kumar V (2022) Antimicrobial resistance in urban river ecosystems. Microbi Res 263:127135. https://doi.org/10.1016/j.micres.2022.127135

Roy R, Tiwari M, Donelli G, Tiwari V (2018) Strategies for combating bacterial biofilms: a focus on anti-biofilm agents and their mechanisms of action. Virulence *9*(1):522–554. https://doi.org/10.1080/21505594.2017.1313372

Santos EL, Freitas PR, Araújo ACJ, Almeida RS, Tintino SR, Paulo CLR, Silva ACA, Silva LE, do Amaral W, Deschamps C, Junior JPS, Filho JMB, de Sousa GR, Ribeiro-Filho J, Coutinho HDM (2021) Enhanced antibacterial effect of antibiotics by the essential oil of Aloysia gratissima (Gillies & Hook.) Tronc. and its major constituent beta-caryophyllene. Phytomed Plus 1(4):1. https://doi.org/10.1016/j.phyplu.2021.100100

Sharifi A, Abdolmajid Mohammadzadeh TZS, Pezhman, Mahmoodi (2018) Antibacterial, antibiofilm and antiquorum sensing effects of Thymus daenensis and Satureja hortensis essential oils against Staphylococcus aureus isolates. J Appl Microbiol 124:379. https://doi.org/10.1111/jam.13639

Sharifi A, Mohammadzadeh A, Salehi TZ (2021) Cuminum cyminum L. essential oil : a promising antibacterial and Antivirulence agent against multidrug-resistant Staphylococcus aureus. Front Microbial 12:1–9. https://doi.org/10.3389/fmicb.2021.667833

Shriram V, Khare T, Bhagwat R, Shukla R, Kumar V (2018) Inhibiting bacterial drug efflux pumps via phyto-therapeutics to combat threatening antimicrobial resistance. Front Microbiol 9:1–18. https://doi.org/10.3389/fmicb.2018.02990

Simpson MG (2019) Diversity and classification of flowering plants: eudicots. In: Plant systematics, vol 366. Columbia University Press, New York, NY. https://doi.org/10.1016/b978-0-12-812628-8.50008-0

Snoussi M, Dehmani A, Noumi E, Flamini G, Papetti A (2016) Chemical composition and antibiofilm activity of Petroselinum crispum and Ocimum basilicum essential oils against vibrio spp. strains. Microb Pathog 90:13–21. https://doi.org/10.1016/j.micpath.2015.11.004

Srivastava A, Kumar V, Sharma D (2023) Data regarding anti-quorum sensing and antimicrobial activity of Melaleuca alternifolia and Salvia sclarea essential oil against Pseudomonas aeruginosa. Data Brief 48:109145. https://doi.org/10.1016/j.dib.2023.109145

Sun J, Deng Z, Yan A (2014) Biochemical and biophysical research communications bacterial multidrug efflux pumps : mechanisms, physiology and pharmacological exploitations. Biochem Biophy Res Commun 453(2):254–267. https://doi.org/10.1016/j.bbrc.2014.05.090

Swamy MK, Akhtar MS, Sinniah UR (2016) Antimicrobial properties of plant essential oils against human pathogens and their mode of action: an updated review. Evid Based Complement Altern Med 2016:1–21. https://doi.org/10.1155/2016/3012462

The Global Risks Report 2018 – Reports—World Economic Forum. (2018). Accessed 11 Feb 2020. httpsreports.weforum/global-risks-2018/anti-microbial-resistance/

Trifan A, Brebu M, Miron A, Aprotosoaie AC (2018) Antibacterial activity of traditional spices against lower respiratory tract pathogens : combinatorial effects of Trachyspermum ammi essential oil with conventional antibiotics. Appl Microbial Int 67(5):449. https://doi.org/10.1111/lam.13069

Vieira M (2017) Chemical composition, antibacterial, antibiofilm and synergistic properties of essential oils from eucalyptus globulus labill. and seven mediterranean aromatic plants. Chem Biodivers 14:1

Vitanza L, Maccelli A, Marazzato M, Scazzocchio F, Comanducci A, Fornarini S, Crestoni ME, Filippi A, Fraschetti C, Rinaldi F, Aleandri M, Goldoni P, Pia M, Ammendolia MG, Longhi C (2018) Satureja montana L. essential oil and its antimicrobial activity alone or in combination with gentamicin. Microbial Pathog 126:323. https://doi.org/10.1016/j.micpath.2018.11.025

Wang J, Jiao H, Meng J, Qiao M, Du H, He M, Ming K, Liu J, Wang D, Wu Y (2019) Baicalin inhibits biofilm formation and the quorum-sensing system by regulating the MsrA drug efflux pump in staphylococcus saprophyticus. Front Microbi 10:1–11. https://doi.org/10.3389/fmicb.2019.02800

Wenqian Yuan H-GY (2019) Effects of sublethal thymol, carvacrol, and trans- cinnamaldehyde adaptation on virulence properties of Escherichia coli O157:H7. Appl Environ Microbiol 85(14):1–11

Yu Z, Tang J, Khare T, Kumar V (2020) The alarming antimicrobial resistance in ESKAPEE pathogens: can essential oils come to the rescue? Fitoterapia 140:104433. https://doi.org/10.1016/j.fitote.2019.104433

Zhang H, Liu Y, Yuan B, Wang W, Jin Y, Wei X (2023) Alkaline tea tree oil nanoemulsion nebulizers for the treatment of pneumonia induced by drug-resistant Acinetobacter baumannii. J Drug Del Sci Technol 79:104047. https://doi.org/10.1016/j.jddst.2022.104047

Zhang F, Ramachandran G, Mothana RA, Noman OM, Alobaid WA, Rajivgandhi G, Manoharan N (2020) Anti-bacterial activity of chitosan loaded plant essential oil against multi drug resistant K. pneumoniae. Saudi J Biol Sci *27*(12):3449–3455. https://doi.org/10.1016/j.sjbs.2020.09.025

Zhang Y, Wang Y, Zhu X, Cao P, Wei S, Lu Y (2017) Antibacterial and antibiofilm activities of eugenol from essential oil of Syzygium aromaticum (L.) Merr. & L. M. Perry (clove) leaf against periodontal pathogen Porphyromonas gingivalis. Microb Pathog 113:396. https://doi.org/10.1016/j.micpath.2017.10.054

Zhao F, Chen YP, Salmaki Y, Drew BT, Wilson TC, Scheen AC, Celep F, Bräuchler C, Bendiksby M, Wang Q, Min DZ, Peng H, Olmstead RG, Li B, Xiang CL (2021) An updated tribal classification of Lamiaceae based on plastome phylogenomics. BMC Biol 19(1):1–27. https://doi.org/10.1186/s12915-020-00931-z

Chapter 3
Phytochemicals as Modulators of Toll-Like Receptors: An Immunopharmacological Perspective

Pritha Chakraborty, Moytrey Chatterjee, Ankita Chakraborty, Somrita Padma, and Suprabhat Mukherjee

Abstract Toll-like receptors (TLRs) are molecular receptors that can "sense" potential foreign antigens to protect the host. Crosstalk between exogenous and endogenous threats with intercellular molecular receptors plays a pivotal role in maintaining the immunological functions of the host. Due to the increased risk of chronic illness and increasing rate of infectious diseases, the immune system has become a significant target of toxicity due to the application of chemotherapeutics. Pathogenic resistance and the adverse side effects of chemotherapeutics can be modulated using phytochemicals. Since ancient times, medicinal plants have been in use as therapeutics due to their immunomodulatory properties. They exhibited anti-inflammatory, antimicrobial, antifilarial, and anticancer actions to ameliorate the cytotoxic effects of chronic illness. They have the ability to interact with molecular receptors like Toll-like receptors (TLRs) and hitherto, studies have shown the use of phytochemicals as a modulator of TLR signaling to revert back immunopathological consequences rising from pathogen and endogenous threats. Herein, in this book chapter, we have discussed about phytochemicals as natural immunomodulators of Toll-like receptors.

Keywords Phytochemicals · Immunomodulation · Toll-like receptors · Inflammation

Abbreviations

BBR	Berberine
CRP	C-reactive protein
DAMP	Danger-associated molecular patterns

P. Chakraborty · M. Chatterjee · A. Chakraborty · S. Padma · S. Mukherjee (✉)
Integrative Biochemistry & Immunology Laboratory, Department of Animal Science, Kazi Nazrul University, Asansol, West Bengal, India
e-mail: suprabhat.mukherjee@knu.ac.in

V. Kumar et al. (eds.), *Medicinal Plants and Antimicrobial Therapies*,
https://doi.org/10.1007/978-981-99-7261-6_3

DSS	Dextran sulfate sodium
Dvl	Dioclea violacea
EA	Ellagic acid
EC	Epicatechin
ECG	Epicatechin-3-gallate
EGC	Epigallocatechin
EGCG	Epigallocatechin-3-gallate
GABA	Gamma-amino butyric acid
HSG	Human submandibular gland
IBD	Inflammatory bowel disease
IG	Iridoid glycosides
IKK-γ	IκB-kinase-γ
iNOS	Inducible nitric oxide synthase
IRAK4	IL-1 receptor-associated kinase 4
LRR	Leucine-rich repeats
MAM	Methylazoxymethanol
MEP	2-C-methyl-D-erythritol-4-phosphate
MVA	Mevalonate
MyD88	Myeloid differentiation primary response protein 88
NCI	National Cancer Institute
NO	Nitric oxide
OBB	Oxyberberine
PAMP	Pathogen-associated molecular patterns
PRR	Pathogen recognition receptors
ROS	Reactive oxygen species
TAB	TAK1/TGF-β-activated kinase
TIR	Toll/IL-1 receptor
TLR	Toll-like receptors
TNF-α	Tumor necrosis factor
TRAF6	TNF receptor-associated factor 6
TREM2	Triggering receptor expressed on myeloid cells 2
WA	Withaferin A

3.1 Introduction

Our immune system showcases two wings of the defense system, viz., innate immunity and adaptive immunity, and has been an integral part of our survival. It maintains the homeostasis between the physiological and immunological functions and the surrounding environment (Zheng et al. 2020). Among the two wings of our defense system, the innate immune system is the primary defense entity of the host to protect against invading pathogens and endogenous threats. Hence, it is thought to be an evolutionarily conserved and phylogenetically ancient arm of the immune system. There are pathogen recognition receptors (PRRs) that initiate innate cellular immune responses and the subsequent adaptive immune responses against infectious and inflammatory disorders. Mammalian Toll-like receptors (TLRs) are members of PRRs that recognize pathogen-associated molecular patterns (PAMPs) and danger-associated molecular patterns (DAMPs) that are not present in the host cells (Akira et al. 2006; Akira 2009). TLRs identify the PAMPs and issue an early warning against the foreign antigen (Chaudhary et al. 1998; Rock et al. 1998). This recognition leads to the activation of TLRs-induced signaling cascades, which induce the release of inflammatory cytokines, chemokines, antigen-presenting molecules, and costimulatory molecules (Vasselon and Detmers 2002; Kawasaki and Kawai 2014).

TLRs are first discovered in *Drosophila melanogaster*. It possesses a Toll-1 protein that specifies the embryonic dorsoventral polarity by directing the activation of the transcription factor NF-κB signaling pathway (Hashimoto et al. 1988; Anderson et al. 1985). *Drosophila* Toll-1 protein interacts with Spätzle to activate downstream TLR-dependent responses (Lindsay and Wasserman 2014). There are eight other TLRs identified in *Drosophila* which also activate NF-κB signaling to raise an immune response against fungi and Gram-positive bacteria (Valanne et al. 2011; Silverman et al. 2009; Norris and Manley 1992). Furthermore, ten TLRs have been identified in humans and 13 in mice that are involved in NF-κB-dependent innate immune signaling (Kawai and Akira 2007). Toll-like receptors (TLRs) are generally categorized as transmembrane proteins of type I, characterized by the presence of three distinctive structural domains: the leucine-rich repeats (LRRs) motif, a transmembrane domain, and the Toll/IL-1 receptor (TIR) domain (Ng and Xavier 2011). TLRs initiate immune signaling through two separate pathways, denoted as the myeloid differentiation primary response protein 88 (MyD88)-dependent pathway and the TIR domain-containing adaptor-inducing IFNβ (TRIF)-dependent pathway. The majority of the TLRs follow the MyD88-dependent pathway, with the exception of TLR3. This pathway involves the recognition of PAMPs or DAMPs, leading to TLR dimerization. Following dimerization, MyD88 associates with the TIR domain of the TLR, and this complex recruits IL-1 receptor-associated kinase 4 (IRAK4) subsequent to autophosphorylation of IL-1 receptor-associated kinase 1 (IRAK1) (Kawasaki and Kawai 2014). The assembled complex facilitates the initiation of (TNF) receptor-associated factor 6 (TRAF6) activation, subsequently inducing the activation of the TAK1/TGF-β-activated kinase (TAB) complex via K-63-linked polyubiquitination of both TAK1 and TRAF6 (El-Zayat et al. 2019). TAB complex mediates phosphorylation of IκB kinase and degrades I kappa B alpha (IκBα). IκBα inhibits the translocation of NF-κB. Hence, degradation of the inhibitor results in translocation of NF-κB into the nucleus for transcription of genes to release inflammatory cytokines resulting in a hallmark of innate immune response known as inflammation. Although inflammation is a result of a series of immune responses, the motive is to restore immune homeostasis in the host body following tissue injury or invasion of pathogens. Proinflammatory cytokines and chemokines result in vasodilation and accumulation of circulating leukocytes and raising the levels of leukocyte adhesion molecules on endothelial cells. However, chronic inflammation can result in dysregulation of cellular response, impaired immunological responses, cell senescence, and tissue dysfunction (Newton and Dixit 2012; Kennedy et al. 2014). Among all the TLRs, TLR4 is mainly involved in immunopathological conditions; hence, it has been considered a potential therapeutic target for drug development (Ain et al. 2020). It has overexpression in diseases; different disease models developed by knockout mice have shown disease resistance. TLR–ligand interaction has exacerbated inflammation in disease models establishing the correlation between TLR4 and the diseases (Hennessy et al. 2010). Studies on genetic polymorphism encoding different TLRs and their signaling molecules have suggested their association with human disease progression, although the compelling link to the diseases is not clear. Since activation of TLRs especially, TLR4, a cascade of events give rise to inflammation; hence, blocking might provide an advantage against the initiation of chronic inflammation (Sutherland and Cook 2005). TLRs can be involved in a positive feedback loop mechanism involving products released from inflamed tissues and the generation of inflammatory products which feed back on TLRs for their further activation (Hennessy et al. 2010).

To counteract inflammation, the anti-inflammatory effects of naturally occurring phytochemicals have been explored. Hitherto, phytochemicals have health benefits, and their use has increased extensively over the years as represented in Fig. 3.1. In this book chapter, we illustrate the signaling cascade of TLRs and their modulation using natural phytochemicals to revert back to immunopathological consequences rising from pathogen and endogenous threats. Therapeutics using different phytochemicals to break the feedback loop might be a strategy to alter the vicious cycle of inflammation.

3.2 Natural Derived Phytochemicals as Toll-like Receptor Modulators

Immune homeostasis is necessary for a healthy organism as various exogenous and endogenous factors influence the immune response and result in immune suppression and immunostimulation. The pathophysiological processes are normalized or modulated by several agents having biological activities also known as immunomodulators (Jantan et al. 2015). They are categorized as immunosuppressants, immunoadjuvants, and immunoadjuvants. Immunoadjuvants modulate the immune system by selecting between cellular and humoral Th1 helper and Th2 helper cells and thus are used to enhance the efficacy of vaccines (Billiau and Matthys 2001). Immunostimulants act non-specifically through innate and adaptive immune response and serve as prophylactic and promoter agents in a healthy individual. Alternatively, these biologically active molecules are used as therapeutics against infection-associated immunopathology, autoimmune disorders, and hypersensitivity reactions (Kumar et al. 2012). There are major challenges using chemically synthesized immunomodulators; hence, natural immunomodulators have the potential to

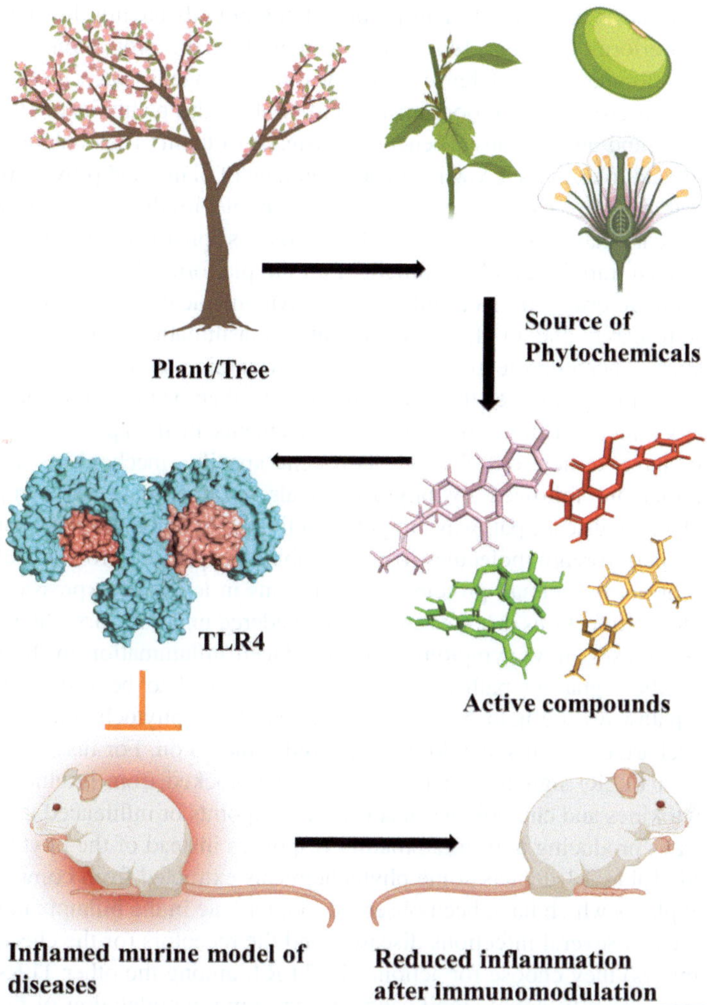

Fig. 3.1 Immunomodulation of TLR4 using phytochemicals: an illustration depicting extracted active compounds from different sources of plant/tree acting directly upon the receptors to block the cascade of inflammation due to immunomodulation of the receptor and/or the downstream signaling pathway

replace the traditionally applied therapeutic regimes. Extensive research has been conducted globally to attain biochemicals, or single compounds to target different diseases, but the success is not competent. Throughout history, the utilization of plant-based remedies for the prevention and management of diverse ailments has been documented. Consequently, exploring the potential of naturally derived immunomodulatory products and their extracts from plants could offer valuable perspectives in mitigating the adverse effects associated with chemotherapeutic interventions.

Phytochemicals are natural plant-produced compounds having distinctive structures and perform different functions for reproduction, protection, pollination, growth, development, and defense from invading pathogens. Such metabolites are produced by a group of microorganisms harbored by the plants inside themselves (endophytes) and are harmless (beneficial instead) to them. These phytochemicals play significant roles in the growth and development of plants and provide protection against factors such as extreme temperatures, ultraviolet irradiations, or even insects. Phytochemicals have been discovered to possess pharmacological properties. Hence, they are rapidly grabbing limelight on the platform of therapeutics, leaving behind synthetic drugs. In ancestral history, phytochemicals have been explored in medicinal research due to their antioxidant, antiasthmatic, antiarrhythmic, anti-inflammatory, hepatoprotective, hypocholesterolemic, antifungal, cardiotonic, diuretic, and other properties that are being used as therapeutic applications against inflammatory and infectious diseases. Advancements in therapeutics with phytochemicals have turned out fruitful. Different specific mechanisms have been observed that are followed by phytochemicals to ameliorate the symptoms of diseases. Human disease pathways might be activated by three different parameters: incorporation of foreign molecules (e.g., microbially expressed proteins), alterations occurring in normal human proteins, or alterations in levels of expression in functional proteins. Infectious diseases are to be considered under the first category. Most infectious diseases show symptoms due to induced inflammation in the body, for which specific signaling pathways have been observed to be responsible. Such signaling pathways are held back by a wide variety of phytochemical types such as polysaccharides, lectins, alkaloids, terpenoids, and so on. For instance, the Toll-like-receptor (TLR) signaling pathways might be blocked from producing inflammatory cytokines and chemokines at several checkpoints or influenced to move into the pathways producing anti-inflammatory responses instead of the phytochemicals concerned. Table 3.1 depicts some phytochemicals extracted from commonly used medicinal plants which have been observed to intervene in the inflammatory signaling crosstalk of several infectious diseases, and the receptors (or the checkpoints in their pathways) they choose for action. The TLR4, among the other TLRs, is to be considered as of substantial significance in the immunomodulation of diseases, as most of the phytochemicals are observed to intervene with the mechanistic pathways of TLR4, an inference that promises a bright future in therapeutics.

The administration of phytochemicals in humans showcases immunomodulatory, anti-inflammatory, and antioxidant activities that support the management of long-term diseases and improve the quality of life. Hitherto, more than 4000 phytochemicals have been identified with different immunomodulatory roles (Behl et al. 2021) which are classified into different classes as depicted in Fig. 3.2.

Table 3.1 Bioactive phytochemicals targeting human TLRs for intervening infectious diseases of human

Name of the plant	common name	Family	Plant type	Phytochemical	Target	References
Acanthopanax senticosus	Siberian ginseng	Araliaceae	Herb	Eleutheroside E	TLR2, TLR4	Han et al. (2003)
Aeginetia indica	Indian broomrape	Orobranchaceae	Herb	AILb-A	TLR4	Chahal et al. (2013), Okamoto et al. (2004)
Aloe barbadensis	Aloe vera	Asphodelaceae	Succulent plant	Emodin	TLR3, TLR4	Ding et al. (2018), Huang et al. (2021)
Withania somnifera	Ashwagandha	Solanaceae	Herb	Withaferin A	TLR 2, TLR 4	Noh et al. (2016)
Allium sativum	Garlic	Amaryllidaceae	Flowering plant	Allicin	TLR 4	Shen et al. (2019)
Andrographis paniculata	Kalmegh	Acanthaceae	Shrub	Andrographolide	TLR4, TLR7, TLR8	Vetvicka and Vannucci (2021)
Angelica sinensis	Female ginseng	Apiaceae	Herb	Polysaccharides	TLR4	Chahal et al. (2013)
Apium graveolens	Celery	Apiaceae	Herb	Quercetin	TLR 2, TLR 4	Yin et al. (2021), Yasui et al. (2015), Arabi et al. (2018)
Camellia sinensis	Green tea	Theaceae	Shrub	Catechin	TLR 2, TLR 4	Hasegawa et al. (2011), Lee et al. (2020)
Carthamus tinctorius	Safflower	Asteraceae	Herb	Daphnoretin	TLR4	Hansur et al. (2022)
Uncaria rhynchophylla	Cat's claw	Rubiaceae	Herb	Ursolic acid	TLR2, TLR4	Chahal et al. (2013), Jung et al. (2010)
Rhodiola imbricata	Sedum	Crassulaceae	Perennial plant	Salidroside	TLR 4	Chahal et al. (2013), Lu et al. (2021)
Chrysanthemum coronarium	Chrysanthemum	Asteraceae	Shrub	Chrysanthenyl acetate	TLR2, TLR4, TLR9	Wang et al. (2021a)
Cinnamomum sp.	Cinnamon	Lauraceae	Tree	Cinnamaldehyde	TLR 4	Wang et al. (2021b)
Curcuma longa	Turmeric	Zingiberaceae	Shrub	Curcumin	TLR2, TLR4, TLR9	

(continued)

Table 3.1 (continued)

Name of the plant	common name	Family	Plant type	Phytochemical	Target	References
						Panaro et al. (2020), Boozari et al. (2019)
Dioscorea alata	Purple yam	Dioscoreaceae	Tubers	Dioscorin	TLR4	Chahal et al. (2013), Fu et al. (2006)
Dioscorea batatas	Yam	Dioscoreaceae	Tubers	Allantoin	TLR4	Chahal et al. (2013), Jeong et al. (2020)
Echinacea purpurea	Purple coneflower	Asteraceae	Flowering plant	Polysaccharides	TLR4	Chahal et al. (2013), Zhang et al. (2020b)
Epimedium brevicornum	Horny goat weed	Berberidaceae	Perennial herb	Icariin	TLR9	Chahal et al. (2013), Li et al. (2011), Xue et al. (2016)
Euphorbia hirta	Garden spurge	Euphorbiaceae	Weed	Quercetin	TLR 2, TLR 4, TLR 7	Yin et al. (2021), Singh and Kumar (2013)
Ginkgo biloba	Maiden hair	Ginkgoaceae	Tree	Ginkgolides	TLR4	Wang and Zhang (2019)
Glycine max	Soybean	Leguminosae	Legumes	Carvacrol	TLR2, TLR4	Chahal et al. (2013), Ghahari et al. (2017)
Glycyrrhiza uralensis	Licorice	Leguminosae	Flowering plant	Glycyrrhetinic acid	TLR4	Chahal et al. (2013), Peng et al. (2011)
Lens culinaris	Lentil	Fabaceae	Legumes	Lectins	TLR2/6, TLR4, TLR5	Chahal et al. (2013), Wang et al. (2021b)
Magnolia officinalis	Magnolia	Magnoliaceae	Tree	Magnolol	TLR 3	Chahal et al. (2013), Wu et al. (2011)
Morus alba	White mulberry	Moraceae	Tree	Rutin	TLR4	Chahal et al. (2013), Han (2009)
Nigella sativa	Black cumin	Ranunculaceae	Flowering plant	Thymoquinone	TLR4	Chahal et al. (2013), Bai et al. (2013)
Panax ginseng	Asian ginseng	Araliaceae	Perennial plant	Ginsenoside	TLR 2, TLR 4, TLR 9	Chahal et al. (2013), Liu et al. (2022), Shaukat et al. (2019)
Platycodon grandiflorum	Chinese bellflower	Campanulaceae	Herb	Platycodin D	TLR 4	Chahal et al. (2013), Wu et al. (2021)

Plantago major	Broadleaf plantain	Plantanginaceae	Flowering plant	Catapol	TLR 4	Chahal et al. (2013), Zhang et al. (2020a)
Polyporus umbellatus	Umbrella polypore	Polyporaceae	Fungus	Polysaccharides	TLR4	Chahal et al. (2013), Li and Xu (2011)
Phytolacca americana	American pokeweed	Phytolaccaceae	Herb	Lectin (PWM)	TLR2	Bekeredjian-Ding et al. (2012)
Vitis vinifera	Grape	Vitaceae	Climber	Resveratrol	TLR 4	Wang et al. (2017), Tabeshpour et al. (2018)
Rheum palmatum	Rhubarb	Polygonaceae	Flowering plant	Emodin, rhein	TLR 2, TLR 3, TLR 4, TLR 7, TLR 9	Dai et al. (2017)
Tinospora cordifolia	Giloy	Menispermaceae	Shrub	G14A	TLR 4	Gupta et al. (2016)
Salvia miltiorrhizae	Sage	Lamiaceae	Herb	Salvianolic acid	TLR 4	Wang et al. (2016)
Zingiber officinale	Ginger	Zingiberaceae	Shrub	Gingerol	TLR 2, TLR 4	Ju et al. (2021)

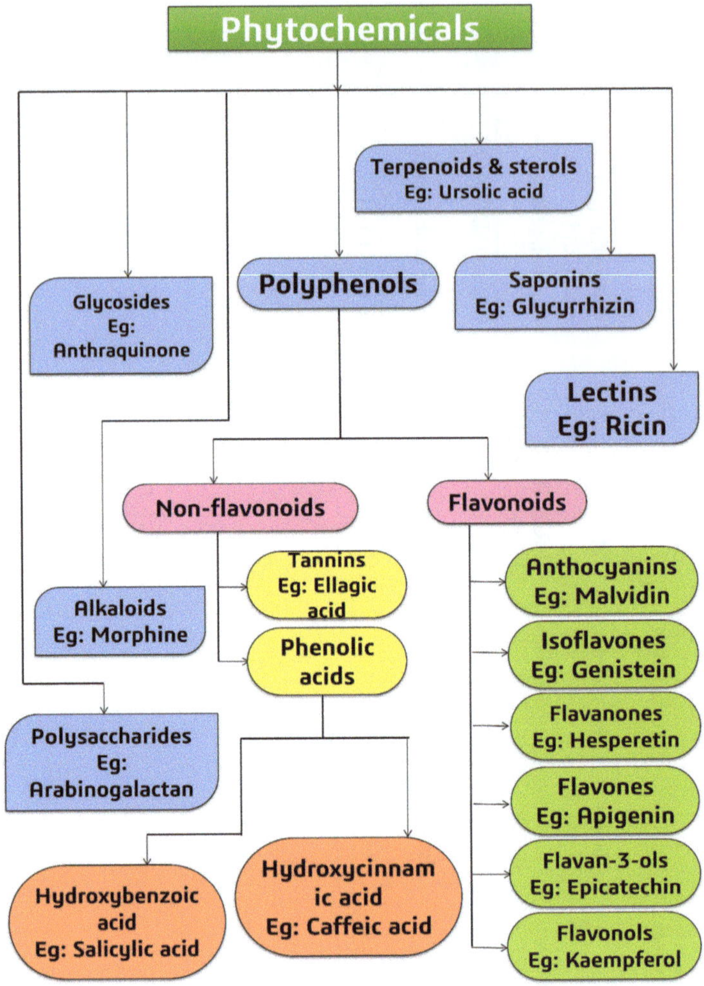

Fig. 3.2 Classification of different phytochemicals

3.2.1 Glycosides

Glycosides represent a significant class of secondary metabolites found in plants, characterized by the presence of a sugar moiety chemically bonded to non-sugar constituents, creating a hemiacetal structure. Within plants, these molecules are often stored in an inert state. However, through enzymatic processes, they can be converted into active forms, enabling them to exert advantageous effects in both human and animal organisms. This activation of glycosides unleashes their potential to contribute to various beneficial activities (Anon 2007). The activated states of glycosides encompass compounds such as anthracin and anthocyanin. These specific glycosides play a pivotal role in stimulating not only the immune system but also the central nervous system and the cardiovascular system. Their influence extends across multiple physiological systems, showcasing their potential to induce positive effects in these vital areas of human and animal well-being.

An important example is found *in Syringa oblata Lindl.* This plant stands out due to its prominent abundance of iridoid glycosides (IG), which are bioactive compounds with diverse therapeutic properties. Throughout its traditional usage, *Syringa oblata Lindl.* has demonstrated efficacy in addressing inflammatory conditions, particularly those affecting the gastrointestinal system. Since historical times, IGs have been in use for the curing of acute enteritis, bacillary dysentery, and even upper respiratory tract infections (Liu and Wang 2011a). IG have high levels of the bioactive substance called syringopicroside, which gives them their distinctive properties. This particular compound exhibits a remarkable capacity to effectively neutralize oxidative free radicals, thereby contributing to potent antioxidant effects. Additionally, syringopicroside demonstrates the ability to mitigate the accumulation of leukocytes, which are central players in inflammatory responses. It also affects how different pro-inflammatory cytokines are modulated, which results in the downregulation of certain cytokines (Liu and Wang 2011b). Research findings have illuminated the intricate molecular pathways involved in dextran sulfate sodium (DSS) induced in a murine colitis model. The IG-treated group, in a dose-dependent manner, exhibited a compelling anti-inflammatory effect, demonstrably repressing both the mRNA and protein expression of TLR2, TLR4, MyD88, and NF-κBp65. This study has provided fundamental insight into treating colonic inflammation as IG has the potential to impede the TLR2/4/MyD88/NF-κB signaling axis (Zhang et al. 2020c). Furthermore, IG has repressed NOX-dependent reactive oxygen species (ROS) production within the DSS-induced colitis model. IG potential in hindering the elevation of both NOX1 and NOX2 expressions was reported by Zhang et al. (2020). The dose-dependent DSS treatment has substantially elevated the presence of IL-2, IL-4, IL-5, and IL-13, whereas IG administration can effectively mitigate the upregulation, thus dampening the inflammatory pathways and cytokine cascades within experimental colitis scenarios (Zhang et al. 2020c).

3.2.2 Alkaloids

Alkaloids, a diverse class of natural compounds, are ubiquitously present in medicinal plants, bacteria, fungi, and animals. Renowned for their defensive attributes against a spectrum of pathogens, alkaloids exert their defensive prowess through inherent toxicity. The extent of this defensive response is contingent upon variables encompassing exposure duration, dosage, organism sensitivity, and targeted physiological sites (Cushnie et al. 2014; Rida et al. 2015). Cepharanthine has exhibited antimicrobial properties, while vincristine has demonstrated potent antitumor effects. Morphine has been recognized for its analgesic attributes and continues to be revered for its pain-relieving capacities. Significantly, alkaloids also embrace anti-inflammatory and immunomodulatory roles, thus underscoring their multifaceted impact on biological systems (Zhao et al. 2015; Souto et al. 2011).

Berberine (BBR), a quaternary isoquinoline alkaloid sourced from diverse traditional Chinese medicinal plants, holds a notable position in therapeutic application. Its historical utilization encompasses the management of gastroenteritis, colitis, as well as conditions characterized by diarrhea and dysentery. This compound's integration into traditional medicine underscores its enduring significance in addressing gastrointestinal ailments (Li et al. 2019). Emerging research has delineated the distinct pharmacological attributes of oxyberberine (OBB), an oxidized derivative of protoberberine alkaloid. OBB exhibits a multifaceted spectrum of effects, encompassing anti-inflammatory, antifungal, antitumor, and anti-arrhythmic properties (Li et al. 2019; Chi et al. 1996). In a recent study conducted by Li et al. in 2020, compelling evidence emerged elucidating the pivotal role of oxyberberine in mitigating intestinal mucosal inflammation and colonic mucosal injury induced by DSS in mice (Li et al. 2020). The study highlighted that oxyberberine's beneficial effects are primarily achieved by reducing the activity of TLR4–MyD88–NF-κB. This discovery provides a clearer understanding of how oxyberberine helps in addressing colitis-related mucosal issues by affecting these molecular pathways. OBB exerts its influence by reducing the expression of TLR4 and MyD88, subsequently restraining the phosphorylation of IκBα and the nuclear translocation of NF-κB p65. This orchestration collectively results in the suppression of the TLR4-MyD88-NF-κB signaling cascade. The investigations conducted by Li et al. in 2006 have revealed berberine's immunomodulatory capabilities through the suppression of key mediators such as IFN-γ, TNF-α, and nitric oxide (Li et al. 2006). Both in vivo and in vitro studies of berberine have showcased its ability to impede cell growth in human colorectal adenocarcinoma through the imposition of cell cycle arrest. This result is consistent with its function in encouraging apoptosis and preventing the growth of colorectal cancer. In this context, berberine has demonstrated the capacity to downregulate PPARγ mRNA expression within LoVo cells, a factor associated with apoptosis induction and colon cancer prevention (Wu et al. 2010; Dionne et al. 2010; Kaur and Sanyal 2010). Berberine's influence also extends to the regulatory factors, as it downregulates the expression of COX-2—a pivotal player in cancer formation—adding to its multifaceted anticancer potential (Zou et al. 2017).

Piperine, identified as 1-piperoyl piperidine, represents a principal alkaloid within the composition of black pepper (*Piper nigrum*) and long pepper (*Piper longum*). Literature indicates piperine's potential efficacy in countering metabolic syndrome alongside its documented anti-inflammatory properties (Diwan et al. 2013). The Piper genus exhibits therapeutic potential within traditional medicine systems, addressing tuberculosis, respiratory tract infections, and arthritic disorders. Of particular significance, piperine, demonstrates a considerable ability to diminish inflammatory mediators involved in inflammatory colitis (Hu et al. 2015). Piperine has been found to have a variety of functions, including lowering depression, defending the liver, inhibiting metastasis, treating thyroid problems, regulating the immune system, and battling cancers (Pathak and Khandelwal 2009). Different in vitro studies conducted on piperine have reported the inhibition of nitric oxide (NO), tumor necrosis factor (TNF-α), and pro-inflammatory gene expression (Kumar et al. 2007). It also acts to mitigate inflammation induced by free fatty acids (FFA) through the TLR-4 pathway, resulting in a decrease in the levels of the pro-inflammatory cytokine TNF-α (Gupta et al. 2015). The study conducted by Gupta et al. in 2015 revealed that piperine exhibited significant effects, including diminishing sub-mucosal edema, reducing cellular infiltration, attenuating hemorrhages, and mitigating ulcerative colitis induced by acetic acid. Additionally, piperine demonstrated the capacity to decrease the secretion of pro-inflammatory mediators such as NO and TNF-α. Additionally, it demonstrated the capacity to inhibit TLR 4-mediated inflammation brought on by FFA (Gupta et al. 2015).

Mahanine, identified as a carbazole alkaloid sourced from *Murraya koenigii*, is a member of the Rutaceae family and has demonstrated anticancer attributes. Roy et al. in 2004 demonstrated effective inhibition of the proliferation of cancer cells and induce apoptosis—a programmed cell death process by mahanine (Roy et al. 2004). In the year 2017, Roy et al. conducted a study revealing the susceptibility of virulent *Leishmania donovani* promastigotes to mahanine treatment. This intervention led to apoptosis, evidenced by phosphatidylserine externalization, DNA fragmentation, and cell cycle arrest. Additionally, their findings indicated the involvement of oxidative stress, triggering the generation of reactive oxygen species (ROS) within macrophages confronting the promastigotes. Oral mahanine administration in a mouse model resulted in an increase in ROS generation and T cell proliferation, which ultimately led to a complete decrease in parasite load (Roy et al. 2017).

3.2.3 Phenolics

Fruits and vegetables contain large amounts of polyphenols, a type of secondary metabolite that is distinguished by its chemical composition. Research indicates that various plant-derived products are rich sources of polyphenols. These compounds feature an aromatic ring with one hydroxyl group attached to either an aromatic or aliphatic ring, resulting in the formation of simple phenolic compounds or more complex, larger molecules (Manach et al. 2004). Flavonoids and non-flavonoids are the two primary categories of polyphenols. Flavonoids have two aromatic rings connected to an oxygen-containing ring structure, while non-flavonoids include phenolic acids made up of benzoic and cinnamic acid units. Polyphenols exhibit various beneficial properties, such as antimicrobial, antifilarial, anti-inflammatory, and antioxidant effects. They have also shown promise in improving conditions such as diabetes, neurodegenerative diseases, cardiovascular issues, and infectious diseases caused by microbes, parasites, and viruses (Williams et al. 2017; Calixto Júnior et al. 2016; Montenegro-Landívar et al. 2021). It has the capacity to modulate the production of NO and impede the mechanisms underlying the action of proinflammatory cytokines (Chuang and McIntosh 2011). The ethanolic extracts derived from *Cajanus scarabaeoides* have demonstrated remarkable efficacy in eradicating all developmental stages of *Setaria cervi*, a parasitic nematode. This effect is achieved through the induction of oxidative stress, leading to a disruption of equilibrium within the nematode and subsequently triggering its progression toward the cell death execution (CED) pathway for parasite elimination. Ray et al., 2018 reported that targeted elimination of worms was achieved without inducing any harmful effects on mammalian cells (Ray et al. 2018). Extracts rich in polyphenols obtained from Zingiber officinale have demonstrated the ability to mitigate hepatoxicity induced by phosphamidon exposure. The administration of ethanolic extracts led to a reduction in markers indicative of oxidative stress, concurrently augmenting the levels of ROS and enhancement in apoptotic markers, which is suggestive of apoptotic processes taking place within the liver were also reported (Mukherjee et al. 2015).

CUR (diferuloylmethane) represents the primary curcuminoid compound within turmeric, an Indian spice obtained from the plant *Curcuma longa Linn* (Zingiberaceae family). This compound displays a diverse array of pharmacological effects, notably encompassing anti-inflammatory and antitumorigenic activities (Duvoix et al. 2005). Research findings indicate that curcumin has the capacity to competitively inhibit the binding of LPS to TLR4, thereby obstructing the MyD88-dependent pathway. This mode of action inhibits LPS-induced inflammation within vascular smooth muscle cells, operating through the TLR4-MAPK/NF-κβ pathways. These effects are mediated by curcumin's ability to impede the production of ROS, consequently curbing the propagation of inflammation (Meng et al. 2013). Curcumin demonstrates a capacity to modulate the TLR4 and MyD88 pathways, subsequently hindering the activation of NF-κβ via inhibition of MAPK and AP-1 transcription factors, thereby yielding an anti-inflammatory milieu(Guimarães et al. 2013). Through the downregulation of TLR4 expression, curcumin achieves a dose-dependent inhibition of M1 macrophage polarization, leading to the suppression of proinflammatory cytokines, including TNF-α, IL-1β, and IL-6 (Azab et al. 2016; Shah et al. 2010). Research findings propose that curcumin holds the potential to enhance the neuroinflammatory response by mitigating microglia/macrophage activation and neuronal apoptosis via modulating the TLR4/MyD88/NF-B signaling pathway (Zhu et al. 2014). The triggering receptor expressed on myeloid cell 2 (TREM2) functions as an effective counteractive modulator of TLR-4 signaling. It orchestrates the repression of inflammatory cytokine activity. Additionally, TREM2 has the capacity to reverse the neuroinflammatory reactions brought on by LPS, including reducing the increased TLR 4 hyperactivity. This reversal is accomplished through the elevation of TREM2 levels (Zhang et al. 2019).

3.2.4 Flavonoids

Plant-based foods and beverages are rich sources of flavonoids, which are frequently characterized by their relatively low molecular weight and water solubility. These compounds exhibit a structural arrangement known as the C6–C3–C6 skeleton and can be encountered both in their unbound forms and as glycosides (Marzocchella et al. 2011). Flavonoids can be categorized into two classes: anthoxanthins and anthocyanins. Their diverse array of health benefits begins from their capacity to exert preventative effects against oxidative cellular stress and their demonstrated anticancer properties. Flavonoids present themselves as promising candidates for therapeutic exploration aimed at reducing the symptomatic manifestations of chronic illnesses such as diabetes, Alzheimer's disease, and atherosclerosis due to their immunomodulatory properties, as well as their roles as antioxidants and anti-inflammatory agents (Panche et al. 2016).

Rutin is a prominent flavanol found abundantly in botanical sources such as tea, apple, passionflower, and buckwheat, which exhibits antioxidant, anti-inflammatory, and anti-carcinogenic properties (Ganeshpurkar and Saluja 2017). The efficacy of treating leishmaniasis has encountered challenges due to the emergence of resistance and associated adverse effects linked to existing therapies. In the study conducted by Chauhan et al. (2018), the anti-promastigote effectiveness of rutin was validated against *Leishmania donovani* strains that had developed resistance to sodium stibogluconate (SSG), also known as a conventional therapeutic agent. The administration of rutin yielded a reduction in parasite burden without inducing renal or hepatotoxic effects. Rutin exhibited the capacity to induce an upregulation in the expression of the inducible nitric oxide synthase (iNOS) and NF-κβ genes, subsequently triggering a T cell response, the release of Th1 cytokines, and the generation of ROS. This orchestrated enhancement in the expression of NF-κβ and iNOS genes culminated in the production of microbicidal nitric oxide (NO), effectively countering the resistance mechanisms exhibited by the parasite and thus reinstating the host's immune responsiveness (Chauhan et al. 2018).

Genistein is characterized chemically as 4,5,7-trihydroxy isoflavone and 5,7-dihydroxy-3-(4-hydroxyphenyl) chrome-4-one. It is sourced from Genista tinctorial, commonly known as Dyer's broom, and is classified as an isoflavone polyphenol (Han et al. 2015). Renowned for its multifaceted properties, genistein is widely acknowledged for its antioxidative, anti-inflammatory, and antiproliferative attributes; hence, it is efficacious as an immunosuppressive agent (Polkowski and Mazurek 2000). Genistein contributes to the recompensating of endothelial inflammatory damage by orchestrating the inhibition of NF-κβ activity, alongside the suppression of the proinflammatory cytokine IL-6 and the intercellular adhesion molecule-1 (ICAM-1) (Han et al. 2015). By inhibiting NF-κβ expression and lowering C-reactive protein (CRP) levels, genistein successfully reduces angiotensin II-induced inflammation in vascular smooth muscle cells (Xu et al. 2019). Genistein amplifies the cascades of MAPK transduction while concurrently suppressing the TLR pathway, a feat achieved through the downregulation of IL-10, IL-1β, IL-6, TNF-α, CSF-3, NF-κβ, and COX-2 expression (Cui and Bilitewski 2014). This intricate regulatory action extends to microglia is elucidated by Jeong et al. (2014), who demonstrated the reduction of NO, IL-1, TNF-α, TLR4, prostaglandin E2 (PGE2), and MyD88 expression upon genistein intervention in LPS-induced BV2 microglial cells (Jeong et al. 2014).

A flavonoid molecule called quercetin, which is derived from fruits and vegetables, has anti-inflammatory, anticancer, and antioxidant properties. Its anti-inflammatory potency resides in the modulation of NF-κβ-orchestrated pro-inflammatory cytokines, a suppressive influence on TLR2 and TLR4 expression as well as the genes encoding inducible iNOS and COX-2. Quercetin inhibits the excessive production of chemokines, notably vascular cell adhesion molecule-1 (VCAM-1), ICAM-1, and MCP-1, in animal models of atherosclerosis, which has a modulatory effect (Bhaskar et al. 2016; Byun et al. 2013). Quercetin along with isorhamnetin demonstrates the ability to downregulate both mRNA and protein levels of proinflammatory mediators, viz., TNF-α, IL-6, and IL-1β. This regulatory result reduces the production of miR-155, which in turn causes NF-κβ to be deactivated and, as a result, lessens the inflammatory cascade (Boesch-Saadatmandi et al. 2011). According to research conducted by Byun et al. (2013), quercetin raises levels of a protein called Toll-interacting protein (Tollip), which inhibits TLR signaling, a process known to be important in immunological responses (Byun et al. 2013). Quercetin also decreases the levels of COX-2, a substance implicated in inflammation, and the production of pro-inflammatory cytokines induced by LPS. Moreover, quercetin promotes the creation of NO by iNOS, again through Tollip's involvement. These effects are achieved through a series of molecular pathways including NF-κβ and MAPK (Byun et al. 2013).

Epigallocatechin-3-gallate (EGCG) is a widely recognized polyphenolic constituent that is found in appropriate quantities within green tea and is also distinct in sources such as onions, apple skin, and plums. EGCG belongs to the larger category of polyphenols, a class of phytochemicals with substantial biological significance. Phenolic extraction of *Camellia sinensis* includes epicatechin-3-gallate (ECG), epigallocatechin (EGC), epigallocatechin-3-gallate (EGCG), and epicatechin (EC), all of which encompass an array of pharmacological effects encompassing antioxidative, anti-inflammatory, and anticancer properties (Cabrera et al. 2006). Research demonstrates that EGCG effectively impedes the MyD88-dependent and TRIF-dependent signaling pathways associated with TLRs in RAW264.7 cells and can suppress the LPS-induced TLR4 signaling ; Youn et al. 2006a, b, Hong Byun et al. 2010). Through its activation of nuclear factor erythroid 2-related factor 2 (Nrf-2), EGCG plays a crucial part in suppressing inflammation-induced oxidative stress. In the setting of the asthmatic mouse model, EGCG appears as a strong mitigator of airway inflammation, significantly decreasing inflammatory cell infiltration and lowering levels of proinflammatory cytokines, TNF-, IL-2, and IL-6 (Shan et al. 2018).

Apigenin is identified as a notable flavonoid encountered in various vegetables including parsley, celery, and chamomile tea. It is acknowledged for exhibiting potency in anti-inflammatory and cardioprotective attributes (Saleh et al. 2021). Apigenin emerges as a substantial modulator of M1/M2 polarization, exerting a preference for M2 polarization and consequently dampening the production of proinflammatory cytokines (Saqib et al. 2018). In vivo experimentation on the inflammation model induced by LPS further demonstrates apigenin's efficacy. Apigenin therapy efficiently reduces NF-κβ, COX-2, and iNOS activity, which causes a decrease in the production of NO, IL-6, IL-1, and TNF-α (Karunaweera et al. 2015). Moreover, apigenin plays a constructive role in mitigating inflammation related to obesity, as validated by a study involving an obese mice model conducted by Saqib et al. (2018). Research undertaken by Zhao et al. highlights apigenin's potential in ameliorating oxidative stress-mediated mitochondrial-driven neuronal apoptosis. This beneficial effect is achieved through the suppression of the TLR4/NF-κβ signaling pathway, thereby culminating in reduced levels of IL-6 (Zhou et al. 2019).

Kaempferol is a naturally occurring flavonoid, which is procured from diverse sources such as tea, and commonplace vegetables including beans, Brussels sprouts, and cabbage, as well as ubiquitous fruits such as grapes and strawberries. It possesses several characteristics that have been linked to anti-inflammatory actions against long-term inflammatory illnesses such as colitis, metabolic disorders, and acute lung damage. Additionally, its anticancer potential is demonstrated in a number of cancers, including ovarian, breast, cervical, and esophageal cancers (Ren et al. 2019). In particular, it has been found that kaempferol reduces the apoptosis brought on by LPS via altering chondrogenic markers including SOX-9, Collagen II, and Aggrecan, which then has an impact on the activity of enzymes that break down the matrix (Zhu et al. 2017). Insightfully, studies have revealed kaempferol's capability to significantly diminish the elevation of TLR4, MyD88, and the phosphorylation levels of MAPKs (Zhang et al. 2017). As indicated by Ming et al. in 2017, kaempferol has been shown to have anti-biofilm capabilities in this situation. By inhibiting sortase A activity and the production of adhesion-related genes, it specifically prevents the development of *Staphylococcus aureus* biofilms, a pathogenic bacteria known to cause severe illnesses (Ming et al. 2017).

3.2.5 Non-Flavonoid Polyphenol

3.2.5.1 Tannins

Tannins, which are water-soluble compounds abundantly present in plant sources, engage in intricate associations with proteins, alkaloids, and polysaccharides. Originating from diverse botanical origins such as apples, grapes, peaches, walnuts, and berries, tannins have demonstrated notable immunomodulatory properties (Serrano et al. 2009). A seminal work by Scarbert in 1991 delineated the antimicrobial attributes of tannins, elucidating their capacity to inhibit extracellular microbial enzymes. Tannins enact microbial growth inhibition by sequestration of metal ions, perturbation of microbial enzyme function, and alteration of cell wall integrity, ultimately leading to augmented membrane permeability (Huang et al. 2018; Scalbert 1991). Tannins exhibit robust efficacy against pathogens such as *Helicobacter pylori*, *Salmonella*, and *Staphylococcus*, primarily due to their oxidative properties that stimulate hydrogen peroxide release (Huang et al. 2018; Funatogawa et al. 2004). Furthermore, tannins manifest anti-parasitic effects, against

Trichostrongylus colubriformis eggs, curbing the development of gastrointestinal and lung worms (Huang et al. 2018; Molan et al. 2000a, b, 2002), particularly the identification of four tannin extracts bearing anthelmintic potential against *T. colubriformis* and *Haemonchus contortus* which can modulate larval exsheathment and migration in a dose-dependent manner, thereby highlighting their intricate influence on parasite developmental processes (Athanasiadou et al. 2001; Molan et al. 2003; Molan et al. 2004).

Resveratrol represents a natural stilbene phytoalexin prevalent in various sources including peanuts, grapes, blueberries, rhubarb, and wine. It plays multifaceted pharmacological roles encompassing anti-inflammatory, chemo-preventive, and hepatoprotective attributes (Elshaer et al. 2018). Resveratrol seems as a viable drug for treating chronic illnesses brought on by TLR activation as well as TLR-mediated inflammatory responses. In LPS-induced macrophages, this regulation may be seen as a reduction in the activity of the protein kinase B (AKT), MAPK, and TLR4-TNF receptor-associated factor 6 (TRAF6) pathways (Jakus et al. 2013). Resveratrol also lowers NO and ROS levels while lowering NF-κβ induced proinflammatory cytokine production (Qureshi et al. 2012). Additionally, resveratrol exerts its anti-inflammatory action by diminishing the expression of markers such as COX2 and inducible iNOS (Cianciulli et al. 2012). Subsequent investigations have substantiated the capacity of resveratrol to attenuate inflammatory responses induced by lipopolysaccharide (LPS) and to diminish the downregulation of NF-κβ activity in human colon cancer cells (Panaro et al. 2012). Its anti-inflammatory effects span diverse experimental models encompassing both MyD88-dependent and independent signaling pathways. Recent findings in TLR4/NF-κβ-mediated cardiac inflammation studies underscore resveratrol's capacity to bolster antioxidant production marked by enhanced glutathione (GSH) and superoxide dismutase (SOD) levels while concurrently dampening TNF-α levels. Furthermore, studies have elucidated the dose-dependent inhibition of TLR4/NF-κβ signaling by resveratrol in an ischemic murine model. This intervention yields downregulation of TLR4 and NF-κβ, consequently curtailing myocardial TNF-α production (Arafa et al. 2014; Li et al. 2014).

3.3 Polysaccharide

3.3.1 Fucoidan

Marine brown algae, specifically *Sargassum swartzii*, serves as a reservoir of sulfated polysaccharides known as fucoidan. Fucoidan, characterized by sulfate groups and intricate structural arrangements, holds considerable demand due to its intricate biological properties. The ability of fucoidan to lessen neutrophilic infiltration in inflamed auricles was shown in a study by Hwang et al. (2015), establishing it as a possible treatment for inflammatory illnesses (Hwang et al. 2015). Moreover, their investigation revealed that fucoidan not only suppresses iNOS production but also significantly diminishes COX-2 levels. In tandem with these observations, while LPS stimulation amplifies the phosphorylation of MAPK proteins, fucoidan exposure induces a marked reduction in these levels. The active component of fucoidan, designated as the active fraction (F4), manifests prominent anti-inflammatory efficacy through the downregulation of MAPK phosphorylation. This intervention is complemented by a substantial reduction in MyD88 expression, as well as diminished Toll-like receptor 2 and 4 (TLR 2/4) expression subsequent to F4 treatment (Jayawardena et al. 2020). In a parallel context, Manikandan et al. (2020) highlighted the ability of fucoidan derived from *Turbinaria decurrens* to reduce the paw edema caused by formalin in a mouse model and to lessen the cytotoxicity caused by LPS in macrophages.

3.3.2 Lectin

Lectins, found in a diversity of sources including plants, yeast, animals, and bacteria, encompass non-enzymatic compounds and glycoproteins that exert an influence on strengthening the immune system. Their multifaceted properties span anticancer and immunomodulatory functions, thereby warranting exploration through experimental and clinical trial settings (de Mejía and Prisecaru 2005). The immunomodulatory capacity of lectins within the context of inflammatory diseases is marked by a degree of controversy, as certain lectins manifest anti-inflammatory effects, while others are implicated in the induction of inflammation (Gong et al. 2017). Certain lectins are thought to be responsible for the genesis of a number of inflammatory illnesses, including rheumatoid arthritis, colitis, celiac disease, and Alzheimer's disease. Con A, a plant-derived lectin, activates the NLRP3 inflammasome, which results in the production of proinflammatory cytokines including IL-1 and IL-18, according to research by Gong et al. (2017). Dioclea violacea (Dvl) is a lectin identified and extracted from the Fabaceae family, which has been documented to exhibit anti-inflammatory properties by impeding neutrophil migration during episodes of inflammation (Nascimento et al. 2018). This phenomenon is corroborated by a study conducted by Clemente-Napimoga et al. (Nunes et al. 2009) that confirms Dvl's ability to reduce inflammation by inhibiting leukocyte infiltration and ICAM-1 levels. Additionally, ConGF, a lectin with specificity for D-glucose/D-mannose, sourced from *Canavalia grandiflora*, has exhibited the ability to dose-dependently suppress the production of pro-inflammatory cytokines, specifically IL-1β and TNF-α.

3.3.3 Saponin

Triterpenes and steroidal saponins are two distinct classes of saponins, glycosides that are naturally present in a variety of plant parts and function as potent immunoadjuvants to stimulate the production of antibodies as well as position them as potential vaccine candidates (Oleszek and Oleszek 2021; Sparg et al. 2004). An exemplary instance is found in glycyrrhizin, the principal constituent of licorice root, which exhibits noteworthy anti-inflammatory and antiviral effects, rendering it efficacious in addressing conditions such as myocardial ischemia, liver injury, and lung injury (Fei et al. 2017). Further investigation reveals that glycyrrhizin exerts a mitigating influence on LPS-induced acute lung injury by reducing TLR2 expression and inhibiting TLRs and NF-κB signaling pathways, thereby acting as a TLR-2 inhibitor (Zhao et al. 2016).

3.3.4 Sterols and Sterolins

Glucocorticoids have gained pharmaceutical significance for addressing chronic inflammatory ailments and immunopathological disorders. Accordingly, sterols and sterolins are assimilated from plants via dietary intake in which the robust evidence from in vivo and in vitro cancer investigations underscores their capacity for immunomodulation, marked by the modulation of T cell proliferation and heightened natural killer (NK) cell activity. It is interesting to note that the dynamic interplay also affects how Th1 and Th2 cytokine profiles are regulated. This carefully controlled factor enhances the host immune system (Bouic 2002; Patel 2008).

Withania somnifera root yields Withaferin A (WA), a highly potent bioactive compound that has been linked to a variety of health benefits, including the reduction of inflammation and the prevention of cancer (Sun et al. 2016). WA intervenes with TLR activity through a direct disruption of the downstream NF-κB pathway. As evidenced across multiple reports, its intervention yields diminished inflammation and a comprehensive amelioration of disease symptoms observed in conditions such as cystic fibrosis, diabetes, and pulmonary fibrosis. This modulation is accomplished through the restraint of NF-κB, signaling kinases, Nrf2, and the inflammasome complex (Logie and vanden Berghe 2020). Grover et al. (2010) postulated a plausible intermolecular interaction between WA and IKK-γ, thereby potentially impeding the assembly of the IKK complex and consequently thwarting IκB degradation. This observed decline in IKK levels induced by WA extends to studies involving lung fibrosis, obesity, and cancer (Kaileh et al. 2007; Sayed et al. 2019). The underlying process may entail a direct interaction between WA with NF-κB, thus hampering p65 dimerization or an indirect influence via IκB inhibition (Kaileh et al. 2007; Sayed et al. 2019; Logie and vanden Berghe 2020).

3.3.5 Terpenoids

Terpenoids, also known as isoprenoids, are isoprene-based chemical compounds that are frequently found in higher plants. Within these routes, diverse categories of terpenoids, spanning hemi-, mono-, di-, tri-, and sesquiterpenoids, are synthesized (Singh and Sharma 2015). Surprisingly, several studies have proven the terpenoids' medicinal potential, demonstrating properties including anti-inflammatory, antiviral, anti-diabetic, and immunomodulatory actions. Their influence encompasses increased T cell suppression alongside heightened antibody production (Singh and Sharma 2015; Ludwiczuk et al. 2017). The pharmaceutical applications also feature an array of terpenoids. A prominent antimalarial drug, Taxol, is a widely embraced anticancer agent, underscoring the significance of terpene-based therapeutic interventions (Wang et al. 2005).

Camphor, a naturally occurring terpenoid predominantly sourced from *Cinnamomum camphora* wood, is recognized for its anti-inflammatory, anticancer, and antioxidant attributes (Silva-Filho et al. 2015; Lee et al. 2006). Recent investigations into camphor's effects have unveiled a propensity to augment catalase and Nrf-2 activities, concomitant with diminished NO and TNFα levels. This pertains to an elevation in neurotransmitters such as serotonin and dopamine, the P190-RHO GTP protein, and gamma-aminobutyric acid (GABA) within neuronal cells of the frontal cortex in a rodent model of ciprofloxacin-induced depression, orchestrated through the modulation of TLR4 (Salama et al. 2021). Menthol is another aromatic terpenoid that has effectiveness against acetic acid-induced acute colitis in an experimental rat model. Followed by the administration of menthol, levels of IL-1β, TNFα, IL-6, and MPO activity have been reduced in the inflamed colons (Ghasemi-Pirbaluti et al. 2017). In-depth exploration by Saini et al. in 2014 delves into the antifilarial effects of ursolic acid derived from *Nyctanthes arbor-tristis*. This compound exhibits micro- and macrofilaricidal activities against *Setaria cervi* by engendering ROS generation and inducing apoptosis (Zaia et al. 2016). In silico analyses highlight the interaction between andrographolide and diverse viral entry points of SARS-CoV-2, simultaneously inhibiting TLR4-MD2 and IL-6 interactions. This strategic intervention is suggestive of its potential to mitigate inflammatory regulators, thereby mitigating the risk of cytokine storms in the host (Saini et al. 2014). In silico studies have suggested the interaction of andrographolide with different viral entry points of SARS-CoV-2 and blocks TLR4-MD2 and IL-6. Inhibition of inflammatory regulators is suggestive of prevention of cytokine storm in the host (Das et al. 2022).

3.4 Advantages and Challenges

At present, due to the high cost and time commitment required for drug discovery, therapeutic repurposing of existing medications was investigated as a treatment option as well and some drugs have no positive outcome but have severe side effects. In this situation, investigating an alternative therapy option that would be both safe and therapeutically beneficial is necessary. The use of phytopharmaceuticals—molecules produced from medicinal plants—as significant resources for immune system modulation against various illnesses might be investigated. Because of their health benefits and economic prospects, functional foods, especially those containing phytochemicals, have increased their market share in recent years. The current therapeutic therapy of inflammatory bowel disease (IBD) is mostly focused on medications and surgeries, although harsh side effects and the uncertain efficacy of drug treatment have limited their use. The significance of phytochemicals in the treatment of IBD has been recognized, owing to their great efficacy and lack of adverse effects. However, low solubility and considerable hazardous side effects are still primary issues; hence, the present focus of research is on mitigating the impact of these issues. In this context, a wide range of chemicals and therapeutic candidates have been created, and methods to enhance water solubility and tumor selectivity have also been established. Some phytochemicals such as aristolochic acid are carcinogenic at low doses and are known as phytotoxin. It is the only known plant toxin that produces spontaneous cases of intestinal or bladder cancer in animals. According to a survey conducted in 2001, Aristolochic acid, which is derived from the Chinese herb *Aristolochia fangchi*, has been linked to renal disease and cancer of the urinary system in obsessive persons who take weight-loss medications. The seed and root of the cycad plant contain cycasin and methylazoxymethanol (MAM), which have the potential to change gene expression associated with cancer. According to the research by the National Cancer Institute (NCI), U.S. MAM could induce a variety of tumors, including liver and renal cells even in nonhuman primates. Normal human submandibular gland (HSG) cells and oral epithelial cells (OSCC-3) experience DNA damage and apoptosis when exposed to bracken fronds, a popular vegetable in portions of China, Korea, and Japan. In addition to this, determining standardized dosages is a significant obstacle to employing phytochemicals as an alternative medicinal medication that must be solved. Therefore, further research and epidemiological studies need to establish the health and therapeutic claim of phytochemicals.

3.5 Future Directions

Phytochemicals are essentially secondary metabolites that are present in plants and provide them with defense against a variety of dangers. By combining a wide range of plants (either as parts or products) into nutrition and medicine, mankind has been attempting to extract the health advantages of these phytochemicals. We often presume that they are safe to consume, placing blind faith in nature's skill. They have been cost-effective, and their bioavailability makes them an alternative therapeutic against chronic illnesses such as arthritis, diabetes, and so on. In addition to treating inflammatory illnesses, phytochemicals can also be used to treat cancer and infectious diseases. However, different studies depict phytochemicals to accentuate, but clinical studies are still lacking as, to evaluate the effects of phytochemicals, laboratory tests and extensive clinical trials are required. To select a drug candidate successfully, phytochemicals need to be evaluated for potential toxicity on the basis of both cellular and molecular levels. Hitherto, companies such as Chromadez Hrozon Science and the American Institute for Cancer Research have been working on different phytochemicals to determine the efficacy of several phytochemicals and their ability to treat chronic diseases (Behl et al. 2021).

3.6 Conclusion

Plant extracts with active phytochemicals have multi-pharmacological activities and have been in use as traditional medicines as depicted in Fig. 3.3. Their health benefits have been known to mankind since ancient times and yet they are explored to achieve essential cures for chronic illness. They have the potential to modulate TLR signaling either by direct activation of receptors or they might act further downstream the TLR pathway although the difference is not clear. Phytochemicals are conglomerations from different sources of plants that are further diversified into glycosides, phenolics, polysaccharides, terpenoids, and proteins. They have been employed as innovative therapies targeting TLRs because they have activated TLRs by mimicry, small compounds, and antibodies to suppress downstream NF-κB signaling and break the positive feedback loop of inflammation. Immune system dysbiosis results in a variety of immunopathological outcomes, including potentially fatal diseases. Phytochemicals have emerged as a glimmer of hope due to their success in treating and/or reducing crippling illnesses. Although mechanistic studies should be performed to determine the action of modulation on different TLRs toward TLR-related conditions. Thus, phytochemicals tend to improve the immunological consequences and restore the homeostasis of the host immune system. Sustainable research in the realm of TLR biology will provide mechanistic insights toward future discoveries and provide aid to the development of TLR immunomodulators as therapeutics.

Different phytochemicals act at different checkpoints for influencing the pathway into producing anti-inflammatory cytokines like IFN or IL-10, instead of the pro-inflammatory cytokines. The PAMPs, on binding with TLRs, activate the MyD88-dependent inflammatory pathway through the activation of IRAKs, followed by activation of AP-1 or NF-κβ. Phytochemicals activate the pathway for the production of anti-inflammatory cytokines by blocking or modulating different steps, thus preventing the expression of inflammation.

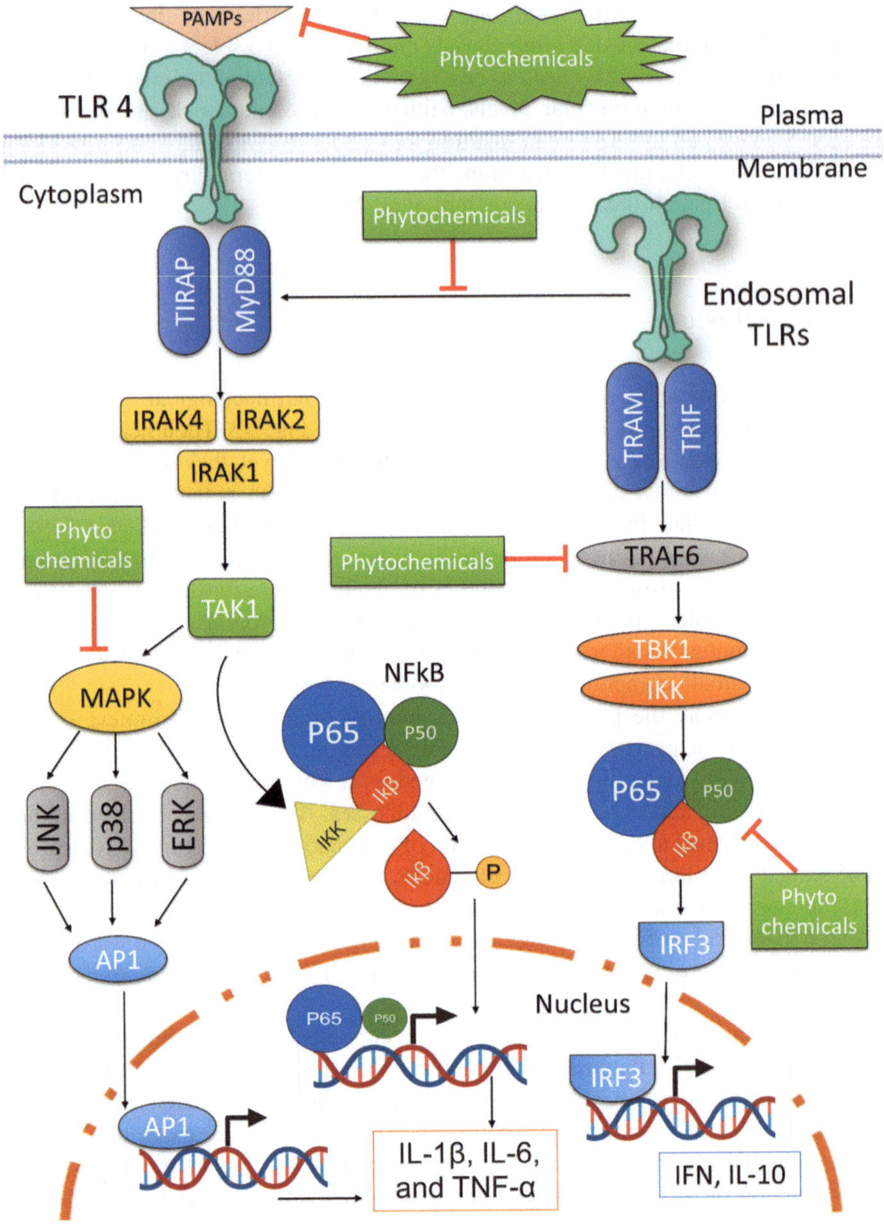

Fig. 3.3 Phytochemicals modulating Toll-like receptors: a generalized illustration of the mechanistic pathway of production of pro-inflammatory and anti-inflammatory cytokines or chemokines, generally initiated by binding of PAMPs on the Toll-like receptors

Acknowledgments SM acknowledges the University Grants Commission (UGC) (Ref no. F.2-12/2019(STRIDE-1) and KNU-UGC-STRIDE (Ref no. KNU/R/STRIDE/1077/21) and Department of Science and Technology—Science and Engineering Research Board (DST-SERB) (Ref no. SRG/2021/002605) for supporting his research activities and laboratory through awarding research grants. PC thanks DST, Govt. of India, for the award of the DST-INSPIRE fellowship (IF190998). SP acknowledges the Department of Higher Education, Government of West Bengal, for her SVMCM fellowship. The authors gratefully acknowledge BioRender.com which is used for drawing the figures.

Conflicts of Interest The authors declare that there is no conflict of interest.

References

Ain Q, Batool M, Choi S (2020) TLR4-targeting therapeutics: structural basis and computer-aided drug discovery approaches. Molecules 25:627. https://doi.org/10.3390/molecules25030627

Akira S (2009) Pathogen recognition by innate immunity and its signaling. Proc Jpn Acad B 85: 143–156. https://doi.org/10.2183/pjab.85.143

Akira S, Uematsu S, Takeuchi O (2006) Pathogen recognition and innate immunity. Cell 124:783–801. https://doi.org/10.1016/j.cell.2006.02.015

Anderson KV, Bokla L, Nüsslein-Volhard C (1985) Establishment of dorsal-ventral polarity in the drosophila embryo: the induction of polarity by the toll gene product. Cell 42:791–798. https://doi.org/10.1016/0092-8674(85)90275-2

Anon (2007) Synthesis and characterization of glycosides. Springer, Boston, MA. https://doi.org/10.1007/978-0-387-70792-1

Arabi M, Ostovan A, Asfaram A, Ghaedi M (2018) Development of an eco-friendly approach based on dispersive liquid–liquid microextraction for the quantitative determination of quercetin in *Nasturtium officinale, Apium graveolens, Spinacia oleracea, Brassica oleracea var. sabellica*, and food samples. New J Chem 42:14340–14348. https://doi.org/10.1039/C8NJ02485E

Arafa MH, Mohammad NS, Atteia HH, Abd-Elaziz HR (2014) Protective effect of resveratrol against doxorubicin-induced cardiac toxicity and fibrosis in male experimental rats. J Physiol Biochem 70:701–711. https://doi.org/10.1007/s13105-014-0339-y

Athanasiadou S, Kyriazakis I, Jackson F, Coop RL (2001) Direct anthelmintic effects of condensed tannins towards different gastrointestinal nematodes of sheep: in vitro and in vivo studies. Vet Parasitol 99:205–219. https://doi.org/10.1016/S0304-4017(01)00467-8

Azab A, Nassar A, Azab A (2016) Anti-inflammatory activity of natural products. Molecules 21: 1321. https://doi.org/10.3390/molecules21101321

Bai T, Lian L-H, Wu Y-L, Wan Y, Nan J-X (2013) Thymoquinone attenuates liver fibrosis via PI3K and TLR4 signaling pathways in activated hepatic stellate cells. Int Immunopharmacol 15:275–281. https://doi.org/10.1016/j.intimp.2012.12.020

Behl T, Kumar K, Brisc C, Rus M, Nistor-Cseppento DC, Bustea C, Aron RAC, Pantis C, Zengin G, Sehgal A, Kaur R, Kumar A, Arora S, Setia D, Chandel D, Bungau S (2021) Exploring the multifocal role of phytochemicals as immunomodulators. Biomed Pharmacother 133:110959. https://doi.org/10.1016/j.biopha.2020.110959

Bekeredjian-Ding I, Foermer S, Kirschning CJ, Parcina M, Heeg K (2012) Poke weed mitogen requires toll-like receptor ligands for proliferative activity in human and murine B lymphocytes. PLoS One 7:e29806. https://doi.org/10.1371/journal.pone.0029806

Bhaskar S, Sudhakaran PR, Helen A (2016) Quercetin attenuates atherosclerotic inflammation and adhesion molecule expression by modulating TLR-NF-κB signaling pathway. Cell Immunol 310:131–140. https://doi.org/10.1016/j.cellimm.2016.08.011

Billiau A, Matthys P (2001) Modes of action of Freund's adjuvants in experimental models of autoimmune diseases. J Leukoc Biol 70:849–860

Boesch-Saadatmandi C, Loboda A, Wagner AE, Stachurska A, Jozkowicz A, Dulak J, Döring F, Wolffram S, Rimbach G (2011) Effect of quercetin and its metabolites isorhamnetin and quercetin-3-glucuronide on inflammatory gene expression: role of miR-155. J Nutr Biochem 22:293–299. https://doi.org/10.1016/j.jnutbio.2010.02.008

Boozari M, Butler AE, Sahebkar A (2019) Impact of curcumin on toll-like receptors. J Cell Physiol 234:12471–12482. https://doi.org/10.1002/jcp.28103

Bouic PJD (2002) Sterols and sterolins: new drugs for the immune system? Drug Discov Today 7: 775–778. https://doi.org/10.1016/S1359-6446(02)02343-7

Byun E-B, Yang M-S, Choi H-G, Sung N-Y, Song D-S, Sin S-J, Byun E-H (2013) Quercetin negatively regulates TLR4 signaling induced by lipopolysaccharide through Tollip expression. Biochem Biophys Res Commun 431:698–705. https://doi.org/10.1016/j.bbrc.2013.01.056

Cabrera C, Artacho R, Giménez R (2006) Beneficial effects of green tea—a review. J Am Coll Nutr 25:79–99. https://doi.org/10.1080/07315724.2006.10719518

Calixto Júnior JT, de Morais SM, Gomez CV, Molas CC, Rolon M, Boligon AA, Athayde ML, de Morais Oliveira CD, Tintino SR, Henrique Douglas MC (2016) Phenolic composition and antiparasitic activity of plants from the Brazilian Northeast "Cerrado". S J Biol Sci 23:434–440. https://doi.org/10.1016/j.sjbs.2015.10.009

Chahal DS, Sivamani RK, Rivkah Isseroff R, Dasu MR (2013) Plant-based modulation of toll-like receptors: an emerging therapeutic model. Phytother Res 27:1423–1438. https://doi.org/10.1002/ptr.4886

Chaudhary PM, Ferguson C, Nguyen V, Nguyen O, Massa HF, Eby M, Jasmin A, Trask BJ, Hood L, Nelson PS (1998) Cloning and characterization of two toll/interleukin-1 receptor–like genes TIL3 and TIL4: evidence for a multi-gene receptor family in humans. Blood 91:4020–4027. https://doi.org/10.1182/blood.V91.11.4020

Chauhan K, Kaur G, Kaur S (2018) Activity of rutin, a potent flavonoid against SSG-sensitive and -resistant leishmania donovani parasites in experimental leishmaniasis. Int Immunopharmacol 64:372–385. https://doi.org/10.1016/j.intimp.2018.09.026

Chi J-F, Chu S-H, Lee C-S, Chou N-K, Su M-J (1996) Mechanical and electrophysiological effects of 8-oxoberberine (JKL1073A) on atrial tissue. Br J Pharmacol 118:503–512. https://doi.org/10.1111/j.1476-5381.1996.tb15431.x

Chuang C-C, McIntosh MK (2011) Potential mechanisms by which polyphenol-rich grapes prevent obesity-mediated inflammation and metabolic diseases. Annu Rev Nutr 31:155–176. https://doi.org/10.1146/annurev-nutr-072610-145149

Cianciulli A, Calvello R, Cavallo P, Dragone T, Carofiglio V, Panaro MA (2012) Modulation of NF-κB activation by resveratrol in LPS treated human intestinal cells results in downregulation of PGE2 production and COX-2 expression. Toxicol In Vitro 26:1122–1128. https://doi.org/10.1016/j.tiv.2012.06.015

Cui S, Bilitewski U (2014) Effect of genistein on the TLR and MAPK transduction cascades in lipopolysaccharide-stimulated macrophages. Chin J Cell Mol Immunol 30:233–236

Cushnie TPT, Cushnie B, Lamb AJ (2014) Alkaloids: An overview of their antibacterial, antibiotic-enhancing and antivirulence activities. Int J Antimicrob Agents 44:377–386. https://doi.org/10.1016/j.ijantimicag.2014.06.001

Dai J-P, Wang Q-W, Su Y, Gu L-M, Zhao Y, Chen X-X, Chen C, Li W-Z, Wang G-F, Li K-S (2017) Emodin inhibition of influenza A virus replication and influenza viral pneumonia via the Nrf 2, TLR4, p 38/JNK and NF-kappa B pathways. Molecules 22:1754. https://doi.org/10.3390/molecules22101754

Das BS, Das NC, Swain SS, Mukherjee S, Bhattacharya D (2022) Andrographolide induces anti-SARS-CoV-2 response through host-directed mechanism: an in silico study. Futur Virol. https://doi.org/10.2217/fvl-2021-0171

de Mejía EG, Prisecaru VI (2005) Lectins as bioactive plant proteins: a potential in cancer treatment. Crit Rev Food Sci Nutr 45:425–445. https://doi.org/10.1080/10408390591034445

Ding Y, Liu P, Chen Z-L, Zhang S-J, Wang Y-Q, Cai X, Luo L, Zhou X, Zhao L (2018) Emodin attenuates lipopolysaccharide-induced acute liver injury via inhibiting the TLR4 signaling pathway in vitro and in vivo. Front Pharmacol 9:962. https://doi.org/10.3389/fphar.2018.00962

Dionne S, Levy E, Levesque D, Seidman EG (2010) PPARgamma ligand 15-deoxy-delta 12, 14-prostaglandin J2 sensitizes human colon carcinoma cells to TWEAK-induced apoptosis. Anticancer Res 30:157–166

Diwan V, Poudyal H, Brown L (2013) Piperine attenuates cardiovascular, liver and metabolic changes in high carbohydrate, high fat-fed rats. Cell Biochem Biophys 67:297–304. https://doi.org/10.1007/s12013-011-9306-1

Duvoix A, Blasius R, Delhalle S, Schnekenburger M, Morceau F, Henry E, Dicato M, Diederich M (2005) Chemopreventive and therapeutic effects of curcumin. Cancer Lett 223:181–190. https://doi.org/10.1016/j.canlet.2004.09.041

Elshaer M, Chen Y, Wang XJ, Tang X (2018) Resveratrol: an overview of its anti-cancer mechanisms. Life Sci 207:340–349. https://doi.org/10.1016/j.lfs.2018.06.028

El-Zayat SR, Sibaii H, Mannaa FA (2019) Toll-like receptors activation, signaling, and targeting: an overview. Bull Natl Res Cent 43:187. https://doi.org/10.1186/s42269-019-0227-2

Fei L, Jifeng F, Tiantian W, Yi H, Linghui P (2017) Glycyrrhizin ameliorate ischemia reperfusion lung injury through downregulate TLR2 signaling Cascade in alveolar macrophages. Front Pharmacol 8:389. https://doi.org/10.3389/fphar.2017.00389

Fu S-L, Hsu Y-H, Lee P-Y, Hou W-C, Hung L-C, Lin C-H, Chen C-M, Huang Y-J (2006) Dioscorin isolated from Dioscorea alata activates TLR4-signaling pathways and induces cytokine expression in macrophages. Biochem Biophys Res Commun 339:137–144. https://doi.org/10.1016/j.bbrc.2005.11.005

Funatogawa K, Hayashi S, Shimomura H, Yoshida T, Hatano T, Ito H, Hirai Y (2004) Antibacterial activity of Hydrolyzable tannins derived from medicinal plants against *Helicobacter pylori*. Microbiol Immunol 48:251–261. https://doi.org/10.1111/j.1348-0421.2004.tb03521.x

Ganeshpurkar A, Saluja AK (2017) The pharmacological potential of Rutin. S Pharm J 25:149–164. https://doi.org/10.1016/j.jsps.2016.04.025

Ghahari S, Alinezhad H, Nematzadeh GA, Tajbakhsh M, Baharfar R (2017) Chemical composition, antioxidant and biological activities of the essential oil and extract of the seeds of Glycine max (soybean) from North Iran. Curr Microbiol 74:522–531. https://doi.org/10.1007/s00284-016-1188-4

Ghasemi-Pirbaluti M, Motaghi E, Bozorgi H (2017) The effect of menthol on acute experimental colitis in rats. Eur J Pharmacol 805:101–107. https://doi.org/10.1016/j.ejphar.2017.03.003

Gong T, Wang X, Yang Y, Yan Y, Yu C, Zhou R, Jiang W (2017) Plant lectins activate the NLRP3 inflammasome to promote inflammatory disorders. J Immunol 198:2082–2092. https://doi.org/10.4049/jimmunol.1600145

Grover A, Shandilya A, Punetha A, Bisaria VS, Sundar D (2010) Inhibition of the NEMO/IKKβ association complex formation, a novel mechanism associated with the NF-κB activation suppression by Withania somnifera's key metabolite Withaferin A. BMC Genomics 11:S25. https://doi.org/10.1186/1471-2164-11-S4-S25

Guimarães MR, Leite FRM, Spolidorio LC, Kirkwood KL, Rossa C (2013) Curcumin abrogates LPS-induced pro-inflammatory cytokines in RAW 264.7 macrophages. Evidence for novel mechanisms involving SOCS-1, −3 and p 38 MAPK. Arch Oral Biol 58:1309–1317. https://doi.org/10.1016/j.archoralbio.2013.07.005

Gupta PK, Chakraborty P, Kumar S, Singh PK, Rajan MGR, Sainis KB, Kulkarni S (2016) G1-4A, a polysaccharide from Tinospora cordifolia inhibits the survival of mycobacterium tuberculosis by modulating host immune responses in TLR4 dependent manner. PLoS One 11:e0154725. https://doi.org/10.1371/journal.pone.0154725

Gupta RA, Motiwala MN, Dumore NG, Danao KR, Ganjare AB (2015) Effect of piperine on inhibition of FFA induced TLR4 mediated inflammation and amelioration of acetic acid induced ulcerative colitis in mice. J Ethnopharmacol 164:239–246. https://doi.org/10.1016/j.jep.2015.01.039

Han Y (2009) Rutin has therapeutic effect on septic arthritis caused by Candida albicans. Int Immunopharmacol 9:207–211. https://doi.org/10.1016/j.intimp.2008.11.002
Han S, Wu H, Li W, Gao P (2015) Protective effects of genistein in homocysteine-induced endothelial cell inflammatory injury. Mol Cell Biochem 403:43–49. https://doi.org/10.1007/s11010-015-2335-0
Han SB, Yoon YD, Ahn HJ, Lee HS, Lee CW, Yoon WK, Park SK, Kim HM (2003) Toll-like receptor-mediated activation of B cells and macrophages by polysaccharide isolated from cell culture of Acanthopanax senticosus. Int Immunopharmacol 3:1301–1312. https://doi.org/10.1016/S1567-5769(03)00118-8
Hansur L, Louisa M, Wuyung PE, Fadilah F (2022) Daphnoretin from Carthamus tinctorius as a potential inflammatory inhibitor in COVID-19 by binding to toll-like Receptor-4: an in silico molecular docking study. Open Access Maced J Med Sci 10:220–227. https://doi.org/10.3889/oamjms.2022.7961
Hasegawa T, Akutsu K, Kishi Y, Nakamura K (2011) Constituents of the green tea seeds of *Camellia sinensis*. Nat Prod Commun 6:1934578X1100600. https://doi.org/10.1177/1934578X1100600314
Hashimoto C, Hudson KL, Anderson K, v. (1988) The toll gene of drosophila, required for dorsal-ventral embryonic polarity, appears to encode a transmembrane protein. Cell 52:269–279. https://doi.org/10.1016/0092-8674(88)90516-8
Hennessy EJ, Parker AE, O'Neill LAJ (2010) Targeting toll-like receptors: emerging therapeutics? Nat Rev Drug Discov 9:293–307. https://doi.org/10.1038/nrd3203
Hong Byun E, Fujimura Y, Yamada K, Tachibana H (2010) TLR4 signaling inhibitory pathway induced by green tea polyphenol epigallocatechin-3-gallate through 67-kDa laminin receptor. J Immunol 185:33–45. https://doi.org/10.4049/jimmunol.0903742
Hu D, Wang Y, Chen Z, Ma Z, You Q, Zhang X, Liang Q, Tan H, Xiao C, Tang X, Gao Y (2015) The protective effect of piperine on dextran sulfate sodium induced inflammatory bowel disease and its relation with pregnane X receptor activation. J Ethnopharmacol 169:109–123. https://doi.org/10.1016/j.jep.2015.04.006
Huang Y, Li X, Pan C, Cheng W, Wang X, Yang Z, Zheng L (2021) The intervention mechanism of emodin on TLR3 pathway in the process of central nervous system injury caused by herpes virus infection. Neurol Res 43:307–313. https://doi.org/10.1080/01616412.2020.1853989
Huang Q, Liu X, Zhao G, Hu T, Wang Y (2018) Potential and challenges of tannins as an alternative to in-feed antibiotics for farm animal production. Anim Nutr 4:137–150. https://doi.org/10.1016/j.aninu.2017.09.004
Hwang P-A, Hung Y-L, Chien S-Y (2015) Inhibitory activity of sargassum hemiphyllum sulfated polysaccharide in arachidonic acid-induced animal models of inflammation. J Food Drug Anal 23:49–56. https://doi.org/10.1016/j.jfda.2014.05.004
Jakus PB, Kalman N, Antus C, Radnai B, Tucsek Z, Gallyas F, Sumegi B, Veres B (2013) TRAF6 is functional in inhibition of TLR4-mediated NF-κB activation by resveratrol. J Nutr Biochem 24:819–823. https://doi.org/10.1016/j.jnutbio.2012.04.017
Jantan I, Ahmad W, Bukhari SNA (2015) Plant-derived immunomodulators: an insight on their preclinical evaluation and clinical trials. Front Plant Sci 6:655. https://doi.org/10.3389/fpls.2015.00655
Jayawardena TU, Sanjeewa KKA, Nagahawatta DP, Lee H-G, Lu Y-A, Vaas APJP, Abeytunga DTU, Nanayakkara CM, Lee D-S, Jeon Y-J (2020) Anti-inflammatory effects of sulfated polysaccharide from sargassum Swartzii in macrophages via blocking TLR/NF-Kb signal transduction. Mar Drugs 18:601. https://doi.org/10.3390/md18120601
Jeong SY, Kim M, Park EK, Kim J-S, Hahn D, Bae J-S (2020) Inhibitory functions of novel compounds from Dioscorea batatas Decne Peel on HMGB1-mediated septic responses. Biotechnol Bioprocess Eng 25:1–8. https://doi.org/10.1007/s12257-019-0382-1
Jeong J-W, Lee HH, Han MH, Kim G-Y, Kim W-J, Choi YH (2014) Anti-inflammatory effects of genistein via suppression of the toll-like receptor 4-mediated signaling pathway in

lipopolysaccharide-stimulated BV2 microglia. Chem Biol Interact 212:30–39. https://doi.org/10.1016/j.cbi.2014.01.012

Ju S-A, Nguyen Q-T, Nguyen T-HT, Suh J-H, An WG, Callaway Z, Joe Y, Chung HT, Kim B-S (2021) Pretreatment with 6-gingerol ameliorates sepsis-induced immune dysfunction by regulating the cytokine balance and reducing lymphocyte apoptosis. Oxidative Med Cell Longev 2021:1–20. https://doi.org/10.1155/2021/5427153

Jung T-Y, Pham TNN, Umeyama A, Shoji N, Hashimoto T, Lee J-J, Takei M (2010) Ursolic acid isolated from Uncaria rhynchophylla activates human dendritic cells via TLR2 and/or TLR4 and induces the production of IFN-γ by CD4+ naïve T cells. Eur J Pharmacol 643:297–303. https://doi.org/10.1016/j.ejphar.2010.06.030

Kaileh M, vanden Berghe W, Heyerick A, Horion J, Piette J, Libert C, de Keukeleire D, Essawi T, Haegeman G (2007) Withaferin A strongly elicits IκB kinase β hyperphosphorylation concomitant with potent inhibition of its kinase activity. J Biol Chem 282:4253–4264. https://doi.org/10.1074/jbc.M606728200

Karunaweera N, Raju R, Gyengesi E, Manch G (2015) Plant polyphenols as inhibitors of NF-Î°B induced cytokine productionâ a potential anti-inflammatory treatment for Alzheimer's disease? Front Mol Neurosci 8:24. https://doi.org/10.3389/fnmol.2015.00024

Kaur J, Sanyal SN (2010) Modulation of inflammatory changes in early stages of colon cancer through activation of PPARγ by diclofenac. Eur J Cancer Prev 19:319–327. https://doi.org/10.1097/CEJ.0b013e32833ad385

Kawai T, Akira S (2007) Signaling to NF-κB by toll-like receptors. Trends Mol Med 13:460–469. https://doi.org/10.1016/j.molmed.2007.09.002

Kawasaki T, Kawai T (2014) Toll-like receptor signaling pathways. Front Immunol 5:461. https://doi.org/10.3389/fimmu.2014.00461

Kennedy BK, Berger SL, Brunet A, Campisi J, Cuervo AM, Epel ES, Franceschi C, Lithgow GJ, Morimoto RI, Pessin JE, Rando TA, Richardson A, Schadt EE, Wyss-Coray T, Sierra F (2014) Geroscience: linking aging to chronic disease. Cell 159:709–713. https://doi.org/10.1016/j.cell.2014.10.039

Kumar D, Arya V, Kaur R, Bhat ZA, Gupta VK, Kumar V (2012) A review of immunomodulators in the Indian traditional health care system. J Microbiol Immunol Infect 45:165–184. https://doi.org/10.1016/j.jmii.2011.09.030

Kumar S, Singhal V, Roshan R, Sharma A, Rembhotkar GW, Ghosh B (2007) Piperine inhibits TNF-α induced adhesion of neutrophils to endothelial monolayer through suppression of NF-κB and IκB kinase activation. Eur J Pharmacol 575:177–186. https://doi.org/10.1016/j.ejphar.2007.07.056

Lee HJ, Hyun E-A, Yoon WJ, Kim BH, Rhee MH, Kang HK, Cho JY, Yoo ES (2006) In vitro anti-inflammatory and anti-oxidative effects of Cinnamomum camphora extracts. J Ethnopharmacol 103:208–216. https://doi.org/10.1016/j.jep.2005.08.009

Lee HA, Song YR, Park MH, Chung H, Na HS, Chung J (2020) Catechin ameliorates *Porphyromonas gingivalis-* induced inflammation via the regulation of TLR2/4 and inflammasome signaling. J Periodontol 91:661–670. https://doi.org/10.1002/JPER.18-0004

Li C, Ai G, Wang Y, Lu Q, Luo C, Tan L, Lin G, Liu Y, Li Y, Zeng H, Chen J, Lin Z, Xian Y, Huang X, Xie J, Su Z (2020) Oxyberberine, a novel gut microbiota-mediated metabolite of berberine, possesses superior anti-colitis effect: impact on intestinal epithelial barrier, gut microbiota profile and TLR4-MyD88-NF-κB pathway. Pharmacol Res 152:104603. https://doi.org/10.1016/j.phrs.2019.104603

Li L, Peng L, Miao J, Qiu Y, Zhou Y, Gao X, Xu Y, Shi Z, Shao D, Ma Z (2011) Icariin induces the expression of toll-like receptor 9 in ana-1 murine macrophages. Phytother Res 25:1732–1735. https://doi.org/10.1002/ptr.3514

Li C-L, Tan L-H, Wang Y-F, Luo C-D, Chen H-B, Lu Q, Li Y-C, Yang X-B, Chen J-N, Liu Y-H, Xie J-H, Su Z-R (2019) Comparison of anti-inflammatory effects of berberine, and its natural oxidative and reduced derivatives from Rhizoma Coptidis in vitro and in vivo. Phytomedicine 52:272–283. https://doi.org/10.1016/j.phymed.2018.09.228

Li F, Wang H, Lu D, Wang Y, Qi R, Fu Y, Li C (2006) Neutral sulfate berberine modulates cytokine secretion and increases survival in endotoxemic mice. Acta Pharmacol Sin 27:1199–1205. https://doi.org/10.1111/j.1745-7254.2006.00368.x

Li J, Xie C, Zhuang J, Li H, Yao Y, Shao C, Wang H (2014) Resveratrol attenuates inflammation in the rat heart subjected to ischemia-reperfusion: role of the TLR4/NF-κB signaling pathway. Mol Med Rep 11:1120. https://doi.org/10.3892/mmr.2014.2955

Li X, Xu W (2011) TLR4-mediated activation of macrophages by the polysaccharide fraction from Polyporus umbellatus (pers.) fries. J Ethnopharmacol 135:1–6. https://doi.org/10.1016/j.jep.2010.06.028

Lindsay SA, Wasserman SA (2014) Conventional and non-conventional drosophila toll signaling. Dev Comp Immunol 42:16–24. https://doi.org/10.1016/j.dci.2013.04.011

Liu Y, Liu N, Liu Y, He H, Luo Z, Liu W, Song N, Ju M (2022) Ginsenoside Rb1 reduces D-gal N/LPS-induced acute liver injury by regulating TLR4/NF-κB signaling and NLRP3 inflammasome. J Clin Transl Hepatol 10:474–485. https://doi.org/10.14218/JCTH.2021.00072

Liu X, Wang J (2011a) Anti-inflammatory effects of iridoid glycosides fraction of folium syringae leaves on TNBS-induced colitis in rats. J Ethnopharmacol 133:780–787. https://doi.org/10.1016/j.jep.2010.11.010

Liu X, Wang JM (2011b) Iridoid glycosides fraction of folium syringae leaves modulates NF-κB signal pathway and intestinal epithelial cells apoptosis in experimental colitis. PLoS One 6: e24740. https://doi.org/10.1371/journal.pone.0024740

Logie E, vanden Berghe W (2020) Tackling chronic inflammation with Withanolide phytochemicals—a Withaferin A perspective. Antioxidants 9:1107. https://doi.org/10.3390/antiox9111107

Lu R, Wu Y, Guo H, Zhang Z, He Y (2021) Salidroside protects against influenza a virus-induced acute lung injury in mice. Dose-Response 19:155932582110113. https://doi.org/10.1177/15593258211011335

Ludwiczuk A, Skalicka-Woźniak K, Georgiev MI (2017) Terpenoids. In: Pharmacognosy. Elsevier, Amsterdam, pp 233–266. https://doi.org/10.1016/B978-0-12-802104-0.00011-1

Manach C, Scalbert A, Morand C, Rémésy C, Jiménez L (2004) Polyphenols: food sources and bioavailability. Am J Clin Nutr 79:727–747. https://doi.org/10.1093/ajcn/79.5.727

Manikandan R, Parimalanandhini D, Mahalakshmi K, Beulaja M, Arumugam M, Janarthanan S, Palanisamy S, You S, Prabhu NM (2020) Studies on isolation, characterization of fucoidan from brown algae Turbinaria decurrens and evaluation of it's in vivo and in vitro anti-inflammatory activities. Int J Biol Macromol 160:1263–1276. https://doi.org/10.1016/j.ijbiomac.2020.05.152

Marzocchella L, Fantini M, Benvenuto M, Masuelli L, Tresoldi I, Modesti A, Bei R (2011) Dietary flavonoids: molecular mechanisms of action as anti- inflammatory agents. Recent Patents Inflamm Allergy Drug Discov 5:200–220. https://doi.org/10.2174/187221311797264937

Meng Z, Yan C, Deng Q, Gao D, Niu X (2013) Curcumin inhibits LPS-induced inflammation in rat vascular smooth muscle cells in vitro via ROS-relative TLR4-MAPK/NF-κB pathways. Acta Pharmacol Sin 34:901–911. https://doi.org/10.1038/aps.2013.24

Ming D, Wang D, Cao F, Xiang H, Mu D, Cao J, Li B, Zhong L, Dong X, Zhong X, Wang L, Wang T (2017) Kaempferol inhibits the primary attachment phase of biofilm formation in Staphylococcus aureus. Front Microbiol 8:2263. https://doi.org/10.3389/fmicb.2017.02263

Molan AL, Duncan AJ, Barry TN, McNabb WC (2003) Effects of condensed tannins and crude sesquiterpene lactones extracted from chicory on the motility of larvae of deer lungworm and gastrointestinal nematodes. Parasitol Int 52:209–218. https://doi.org/10.1016/S1383-5769(03)00011-4

Molan AL, McNabb WC, Hoskin SO, Barry TN (2000a) Effect of condensed tannins extracted from four forages on the viability of the larvae of deer lungworms and gastrointestinal nematodes. Vet Rec 147:44–48. https://doi.org/10.1136/vr.147.2.44

Molan AL, Sivakumaran S, Spencer PA, Meagher LP (2004) Green tea flavan-3-ols and oligomeric proanthocyanidins inhibit the motility of infective larvae of Teladorsagia circumcincta and Trichostrongylus colubriformis in vitro. Res Vet Sci 77:239–243. https://doi.org/10.1016/j.rvsc.2004.04.010

Molan AL, Waghorn GC, McNabb WC (2002) Effect of condensed tannins on egg hatching and larval development of *Trichostrongylus colubriformis* in vitro. Vet Rec 150:65–69. https://doi.org/10.1136/vr.150.3.65

Molan AL, Waghorn GC, Min BR, McNabb WC (2000b) The effect of condensed tannins from seven herbages on Trichostrongylus colubriformis larval migration in vitro. Folia Parasitol 47: 39–44. https://doi.org/10.14411/fp.2000.007

Montenegro-Landívar MF, Tapia-Quirós P, Vecino X, Reig M, Valderrama C, Granados M, Cortina JL, Saurina J (2021) Polyphenols and their potential role to fight viral diseases: an overview. Sci Total Environ 801:149719. https://doi.org/10.1016/j.scitotenv.2021.149719

Mukherjee S, Mukherjee N, Saini P, Roy P, SPS B (2015) Ginger extract ameliorates phosphamidon induced hepatotoxicity. Indian J Exp Biol 53:574

Nascimento APM, Knaut JL, Rieger DK, Wolin IAV, Heinrich IA, Mann J, Juarez AV, del Sosa LV, de Paul AL, Moreira CG, Silva IB, Nobre CS, Osterne VJS, Nascimento KS, Cavada BS, Leal RB (2018) Anti-glioma properties of DVL, a lectin purified from Dioclea violacea. Int J Biol Macromol 120:566–577. https://doi.org/10.1016/j.ijbiomac.2018.08.106

Newton K, Dixit VM (2012) Signaling in innate immunity and inflammation. Cold Spring Harb Perspect Biol 4:a006049–a006049. https://doi.org/10.1101/cshperspect.a006049

Ng A, Xavier RJ (2011) Leucine-rich repeat (LRR) proteins: integrators of pattern recognition and signaling in immunity. Autophagy 7:1082–1084. https://doi.org/10.4161/auto.7.9.16464

Noh E-J, Kang M-J, Jeong Y-J, Lee J-Y, Park J-H, Choi H-J, Oh S-M, Lee K-B, Kim D-J, Shin J-A, Cho S-D, Park J-H (2016) Withaferin A inhibits inflammatory responses induced by fusobacterium nucleatum and Aggregatibacter actinomycetemcomitans in macrophages. Mol Med Rep 14:983–988. https://doi.org/10.3892/mmr.2016.5326

Norris JL, Manley JL (1992) Selective nuclear transport of the drosophila morphogen dorsal can be established by a signaling pathway involving the transmembrane protein toll and protein kinase A. Genes Dev 6:1654–1667. https://doi.org/10.1101/gad.6.9.1654

Nunes BS, Rensonnet NS, Dal-Secco D, Vieira SM, Cavada BS, Teixeira EH, Moura TR, Teixeira CS, Clemente-Napimoga JT, Cunha FQ, Napimoga MH (2009) Lectin extracted from Canavalia grandiflora seeds presents potential anti-inflammatory and analgesic effects. Naunyn Schmiedeberg's Arch Pharmacol 379:609–616. https://doi.org/10.1007/s00210-009-0397-9

Okamoto M, Oh-e G, Oshikawa T, Furuichi S, Tano T, Ahmed SU, Akashi S, Miyake K, Takeuchi O, Akira S, Himeno K, Sato M, Ohkubo S (2004) Toll-like receptor 4 mediates the antitumor host response induced by a 55-kilodalton protein isolated from *Aeginetia indica* L., a parasitic plant. Clin Vaccine Immunol 11:483–495. https://doi.org/10.1128/CDLI.11.3.483-495.2004

Oleszek M, Oleszek W (2021) Saponins in food. In: Handbook of dietary phytochemicals. Springer, Singapore, pp 1501–1540. https://doi.org/10.1007/978-981-15-4148-3_34

Panaro MA, Carofiglio V, Acquafredda A, Cavallo P, Cianciulli A (2012) Anti-inflammatory effects of resveratrol occur via inhibition of lipopolysaccharide-induced NF-κB activation in Caco-2 and SW480 human colon cancer cells. Br J Nutr 108:1623–1632. https://doi.org/10.1017/S0007114511007227

Panaro MA, Corrado A, Benameur T, Paolo CF, Cici D, Porro C (2020) The emerging role of curcumin in the modulation of TLR-4 signaling pathway: focus on neuroprotective and anti-rheumatic properties. Int J Mol Sci 21:2299. https://doi.org/10.3390/ijms21072299

Panche AN, Diwan AD, Chandra SR (2016) Flavonoids: an overview. J Nutr Sci 5:e47. https://doi.org/10.1017/jns.2016.41

Patel SB (2008) Plant sterols and stanols: their role in health and disease. J Clin Lipidol 2:S11–S19. https://doi.org/10.1016/j.jacl.2008.01.007

Pathak N, Khandelwal S (2009) Immunomodulatory role of piperine in cadmium induced thymic atrophy and splenomegaly in mice. Environ Toxicol Pharmacol 28:52–60. https://doi.org/10.1016/j.etap.2009.02.003

Peng L-N, Li L, Qiu Y-F, Miao J-H, Gao X-Q, Zhou Y, Shi Z-X, Xu Y-L, Shao D-H, Wei J-C, Ma Z-Y (2011) Glycyrrhetinic acid extracted from Glycyrrhiza uralensis Fisch. Induces the expression of toll-like receptor 4 in ana-1 murine macrophages. J Asian Nat Prod Res 13:942–950. https://doi.org/10.1080/10286020.2011.603305

Polkowski K, Mazurek AP (2000) Biological properties of genistein. A review of in vitro and in vivo data. Acta Pol Pharm 57:135–155

Qureshi AA, Guan X, Reis JC, Papasian CJ, Jabre S, Morrison DC, Qureshi N (2012) Inhibition of nitric oxide and inflammatory cytokines in LPS-stimulated murine macrophages by resveratrol, a potent proteasome inhibitor. Lipids Health Dis 11:76. https://doi.org/10.1186/1476-511X-11-76

Ray AS, Joardar N, Mukherjee S, Rahaman CH, Babu S (2018) S.P.: polyphenol enriched ethanolic extract of Cajanus scarabaeoides (L.) Thouars exerts potential antifilarial activity by inducing oxidative stress and programmed cell death. PLoS One 13:e0208201. https://doi.org/10.1371/journal.pone.0208201

Ren J, Lu Y, Qian Y, Chen B, Wu T, Ji G (2019) Recent progress regarding kaempferol for the treatment of various diseases (review). Exp Ther Med 18:2759. https://doi.org/10.3892/etm.2019.7886

Rida PCG, LiVecche D, Ogden A, Zhou J, Aneja R (2015) The noscapine chronicle: a Pharmaco-historic biography of the opiate alkaloid family and its clinical applications. Med Res Rev 35:1072–1096. https://doi.org/10.1002/med.21357

Rock FL, Hardiman G, Timans JC, Kastelein RA, Bazan JF (1998) A family of human receptors structurally related to *drosophila* toll. Proc Natl Acad Sci 95:588–593. https://doi.org/10.1073/pnas.95.2.588

Roy S, Dutta D, Satyavarapu EM, Yadav PK, Mandal C, Kar S, Mandal C (2017) Mahanine exerts in vitro and in vivo antileishmanial activity by modulation of redox homeostasis. Sci Rep 7:4141. https://doi.org/10.1038/s41598-017-03943-y

Roy MK, Thalang VN, Trakoontivakorn G, Nakahara K (2004) Mechanism of mahanine-induced apoptosis in human leukemia cells (HL-60). Biochem Pharmacol 67:41–51. https://doi.org/10.1016/j.bcp.2003.07.021

Saini P, Gayen P, Kumar D, Nayak A, Mukherjee N, Mukherjee S, Pal BC, Babu SPS (2014) Antifilarial effect of ursolic acid from Nyctanthes arbortristis: molecular and biochemical evidences. Parasitol Int 63:717–728. https://doi.org/10.1016/j.parint.2014.06.008

Salama A, Mahmoud HA-A, Kandeil MA, Khalaf MM (2021) Neuroprotective role of camphor against ciprofloxacin induced depression in rats: modulation of Nrf-2 and TLR4. Immunopharmacol Immunotoxicol 43:309–318. https://doi.org/10.1080/08923973.2021.1905658

Saleh HA, Yousef MH, Abdelnaser A (2021) The anti-inflammatory properties of phytochemicals and their effects on epigenetic mechanisms involved in TLR4/NF-κB-mediated inflammation. Front Immunol 12:606069. https://doi.org/10.3389/fimmu.2021.606069

Saqib U, Sarkar S, Suk K, Mohammad O, Baig MS, Savai R (2018) Phytochemicals as modulators of M1-M2 macrophages in inflammation. Oncotarget 9:17937–17950. https://doi.org/10.18632/oncotarget.24788

Sayed N, Khurana A, Saifi MA, Singh M, Godugu C (2019) Withaferin A reverses bile duct ligation-induced liver fibrosis by modulating extracellular matrix deposition: role of LOXL2/snail 1, vimentin, and NFκB signaling. Bio Factors 45:959–974. https://doi.org/10.1002/biof.1546

Scalbert A (1991) Antimicrobial properties of tannins. Phytochemistry 30:3875–3883
Serrano J, Puupponen-Pimiä R, Dauer A, Aura A-M, Saura-Calixto F (2009) Tannins: current knowledge of food sources, intake, bioavailability and biological effects. Mol Nutr Food Res 53: S310–S329. https://doi.org/10.1002/mnfr.200900039
Shah VO, Ferguson JE, Hunsaker LA, Deck LM, vander Jagt, D.L. (2010) Natural products inhibit LPS-induced activation of pro-inflammatory cytokines in peripheral blood mononuclear cells. Nat Prod Res 24:1177–1188. https://doi.org/10.1080/14786410903112680
Shan L, Kang X, Liu F, Cai X, Han X, Shang Y (2018) Epigallocatechin gallate improves airway inflammation through TGF-β1 signaling pathway in asthmatic mice. Mol Med Rep 18:2088. https://doi.org/10.3892/mmr.2018.9183
Shaukat A, Guo Y, Jiang K, Zhao G, Wu H, Zhang T, Yang Y, Guo S, Yang C, Zahoor A, Akhtar M, Umar T, Shaukat I, Rajput SA, Hassan M, Deng G (2019) Ginsenoside Rb1 ameliorates Staphylococcus aureus-induced acute lung injury through attenuating NF-κB and MAPK activation. Microb Pathog 132:302–312. https://doi.org/10.1016/j.micpath.2019.05.003
Shen N, Cheng A, Qiu M, Zang G (2019) Allicin improves lung injury induced by sepsis via regulation of the toll-like receptor 4 (TLR4)/myeloid differentiation primary response 88 (MYD88)/nuclear factor kappa B (NF-κB) pathway. Med Sci Monit 25:2567–2576. https://doi.org/10.12659/MSM.914114
Silva-Filho S, de Souza Silva-Comar F, Wiirzler L, do Pinho R, Grespan R, Bersani-Amado C, Cuman R (2015) Effect of camphor on the behavior of leukocytes: in vitro and in vivo in acute inflammatory response. Trop J Pharm Res 13:2031. https://doi.org/10.4314/tjpr.v13i12.13
Silverman N, Paquette N, Aggarwal K (2009) Specificity and signaling in the drosophila immune response. Invertebrate Surviv J 6:163–174
Singh G, Kumar P (2013) Phytochemical study and screening for antimicrobial activity of flavonoids of Euphorbia hirta. Int J Appl Basic Med Research 3:111. https://doi.org/10.4103/2229-516X.117082
Singh B, Sharma RA (2015) Plant terpenes: defense responses, phylogenetic analysis, regulation and clinical applications. 3 Biotech 5:129–151. https://doi.org/10.1007/s13205-014-0220-2
Souto AL, Tavares JF, da Silva MS, Diniz M, de Athayde-Filho PF, Barbosa Filho JM (2011) Anti-inflammatory activity of alkaloids: an update from 2000 to 2010. Molecules 16:8515–8534. https://doi.org/10.3390/molecules16108515
Sparg SG, Light ME, van Staden J (2004) Biological activities and distribution of plant saponins. J Ethnopharmacol 94:219–243. https://doi.org/10.1016/j.jep.2004.05.016
Sun GY, Li R, Cui J, Hannink M, Gu Z, Fritsche KL, Lubahn DB, Simonyi A (2016) Withania somnifera and its Withanolides attenuate oxidative and inflammatory responses and up-regulate antioxidant responses in BV-2 microglial cells. Neuro Mol Med 18:241–252. https://doi.org/10.1007/s12017-016-8411-0
Sutherland AM, Cook DN (2005) Polymorphisms of the toll-like receptors and human disease. Clin Infect Dis 41:S403–S407. https://doi.org/10.1086/431985
Tabeshpour J, Mehri S, Shaebani Behbahani F, Hosseinzadeh H (2018) Protective effects of Vitis vinifera (grapes) and one of its biologically active constituents, resveratrol, against natural and chemical toxicities: a comprehensive review. Phytother Res 32:2164–2190. https://doi.org/10.1002/ptr.6168
Valanne S, Wang J-H, Rämet M (2011) The drosophila toll signaling pathway. J Immunol 186:649–656. https://doi.org/10.4049/jimmunol.1002302
Vasselon T, Detmers PA (2002) Toll receptors: a central element in innate immune responses. Infect Immun 70:1033–1041. https://doi.org/10.1128/IAI.70.3.1033-1041.2002
Vetvicka V, Vannucci L (2021) Biological properties of andrographolide, an active ingredient of Andrographis Paniculata: a narrative review. Ann Transl Med 9:1186–1186. https://doi.org/10.21037/atm-20-7830

Wang Y, Chen G, Yu X, Li Y, Zhang L, He Z, Zhang N, Yang X, Zhao Y, Li N, Qiu H (2016) Salvianolic acid B ameliorates cerebral ischemia/reperfusion injury through inhibiting TLR4/MyD88 signaling pathway. Inflammation 39:1503–1513. https://doi.org/10.1007/s10753-016-0384-5

Wang G, Hu Z, Fu Q, Song X, Cui Q, Jia R, Zou Y, He C, Li L, Yin Z (2017) Resveratrol mitigates lipopolysaccharide-mediated acute inflammation in rats by inhibiting the TLR4/NF-κBp65/MAPKs signaling cascade. Sci Rep 7:45006. https://doi.org/10.1038/srep45006

Wang R, Li S, Jia H, Si X, Lei Y, Lyu J, Dai Z, Wu Z (2021a) Protective effects of cinnamaldehyde on the inflammatory response, oxidative stress, and apoptosis in liver of salmonella typhimurium-challenged mice. Molecules 26:2309. https://doi.org/10.3390/molecules26082309

Wang W, Li Q, Wu J, Hu Y, Wu G, Yu C, Xu K, Liu X, Wang Q, Huang W, Wang L, Wang Y (2021b) Lentil lectin derived from *Lens culinaris* exhibit broad antiviral activities against SARS-CoV-2 variants. Emerg Microbes Infect 10:1519–1529. https://doi.org/10.1080/22221751.2021.1957720

Wang G, Tang W, Bidigare RR (2005) Terpenoids as therapeutic drugs and pharmaceutical agents. In: Natural products. Humana Press, Totowa, NJ, pp 197–227. https://doi.org/10.1007/978-1-59259-976-9_9

Wang H-Y, Zhang Y-Q (2019) The main active constituents and detoxification process of Ginkgo biloba seeds and their potential use in functional health foods. J Food Compos Anal 83:103247. https://doi.org/10.1016/j.jfca.2019.103247

Williams AR, Krych L, Fauzan Ahmad H, Nejsum P, Skovgaard K, Nielsen DS, Thamsborg SM (2017) A polyphenol-enriched diet and ascaris suum infection modulate mucosal immune responses and gut microbiota composition in pigs. PLoS One 12:e0186546. https://doi.org/10.1371/journal.pone.0186546

Wu Y, Huang D, Wang X, Pei C, Xiao W, Wang F, Wang Z (2021) Suppression of NLRP3 inflammasome by Platycodin D via the TLR4/MyD88/NF-κB pathway contributes to attenuation of lipopolysaccharide induced acute lung injury in rats. Int Immunopharmacol 96:107621. https://doi.org/10.1016/j.intimp.2021.107621

Wu K, Yang J, Zhou Q (2010) Preventive effects of berberine on experimental colon cancer and relationship with cyclooxygenase-2 expression. Zhongguo Zhong Yao Za Zhi 35:2768–2773

Wu X, Yu C, Cai W, Hua J, Li S, Wang W (2011) Protective effect of a polyphenolic rich extract from Magnolia officinalis bark on influenza virus-induced pneumonia in mice. J Ethnopharmacol 134:191–194. https://doi.org/10.1016/j.jep.2010.11.074

Xu L, Liu J-T, Li K, Wang S-Y, Xu S (2019) Genistein inhibits ang II-induced CRP and MMP-9 generations via the ER-p 38/ERK1/2-PPARγ-NF-κB signaling pathway in rat vascular smooth muscle cells. Life Sci 216:140–146. https://doi.org/10.1016/j.lfs.2018.11.036

Xue L, Jiang Y, Han T, Zhang N, Qin L, Xin H, Zhang Q (2016) Comparative proteomic and metabolomic analysis reveal the antiosteoporotic molecular mechanism of icariin from Epimedium brevicornu maxim. J Ethnopharmacol 192:370–381. https://doi.org/10.1016/j.jep.2016.07.037

Yasui M, Matsushima M, Omura A, Mori K, Ogasawara N, Kodera Y, Shiga M, Ito K, Kojima S, Kawabe T (2015) The suppressive effect of quercetin on toll-like receptor 7-mediated activation in alveolar macrophages. Pharmacology 96:201–209. https://doi.org/10.1159/000438993

Yin J, Peng X, Lin J, Zhang Y, Zhang J, Gao H, Tian X, Zhang R, Zhao G (2021) Quercetin ameliorates aspergillus fumigatuskeratitis by inhibiting fungal growth, toll-like receptors and inflammatory cytokines. Int Immunopharmacol 93:107435. https://doi.org/10.1016/j.intimp.2021.107435

Youn HS, Lee JY, Saitoh SI, Miyake K, Kang KW, Choi YJ, Hwang DH (2006b) Suppression of MyD88- and TRIF-dependent signaling pathways of toll-like receptor by (−)-epigallocatechin-3-gallate, a polyphenol component of green tea. Biochem Pharmacol 72:850–859. https://doi.org/10.1016/j.bcp.2006.06.021

Youn HS, Saitoh SI, Miyake K, Hwang DH (2006a) Inhibition of homodimerization of toll-like receptor 4 by curcumin. Biochem Pharmacol 72:62–69. https://doi.org/10.1016/j.bcp.2006.03.022

Zaia MG, di Cagnazzo TO, Feitosa KA, Soares EG, Faccioli LH, Allegretti SM, Afonso A, de Anibal F (2016) Anti-inflammatory properties of menthol and menthone in Schistosoma mansoni infection. Front Pharmacol 7:170. https://doi.org/10.3389/fphar.2016.00170

Zhang R, Ai X, Duan Y, Xue M, He W, Wang C, Xu T, Xu M, Liu B, Li C, Wang Z, Zhang R, Wang G, Tian S, Liu H (2017) Kaempferol ameliorates H9N2 swine influenza virus-induced acute lung injury by inactivation of TLR4/MyD88-mediated NF-κB and MAPK signaling pathways. Biomed Pharmacother 89:660–672. https://doi.org/10.1016/j.biopha.2017.02.081

Zhang S, Feng Z, Gao W, Duan Y, Fan G, Geng X, Wu B, Li K, Liu K, Peng C (2020a) Aucubin attenuates liver ischemia-reperfusion injury by inhibiting the HMGB1/TLR-4/NF-κB signaling pathway, oxidative stress, and apoptosis. Front Pharmacol 11:544124. https://doi.org/10.3389/fphar.2020.544124

Zhang Y, Han D, Yu S, An C, Liu X, Zhong H, Xu Y, Jiang L, Wang Z (2020c) Protective effect of iridoid glycosides of the leaves of syringa oblata lindl. on dextran sulfate sodium-induced ulcerative colitis by inhibition of the TLR2/4/MyD88/NF- κ B signaling pathway. Bio Med Res Int 2020:1–13. https://doi.org/10.1155/2020/7650123

Zhang H, Lang W, Wang S, Li B, Li G, Shi Q (2020b) Echinacea polysaccharide alleviates LPS-induced lung injury via inhibiting inflammation, apoptosis and activation of the TLR4/NF-κB signal pathway. Int Immunopharmacol 88:106974. https://doi.org/10.1016/j.intimp.2020.106974

Zhang J, Zheng Y, Luo Y, Du Y, Zhang X, Fu J (2019) Curcumin inhibits LPS-induced neuroinflammation by promoting microglial M2 polarization via TREM2/TLR4/NF-κB pathways in BV2 cells. Mol Immunol 116:29–37. https://doi.org/10.1016/j.molimm.2019.09.020

Zhao Z, Xiao J, Wang J, Dong W, Peng Z, An D (2015) Anti-inflammatory effects of novel sinomenine derivatives. Int Immunopharmacol 29:354–360. https://doi.org/10.1016/j.intimp.2015.10.030

Zhao H, Zhao M, Wang Y, Li F, Zhang Z (2016) Glycyrrhizic acid prevents sepsis-induced acute lung injury and mortality in rats. J Histochem Cytochem 64:125–137. https://doi.org/10.1369/0022155415610168

Zheng D, Liwinski T, Elinav E (2020) Interaction between microbiota and immunity in health and disease. Cell Res 30:492–506. https://doi.org/10.1038/s41422-020-0332-7

Zhou J, Yu W, Zhang M, Tian X, Li Y, Lü Y (2019) Imbalance of microglial TLR4/TREM2 in LPS-treated APP/PS1 transgenic mice: a potential link between Alzheimer's disease and systemic inflammation. Neurochem Res 44:1138–1151. https://doi.org/10.1007/s11064-019-02748-x

Zhu H, Bian C, Yuan J, Chu W, Xiang X, Chen F, Wang C, Feng H, Lin J (2014) Curcumin attenuates acute inflammatory injury by inhibiting the TLR4/MyD88/NF-κB signaling pathway in experimental traumatic brain injury. J Neuroinflammation 11:59. https://doi.org/10.1186/1742-2094-11-59

Zhu J, Tang H, Zhang Z, Zhang Y, Qiu C, Zhang L, Huang P, Li F (2017) Kaempferol slows intervertebral disc degeneration by modifying LPS-induced osteogenesis/adipogenesis imbalance and inflammation response in BMSCs. Int Immunopharmacol 43:236–242. https://doi.org/10.1016/j.intimp.2016.12.020

Zou K, Li Z, Zhang Y, Zhang H, Li B, Zhu W, Shi J, Jia Q, Li Y (2017) Advances in the study of berberine and its derivatives: a focus on anti-inflammatory and anti-tumor effects in the digestive system. Acta Pharmacol Sin 38:157–167. https://doi.org/10.1038/aps.2016.125

Chapter 4
Rejuvenating the Potential of Antimicrobials Via Targeted Therapy of Efflux Pumps: The Advent of Phytotherapeutics

Tannishtha Biswas, Mehnaz Ahmed, and Susmita Mondal

Abstract Antimicrobial resistance (AMR) and the emergence of multidrug-resistant organisms are one of the most eminent impediments of recent times. The intrinsic resistance mechanisms of the pathogenic organisms, especially the efflux pumps, play a significant part in the development of AMR and have to date resulted in the failure of a wide range of antimicrobial drugs. The efflux mechanisms exude the drugs from the cells and prevent their accumulation, thereby rendering the drug futile against the pathogenic microbe. Therefore, in order to combat with these evolving MDR microorganisms, their intrinsic resistance processes like the efflux pumps need to be targeted. In this regard, the advent of phytotherapeutic molecules isolated from specific medicinal plants has shown promising potential as efflux pump inhibitors (EPIs) with their capability of modifying the resistance mechanisms of pathogens. These botanicals show structural similarity or mimic certain peptide structure that increases their binding affinity to the efflux pump proteins. As a result, either the EPI binds to the drug binding pocket or the configuration of the efflux pumps is altered as a result of the binding, thus preventing the drug efflux. This review primarily focuses on the mode of action of the microbial efflux pumps, their contribution to enhancing microbial resistance, the emerging role of phytotherapeutics, and their potential to act as efflux pump inhibitors (EPIs).

Keywords Antimicrobial resistance (AMR) · Multidrug resistance · Efflux pumps · Efflux pump inhibitors · Phytotherapeutics · Medicinal plants

Abbreviations

ABC	ATP-binding cassette
AceI	Acinetobacter chlorhexidine efflux I

T. Biswas · M. Ahmed · S. Mondal (✉)
Department of Life Sciences, Presidency University, Kolkata, India
e-mail: susmita.dbs@presiuniv.ac.in

V. Kumar et al. (eds.), *Medicinal Plants and Antimicrobial Therapies*,
https://doi.org/10.1007/978-981-99-7261-6_4

AMR	Antimicrobial resistance
ATP	Adenosine triphosphate
EmrE	Efflux multidrug resistance E
EP	Efflux pump
EPI	Efflux pump inhibitor
MATE	Multidrug and toxic compound extrusion
MDR	Multidrug resistant
MFS	Major facilitator superfamily
MHC	Methoxyhydnocarpin
MsrA	Methionine sulfoxide reductase A
PACE	Proteobacterial antimicrobial compound efflux
PAβN	Phe-Arg-β-naphthylamine
PMF	Proton motive force
RND	Resistance nodulation division
SMR	Small multidrug resistant
TMS	Transmembrane segments

4.1 Introduction

One of the most significant obstacles faced by the contemporary world of research is the aspect of antimicrobial resistance (AMR). The vast range of infections caused by microorganisms such as bacteria, fungi, viruses, and parasites, no longer remain susceptible to the standardized therapeutics developed against them, primarily owing to AMR (Abushaheen et al. 2020). The evolving capacity of these microorganisms inculcates them with the power to develop resistance mechanisms, thereby allowing them to easily bypass the action of the applied drugs and thrive in their very presence. Resistant organisms therefore have the power to wreak havoc in the host system without getting affected by any of the administered drugs (Hughes and Andersson 2017). Furthermore, the rapid development of resistant mechanisms of microbes has almost rendered them imperishable, and this has turned out to be a major challenge in the times of modern research. The advent of microbial resistance to multiple drugs also known as multidrug resistance (MDR) has increased the threat by multiple folds (Piddock 2006a). The mechanisms via which the organisms develop resistance can be broadly distributed into three categories, namely, adaptive resistance, acquired resistance, and intrinsic resistance (Kabra et al. 2019). Gradual exposure to increasing doses of drugs often results in an onset of sudden inception of resistance in certain specific microorganisms that is to some extent associated with the diverseness of patterns of gene expression and might require phenotypic and epigenetic inheritance (Kabra et al. 2019). This form of AMR is known as adaptive resistance. On the contrary, some microorganisms develop acquired resistance primarily because of resistant specific horizontal gene transfer or due to some form of genetic mutation. Intrinsic resistance, however, is mainly associated with the internal mechanisms of microorganisms, which are predominantly concerned with reducing the uptake of drugs by the cell with the help of efflux pumps or with decreasing the cellular concentration of the drugs that have already been taken up by the cell (Reygaert et al. 2018). Therefore, with respect to AMR, the only probable option to deter the threat posed by the resistant superbugs is to develop therapeutic strategies that target the very core of resistant mechanisms in the microbial cells (Motta et al. 2015). Targeting the resistant developing mechanisms in the microbes would provide an opportunity to increase the susceptibility of the organisms to a wide range of drugs (Golparian et al. 2014). The aspect of targeting the AMR mechanisms mainly efflux pumps with medicinal plant extracts has come to the forefront and is being considered a prospective option of therapy (Li and Nikaido 2009; Poole 2001). This study mainly deals with the efflux pump mechanisms highlighting their association with the development of antimicrobial resistance, the advent of phytotherapeutics, and its role in efflux pump inhibition.

4.2 Efflux Pumps: Antimicrobial Resistance Mechanisms

Even after 94 years of the advent of penicillin, the aspect of microbial infections and manifested diseases still remains an unfinished battle (Poole 2001). At present, although an arsenal of treatment options is available for these infections, the microbes stand undefeated powered by their developed AMR (Prestinaci et al. 2015). One of the strongest components of the intrinsic resistance mechanisms is the efflux pumps (Poole 2007). It was the last AMR mechanism discovered and was first observed in *Escherichia coli*, where it provided resistance against tetracycline. According to the initial discovery of efflux pumps related to Tetracycline, four different plasmids, namely, R144, RP1, RA1, and R222, coded for tetracycline-related efflux proteins in several strains of *E. coli* and these efflux machineries were significantly responsible for reducing the cellular accumulation of tetracycline in the *E. coli* cells (Poole 2007). In the later years, a number of essential efflux mechanisms have come to the forefront, which are either chromosome or plasmid-encoded, and have been categorized into various classes on the basis of their protein family as well as on the basis of the antimicrobial resistance features shown by them (Lomovskaya and Watkins 2001). Structurally, efflux pumps can be predominantly defined as transmembrane proteins, whose presence is mostly ubiquitous in all living organisms. Functionally, they have been majorly divided into six protein families (Borges-Walmsley et al. 2003), namely, Major Facilitator Superfamily (MFS), Resistance Nodulation Division (RND) family, Small Multidrug Resistant (SMR) Superfamily, Proteobacterial Antimicrobial Compound Efflux (PACE) Superfamily, Multidrug and Toxic Compound Extrusion (MATE) Superfamily, and, finally, the ATP Binding Cassette (ABC) Superfamily (Du et al. 2018; Eswaran et al. 2004). For most of these efflux pumps, viz., MFS, RND, SMR, MATE, PACE, etc., the source of energy is either the proton motive force (PMF) or the sodium ion gradient, and therefore, they primarily function as secondary active transporters (Marquez 2005;

Nikaido 2011). However, for ABC family proteins, the energy source is ATP, and therefore they act as the primary active transporter (Hollenstein et al. 2007). Each of these efflux pumps has separate mechanisms of action, thereby providing resistance against a wide range of microbial infections.

4.3 Small Multidrug Resistant (SMR) Superfamily

4.3.1 Structure

The SMR efflux family is mainly driven by the energy provided by the proton motive force (PMF) (Paulsen et al. 1996). They are hydrophobic by nature and are primarily lipophilic cations that render their range of molecular recognition significantly lower, compared with other efflux pumps (Bay and Turner 2009). Structurally SMR family proteins are homotetramers, being constituted of four transmembrane segments (TMS) (Poulsen et al. 2009). SMR pump genes have been reported to be expressed in the chromosomal DNA and in certain cases on transposable elements and plasmids as well (Bay and Turner 2009). Although the substrate range of SMR efflux pumps is considerably lower than others, it confers resistance to several drugs that are mostly β-lactams and aminoglycosides in certain cases (Kornelsen and Kumar 2021). For example, the SMR pump of *Escherichia coli* (EmrE pump) reportedly transports drugs such as erythromycin, tetracycline, and vancomycin, thereby rendering the organism invincible against these drugs (Bay et al. 2008; Yerushalmi et al. 1996). On the contrary, *Staphylococcus epidermidis* has also shown resistance provided by SMR pumps that have effectively transported tetracycline, erythromycin, as well as ampicillin (Blair et al. 2014; Kumar and Schweizer 2005).

4.3.2 Mechanism

The SMR pumps comprise oligomeric functional complexes and a hydrophobic core. The pump pathway lying in the transmembrane region, through which the toxic molecules and drugs are effluxed out of the cell, comprises the polar countenance of the amphipathic membrane molecules, along with the residues of serine, tyrosine, tryptophan, as well as glutamate (Rath et al. 2006). The substrates generally bind to the transmembrane during the process of efflux with the help of the conserved glutamate residues, being specifically cationic in nature (Kolbusz et al. 2010). Moreover, as SMRs are associated with the efflux of multiple drugs and hence the name, the versatility and pliability are achieved owing to the presence of conserved sequences of proline and glycine, the active sites of which allow the conformational affability (Jack et al. 2000). Initially, the drug interacts with the transmembrane glutamate residues of the SMR efflux pumps, followed by the exchange of the drug cation with a proton. It then forms a complex with the carrier moiety via a carboxyl group and is regarded as the drug carrier complex, which crosses the hydrophobic internal region of the membrane of the cell. The translocation across the membrane occurs on the basis of several intricate reversible alterations in the protein structure. Following this step, protons acquired from the medium come in contact with the drug carrier complex and replace the drug with the help of competitive binding. Therefore, the drug is released and the efflux proteins are once again complexed with the proton, which results in the reversal of the protein's confirmation, bringing it back to its original form (Spengler et al. 2017).

4.4 Proteobacterial Antimicrobial Compound Efflux (PACE) Superfamily

4.4.1 Structure

The PACE superfamily is so named owing to the presence of these pump proteins in a wide range of proteobacteria. These pumps have four transmembrane segments and comprise alpha helices (Zgurskaya and Introduction 2021). In PACE pumps, four amino acid residues are universally conserved. They are, namely, a residue of glutamic acid that is present within the first transmembrane segment, a residue of asparagine present in the second transmembrane segment, a residue of alanine present within the periplasmic space of the fourth transmembrane section, and, finally, a residue of aspartic acid, present in the boundary of the cytoplasmic membrane of the fourth transmembrane segment (Ahmad et al. 2018). These pumps are mainly responsible for the efflux of drugs such as chlorhexidine, benzalkonium, acriflavine, dequalinium, proflavine, and so on (Spengler et al. 2017).

4.4.2 Mechanism

Proteobacterial antimicrobial compound efflux superfamily proteins are mainly transporters of biocides such as acriflavine and chlorhexidine. The Acinetobacter chlorhexidine efflux pumps (AceI) are one of the most significant prototypes of PACE efflux pumps (Abdi et al. 2020). The AceI proteins primarily use the electrochemical proton gradient as the prominent source of energy, utilizing which they transport the substrates. AceI pumps are to some extent similar to the SMR family efflux pumps, as they are both small in size and share a similar secondary structure; however, they do not have any common functional attributes. The mechanism of transport mainly comprises the recognition of the substrates with the help of the glutamic acid residue lying in the core of the first transmembrane segment (Reddy Bolla et al. 2020). Post recognition, the substrate forms a complex with the efflux protein machinery, thereby resulting in a conformational change. Finally, translocation takes place and the drug molecules are released outside the cell (Hassan et al. 2019). The intricate details of the mechanism of action of the PACE efflux pumps other than AceI have not yet been thoroughly researched to date.

4.5 Major Facilitator Superfamily (MFS)

4.5.1 Structure

The MFS efflux pumps function on the basis of the energy derived from the transport of NA^+ or H^+ ions. The MFS fold can be defined as a canonical section comprising 12 transmembrane segments, which can be further subdivided into two domains, namely, the N-terminal domain and the C-terminal domain. Each of the two domains consists of six transmembrane segments, which are interconnected with the help of a cytoplasmic loop and might also sometimes bear the presence of a helix that is amphipathic by nature (Varela and Mukherjee n.d.). These pumps are mainly associated with the transport of metabolites, drugs, anions, and so on. They have the capacity to efflux out drugs such as tetracycline, macrolides, and so on, thereby rendering them inactive (Kumar et al. 2020). These pumps show a great amount of variation with respect to their substrates. For example, organisms like *Acinetobacter baumannii* express distinctly separate MFS efflux pumps, i.e., the SmvA pump specific for erythromycin and the CraA pump for chloramphenicol (Pasqua et al. 2019). *E. coli* also expresses separate MFS pumps such as MefB, QepA, and Fsr; MefB shows affinity toward macrolides, Qep toward fluoroquinolones, and Fsr toward trimethoprim (Kumar et al. 2013).

4.5.2 Mechanism

The MFS superfamily proteins are ubiquitously present and are susceptible to a wide range of molecules or substrates such as oligosaccharides, small metabolites, amino acids, ions, as well as several other antimicrobial molecules, including a variety of drugs (Kumar et al. 2013). The structure and organization of the MFS efflux pumps are more or less reserved and are termed the MFS fold. The mechanism of transport is carried out by continuous changing of the binding site of the substrate between the extracellular side of the cell membrane and the intracellular or periplasmic side of the cell membrane (Ranaweera et al. 2015). The models developed to demonstrate the mode of action of the MFS efflux pumps are termed the "clamp switch model" and the "'rocker switch model." According to these models, the metamorphosis between different states includes both structural modifications and rotations of fixed and rigid sections. The fifth transmembrane segment is technically responsible for the transition between the states of inwardly open to states of outwardly open, along with the reassembly of the hydrophobic centers present in the N terminal domain and the C terminal domain's cytoplasmic loop, in order to render the cytoplasmic entrance as closed (Du et al. 2018).

4.6 Multidrug and Toxic Compound Extrusion (MATE) Superfamily

4.6.1 Structure

The MATE family efflux pumps are mostly related to the efflux of fluoroquinolones. However, some MATE pumps are also known to be associated with aminoglycosides (Piddock 2006b). Structurally, it is made up of 12 transmembrane segments or helices, along with the amino acid residues existing within a range of 400–700 (Radchenko et al. 2015). Most of the MATE pumps that are found in bacteria have been observed in Gram-negative organisms. One of the first MATE pumps was observed in the organism *Vibrio parahaemolyticus*, and the pump was named NorM. Other organisms such as *Neisseria meningitidis* and *Neisseria gonorrhoeae* also express the NorM MATE efflux pump (Rouquette-Loughlin et al. 2003).

4.6.2 Mechanism

MATE efflux pumps function as proton antiporters, sodium antiporters, and drug antiporters, being present in archaea, bacteria, as well as eukaryotic cells. Each MATE transporter protein comprises two domains, the N- and C-terminal domains (Sun et al. 2014). These domains are so placed inside the MATE transporters, that they form a central cavity that is shaped like a V. The lower section of this cavity ends halfway in the bilayer of the cellular membrane, therefore making the structure acquire an outwardly open stable state. When the efflux mechanism of the substrate is initiated, the substrate and the cation enter into a competition as both of them target to interact with the same groups of amino acids. In most cases, the drug-binding sites of the MATE efflux proteins are the same as the cation-binding sites; however, separate drug-affiliated pockets are also present in certain MATE efflux pumps, like NorM, where both the cation and the substrate have the capacity to bind simultaneously to the protein owing to the presence of distinct and separate binding sites. After drug binding is achieved, the proton undergoes a conformational change in order to switch between the inwardly open state and the outwardly open state, thereby resulting in the efflux of the drug. The MATE family recognizes a vast range of drugs, viz., kanamycin, cimetidine, ampicillin, ethidium, norfloxacin, triethylammonium, chloramphenicol, metformin, ciprofloxacin, and so on, thereby proving protection against the microorganisms against such potent antimicrobial agents (Du et al. 2018; Kuroda and Tsuchiya 2009).

4.7 Resistance Nodulation Division (RND) Family

4.7.1 Structure

RND efflux pumps are mainly localized in Gram-negative bacteria. They are responsible for the efflux of a variety of compounds such as detergents, heavy metals, yes, antibiotics, and so on and thus functionally diverse. Some RND pumps such as the Tet and Mef pumps are substrate-specific and are only associated with the transport of tetracycline and macrolides, respectively (Blair et al. 2015). On the contrary, RND pumps like MexAB-OprM found in *Pseudomonas aeruginosa* provide protection to the organism against a wide variety of substances ranging from beta-lactams, tetracycline, chloramphenicol, sulfamethoxazole, and trimethoprim to fluoroquinolones as well as (Kourtesi et al. 2013) eighth transmembrane segments.

4.7.2 Mechanism

The RND Superfamily of efflux pumps also functions as antiporters of proton and/or drugs and are found in bacteria, eukaryotes, and archaea, although RND pumps are primarily present in Gram-negative bacteria (Routh et al. 2011). With respect to AcrB, an established prototype of RND efflux pumps, the structure of RND proteins constitutes of three states that are not equivalent to each other, namely, the loose (L) state or access state, the tight (T) state or bonding state, and the open (O) state or extrusion state. In other words, the RND transporters show stable interchangeable and reversible shifts between three stages—L, T, and, O, that is loose (access), tight (bonding), and open (extrusion) states, respectively. According to this model, the drug enters the binding site or drug pocket that usually lies in the periplasmic domain, when the RND transporter molecules are in the loose or access state (Du et al. 2018). Post entry, the drug moves ahead in the drug-associated pocket, and then the transporter shifts its state from access or loose to binding or tight state, thereby allowing the drug to bind to the transporter protein and form a complex. Post binding, the drug is further exuded into the docking domain of the extrusion or open state (Shafer et al. 1998). The side chains present in the transmembrane segments of the RND transporter proteins undergo protonation, which in turn allows the change in the states to take place (Fernando and Kumar 2013).

4.8 ATP-Binding Cassette (ABC) Superfamily

4.8.1 Structure

The efflux pumps of the ABC superfamily obtain their energy from the hydrolysis of ATP. They have a wide range of substrates starting from ions, amino acids, proteins, and polysaccharides to drugs (Davidson and Chen 2004). Structurally, these pumps comprise six transmembrane segments, each made up of alpha helices. They generally are active when present in pairs, either in the homodimeric state or in the heterodimeric state (Cole and Deeley 1998). They generally show concomitance with cytoplasmic ATPases. For example, the VcaM efflux pump present in *Vibrio cholerae* is a noteworthy ABC family efflux pump and is responsible for the efflux of drugs like tetracycline and fluoroquinolones (Blair et al. 2015; Davidson and Chen 2004).

4.8.2 Mechanism

ABC efflux pumps of pathogenic microorganisms are one of the principal contributors of intrinsic antimicrobial resistance. The range of molecules exuded by the ABC cascade is extremely vast and huge, and these proteins have therefore been termed orphan proteins. The ABC transporters are constituted of two separate domains, one is the transmembrane domains containing binding sites for substrates, and the other is the nucleotide-binding domain, where, upon binding of ATP, it is hydrolyzed, thereby providing the energy required to initiate and complete the process of transport (Prasad and Goffeau 2012). ABC transporters show two forms, the homodimeric and heterodimeric forms. The homodimeric form falls under intrinsic resistance and possesses a separate binding site for substrate and ATP, thereby allowing ATP hydrolysis to fuel the process of drug exudation. However, the heterodimeric form is related to acquired AMR and often possesses a degenerative site that does not allow hydrolysis of ATP. With respect to the homodimeric form, it too shows the reversible changes and alterations between two states, that is, the switch between an occluded state that is inward open and the outward open state, in order to ensure the translocation of substrates across the bilayer of the cell membrane, and this process is referred to as the process of alternating access mechanism. In this regard, another crucial factor is the changes taking place in the nucleotide-binding domain. The structural and conformational changes associated with the ABC transporters are directly linked with the dissociation as well as dimerization of the nucleotide-binding domain, owing to the binding and hydrolysis of ATP. The change between these two states is also facilitated by the presence of certain enzymes like flippase that facilitate the shift of lipid precursors from the cytoplasmic leaflet associated with the inner membrane to the periplasmic leaflet (Du et al. 2018).

4.9 Challenges of Targeting Efflux Pumps

The aspect of targeting efflux pumps was considered a point of significant concern till a few years back, prior to the advent of phytotherapeutics, primarily due to the wide varieties in the structure and function of the efflux machineries, that make them immensely difficult drug targets. Apart from that, several key factors need to be taken into account while designing drugs against efflux pumps, the most essential one being, the efflux pump inhibitors (EPIs) which should either mimic the substrate structure or should be closely related to the substrate configuration of that respective efflux pump. Only then, it would be possible for the molecules to interfere with the functioning mechanisms of the pumps, thereby preventing drug efflux in the long run. Other processes of targeting the efflux proteins also exist, such as targeting the glutamate residues in certain efflux proteins (Dashtbani-Roozbehani and Brown 2021). This might result in effective modification of the binding site of the transporter moiety, thereby resulting in the loss of efflux function. Therefore, the specificity of the drugs is immensely necessary. However, several challenges arise when efflux pumps of the pathogen are being targeted as a potential option for therapy in order to increase the susceptibility of the microbial cells to the pre-administered drugs (Verma et al. 2021). One of the most significant hindrances can be the ubiquitous presence of efflux pumps. Efflux pump mechanisms are required in all living organisms, in order to maintain the functional stability of the cellular metabolism. Hence, efflux pumps of the same family might be present in both the host and the pathogen. Although the administration of established drugs targeting the pathogenic cells would not pose a problem, the administration of inhibitors targeting the efflux mechanisms might give rise to a major challenge (Pagès and Amaral 2009). The EPIs showing affinity toward both the host and pathogen efflux machineries will inhibit the host efflux functions as well, ultimately resulting in toxic and detrimental outcomes.

4.10 Emergence of Phytotherapeutics Against AMR: Its Potential as a Therapeutic Option

The development of MDR in microbes owing to the significant efflux of drugs from the pathogens is a cause of growing concern, and the formation of effective EPIs is of utmost importance (Pagès and Amaral 2009). Moreover, treatment with antibiotics has taken a backseat and the symptoms and manifestations of the microbial diseases have taken a turn toward fatal outcomes. Therefore, the emergence of phytotherapeutics has acted as a ray of hope for researchers in recent times. Efflux pump inhibitors (EPIs) from plant sources have reinstated the hope of vanquishing MDR organisms to a great extent (Sana et al. 2015). The primary reason why medicinal plants can act as propitious sources of efflux pump inhibitors is the abundance of secondary metabolites in these plants that are diversified both structurally and chemically and are aided with several known medicinal properties. In recent years, numerous research works have been carried out on these medicinal plants and the efficacy of their extracts to act as potent EPIs (Seukep et al. 2020). One significant obstacle is that the activity of these efflux pumps is often diverse and multiple molecular mechanisms are associated with them, because of which it becomes difficult to target these ubiquitous machineries in the microorganisms. Certain plant extracts possess molecules that have the ability to block and stop the functioning of efflux pumps of microorganisms, especially Gram-positive and Gram-negative bacteria (Fig. 4.1). This action of the plant-derived EPIs reduces the resistance of the pathogenic organisms against the proposed therapeutics and in a way re-establishes the efficiency of the initial treatment methods. Accumulation of the drugs inside the microbial cells is quite easily possible, thereby allowing the drug to exhibit its true antimicrobial effect. The plant families that have been thoroughly researched with respect to the presence of EPs or EPI-associated molecules are, namely, Cucurbitaceae, Berberidaceae, Zingiberaceae, Lamiaceae, Apocynaceae, Fabaceae, Convolvulaceae, and so on (Stavri et al. 2007). The main reason behind zoning in upon the above-mentioned plant families is because of the fact that our daily dietary elements come from plants belonging to these families. In fact, potent EPIs from plant sources were identified from simple plant products of our diet, such as tea leaves, lemon grass, pepper, grapes, pomegranates, and so on. Some of the isolated compounds from these sources are geraniol, theobromine, piperine, resveratrol, farnesol, and so on, and their potentiality has been observed in a wide variety of pathogens, including Gram-positive and Gram-negative bacteria.

One of the earliest discoveries of phytotherapeutic molecules against bacterial efflux pumps identified and analyzed a cluster of molecules that mimic the peptide structure and proved that these molecules were highly effective against certain strains of *P. aeruginosa* that show heightened expression of efflux pumps like MexAB (Pagès and Amaral 2009; Stavri et al. 2007). The primary compound that showed the closest structural similarity with the peptide structure was extensively studied, and its associated and derivative molecules were also taken into account, in order to ascertain the effective relationship between conformation and function. It was also shown that these compounds have the ability to act against the efflux of fluoroquinolone, the drug levofloxacin playing the role of a marker. After analyzing the group of peptide analogs, the most significant compound playing the lead role was deduced to be Phe-Arg-β-naphthylamine or PAβN (Thomas et al. 1999). The primary activity profile of PAβN was found to be the capability to inhibit the function of the efflux pumps that majorly efflux out quinolones, in this case, fluoroquinolones like levofloxacin.

The one fact that made the discovery of PAβN all the more interesting was that not only did it structurally mimic the substrate of the MexB efflux pump but was also found to affect the activity of efflux mechanisms of *Klebsiella pneumoniae*, *E. coli*, *Salmonella enterica*, *Enterobacter aerogenes*, and so on (Lomovskaya et al. 2001). The mechanism of action of PAβN indicates the fact that it shows prominent competition with the antibiotic substrates with respect to binding with the efflux proteins. As a result, PAβN binds with the efflux proteins of pumps like MexB,

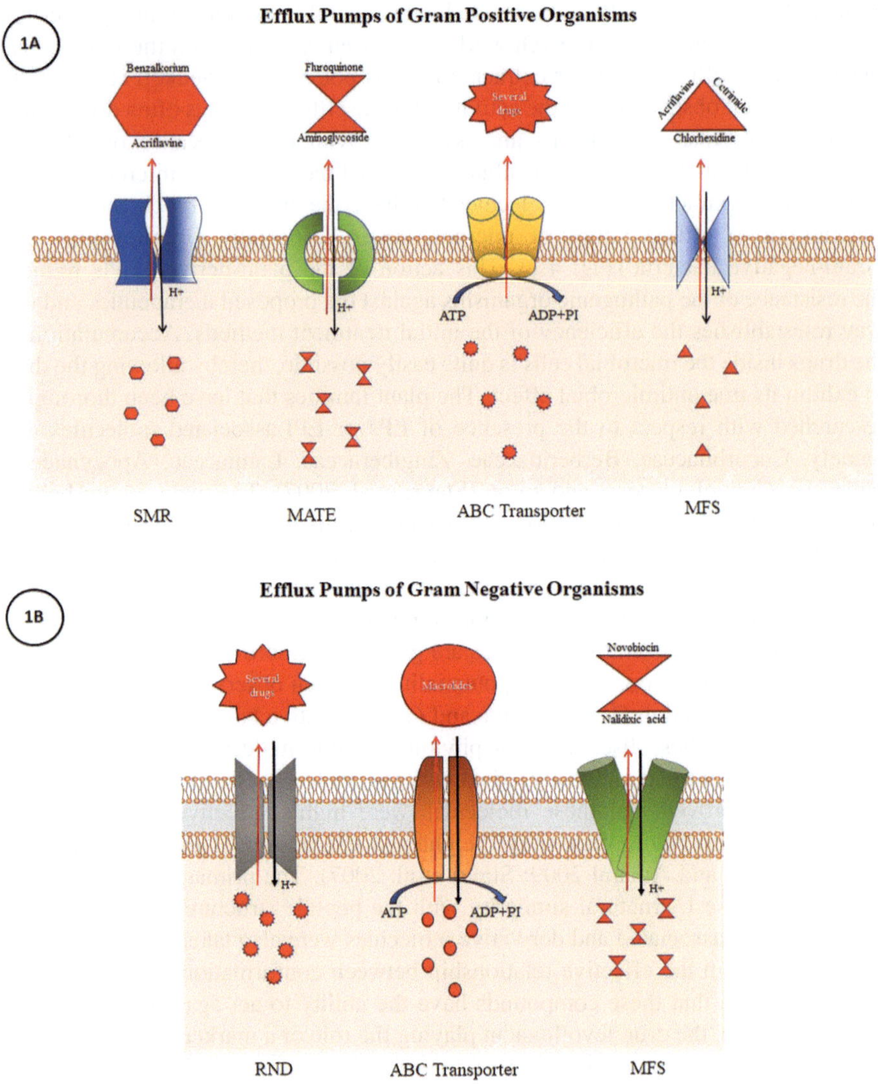

Fig. 4.1 The types of efflux pumps present in Gram-positive and Gram-negative bacteria. (**a**) Different types of efflux pumps including SMR, MATE, ABC transporters, and MFS present in Gram-positive organisms, and the drugs against which they provide resistance. (**b**) The types of efflux pumps including RND, ABC transporters, MFS present in Gram-negative organisms, and the drugs against which they provide resistance

thereby allowing the drug to be retained within the cell and consequently resulting in an increase of cellular drug concentration, ultimately reaching the state required for the drug to exhibit its action on its respective target within the cell. Apart from this, PAβN also shows the capacity to either reduce the resistance of the microorganism toward that particular drug or permanently reverse the resistance mechanism of the microbe, thereby indicating the emergence of a long-term solution against AMR (Murakami et al. 2006). However, the functional activity of PAβN has certain limitations as well, giving rise to specific drawbacks. Although PAβN shows structural similarity with some specific antibiotics, its range leaves out several important ones, and as a result of this, it shows partial inhibition with respect to certain efflux pumps. Thus, the relationship between PAβN and antibiotics can be regarded as a disparate one, the eminent regulatory factors being the presence of the substrate or drug binding domain with affinity toward PaβN and the structural characteristics of the efflux pumps. Another disadvantage regarding PAβN might be its lack of affinity toward the antibiotic-specific site of the drug-binding domain of the efflux pumps. If PAβN binds to any site apart from the drug binding site, it might not be able to show the inhibitory actions as mentioned before. Although alteration of the structural and functional properties of the efflux pumps can be initiated in this regard, complete inhibition cannot be guaranteed. In order to overcome these limitations, several strategic schemes were considered and the synthesis of more potent derivatives of PAβN turned out to be the most effective option. The synthesis of derivative molecules of PAβN therefore led to the stability of the EPI functions and also contributed to establishing phytotherapeutics as a promising solution. The derived molecule was named MC-04,124, which showed heightened stability and features exhibiting a reduced range of toxicity and side effects. Moreover, compared with PAβN, MC-04,124 also showed greater stability in the biological system fluids (Watkins et al. 2003). To date, several derivatives of PAβN have been synthesized, and all of them have shown gradual refinement in their pharmacokinetic behavior as efflux pump inhibitors.

Although EPIs from natural sources working on the majority of the microbial population have been identified, the number of these molecules effectively showing their inhibitory activities against Gram-negative bacteria is relatively on the lower scale. The intricate and complex arrangement of the efflux pumps in Gram-negative organisms that is the tripartite pump system can be considered responsible for the lower range of availability of EPIs against them. Despite the daunting tripartite structure of these pumps in Gram-negative organisms, several molecules have been discovered from plant sources that effectively target these pumps (Prasch and Bucar 2015). For example, karavilagenin isolated from *Momordica balsamnia* and gallotannin isolated from *Terminalia chebula* have been reported to show their inhibitory action against the ArcAB-TolC tripartite efflux pump present in *E. coli* strains. Furthermore, MexAB-OprM tripartite pumps in *P. aeruginosa* have been reported to be inhibited by compounds such as falcarindiol, palmatine, berberine, conessine, catharanthine, osthol, curcumin, and resveratrol isolated from *Levisticum officinale*, *Berberis vulgaris*, *Holarrhena antidysenterica*, *Catharanthus roseus*, *Cnidii monnieri*, *Curcuma longa*, and *Nauclea pobeguinii*, respectively (Fig. 4.2).

Phytotherapeutic botanicals isolated from various plant sources inhibiting drug efflux by Eps such as TetK, MdeA, MsrA, NorA, LmrS, Bmr MDR, and so on

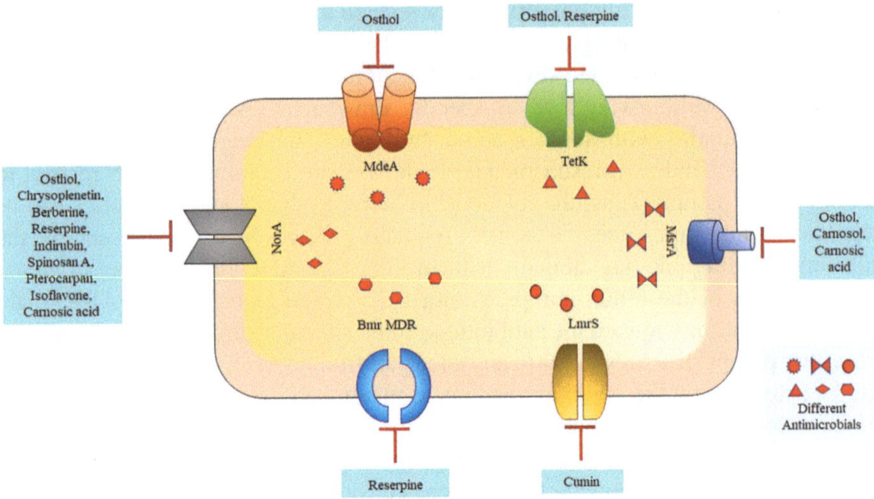

Fig. 4.2 Inhibition of various efflux pumps by phytotherapeutic molecules

On the contrary, theobromine isolated from *Theobroma cacao* works equally effectively against both ArcAB-TolC as well as MexAB-OprM tripartite efflux pumps in Gram-negative organisms (Seukep et al. 2020) (Table 4.1).

4.11 Strategies to Overcome Intrinsic Resistance of Efflux Pumps Using Phytotherapeutics

The rapid surge in the number of multiple drug-resistant microbes urgently calls for the initiation of an effective solution. Emergence of the phytological molecules and their efflux pump inhibitory capabilities have opened up a new field of research, wherein these molecules can be considered potential drug candidates. A prime example of plant-derived EPI is 5-MHC or 5′-methoxyhydnocarpin isolated from plants such as *Hydnocarpus wightianus* and *Berberis* sp.; this chemical compound has reinstated the function of the antimicrobial berberine. In other words, 5-MHC promotes the accumulation of berberine in the pathogens, thereby allowing it to demonstrate its antimicrobial properties after reaching the desired concentration. This clearly demonstrates that berberine can effectively reverse the berberine resistance developed by pathogenic microorganisms. 5-MHC, however, does not exhibit any form of antimicrobial property by itself (Stavri et al. 2007). Therefore, in order to achieve reversal of resistance toward eminent drugs amongst the pathogenic microflora, strategies need to be developed, wherein, with the use of phytotherapeutic molecules, the intrinsic resistance mechanisms like efflux pumps of microbes can be targeted.

Table 4.1 Characterization of medicinal plant resources—action against efflux pumps

Sl. no.	Plant source	Compounds isolated	Efflux pumps inhibited	References
1.	*Holarrhena antidysenterica*	Conessine	MexAB-OprM pump of *P. aeruginosa.*	Siriyong et al. (2017)
2.	*Carpobrotus edulis*	Uvaol	MRSA COLOXA efflux	Anon (n.-d. Seukep et al. (2020)
3.	*Cnidii monnieri*	Osthol	MdeA, MsrA, TetK, NorA of *S. aureus*; efflux pumps of *P. aeruginosa*	Negi et al. (2014), Joshi et al. (2014)
4.	*Rauwolfia serpentina*	Reserpine	Bmr MDR in *Bacillus subtilis*	Mahamoud et al. (2007)
5.	*Cuminum cyminum*	Cumin	LmrS efflux pump	Kakarla et al. (2017)
6.	*Catharanthus roseus*	Catharanthine	Effective against efflux pumps of *P. aeruginosa*	Dwivedi et al. (2018)
7.	*Vernonia adoensis*	Chrysoplenetin	Efflux pumps of *S. aureus* and *Enterococcus faecalis*	Fiamegos et al. (2011)
8.	*Artemisia annua*	Chrysoplenetin	NorA efflux pump of *S. aureus*	Stermitz et al. (n.d.)
9.	*Berberis* spp.	Berberine	NorA present in *S. aureus* and MexABOprM present in *P. aeruginosa*	Sajjad Aghayan et al. (2017)
10.	*Levisticum officinale*	Falcarindiol	Effective against efflux pumps of Gram-negative bacteria	Garvey et al. (2011)
11.	*Rauwolfia vomitoria*	Reserpine	Effective against efflux pumps of *S. aureus* and *S. pneumoniae*, NorA and TetK-mediated MDR in MRSA, and LmrA of *Lactococcus lactis*	Siriyong et al. (2017)
12.	*Wrightia tinctoria*	Indirubin	NorA present in S*taphylococcus epidermidis and S. aureus*	Garvey et al. (2011)
13.	*Silybum marianum*	Silybin	Effective against efflux pumps present in *S. aureus*	Garvey et al. (2011)
14.	*Dalea spinosa*	Arylbenzofuran aldehyde (Spinosan A)	NorA efflux pump of *S. aureus*	Mahamoud et al. (2007)
15.	*Dalea spinosa*	Pterocarpan	NorA efflux pump of *S. aureus*	Mahamoud et al. (2007)
16.	*Dalea spinosa*	Isoflavone	NorA efflux pump of *S. aureus*	Mahamoud et al. (2007)
17.	*Rosmarinus officinalis*	Carnosol	MsrA and NorA efflux pump of *S. aureus*	Oluwatuyi et al. (2004)
18.	*Rosmarinus officinalis*	Carnosic acid	MsrA and NorA efflux pump of *S. aureus*	Oluwatuyi et al. (2004)
19.	*Mesua ferrea*	Coumarins	NorA efflux pump of *S. aureus*	Roy et al. (2013)
20.				

(continued)

Table 4.1 (continued)

Sl. no.	Plant source	Compounds isolated	Efflux pumps inhibited	References
	Alpinia galanga	Phenylpropanoids (10-S-10-acetoxyeugenol acetate)	Effective against efflux pumps of *Mycobacterium smegmatis*	Roy et al. (2012)

In general, the efflux mechanisms in pathogens can be inhibited via the following strategies:

1. Inducing suppression of the genes that are responsible for the structural and functional expression of the efflux pumps
2. Preventing appropriate assembly of the efflux proteins in the pumps, thereby restricting the development of efflux pump structure, making it functionally inactive
3. Blocking the outer and/or inner membrane channel openings, in order to prevent the efflux of drugs
4. Using antiporter directly or indirectly to prevent the pumps from using the proton motive force, thereby resulting in their collapse by cutting off the energy source
5. Introducing certain chemical molecules that can exhibit either competitive or non-competitive inhibition with respect to the administered drug
6. Altering the structural design of the drugs used for treatment, so as to reduce its binding affinity with the efflux pump proteins (Pagès and Amaral 2009)

Of the above-mentioned strategies, plant-derived EPIs can be used to effectively demonstrate at least two of them (Fig. 4.3). Most EPIs generally mimic the drug or substrate structure, because of which their affinity toward the efflux proteins is high. As a result, they can show significant competition with the drug, therefore, getting exuded itself, in place of the administered drug and allowing it to get accumulated within the cell. Once the drug is allowed to reach the desired concentration within the pathogen, it shows its antimicrobial activity. Thus, plant-derived EPIs can execute competitive inhibition, thereby reducing the drug-resistant capacity of pathogenic organisms (Pulingam et al. 2022). On the contrary, EPIs can also show non-competitive binding. In this case, the EPI binds to a site other than the drug binding site, and as a result of this binding, the EP protein undergoes a significant conformational change, thereby rendering the efflux pumps incapable of transporting the drugs outside the cell. Another strategy that the plant-derived chemical compounds can execute is the aspect of blocking the entry and exit channels of the efflux pumps, thus preventing the drug from getting exuded out of the cells.

Therefore, keeping these strategies in mind, chemical compounds from plant sources can be used to design accessory drugs, that would reverse, remove, or reduce the efflux pump-induced intrinsic resistance mechanisms of the pathogenic microflora and improve the efficacy of the administered drugs (Pulingam et al. 2022; Annunziato 2019).

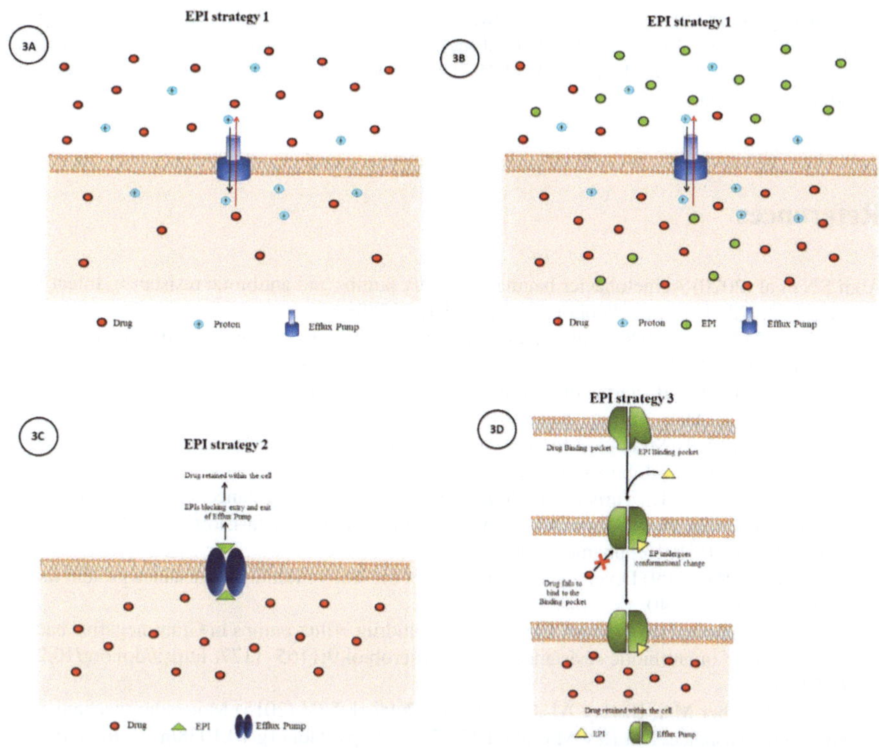

Fig. 4.3 Efflux pump inhibition strategies. (**a**) *Competitive inhibition***:** the mode of action of efflux pumps, exuding the drug from the cells, thereby enhancing the antimicrobial resistance properties of the pathogenic microbes. (**b**) Competitive binding of efflux pump inhibitors (EPIs) and exclusion of EPIs from the cell in place of the drugs, resulting in the retention and accumulation of the drug within the cell. (**c**) *Blocking the EP channels***:** Drug is retained within the pathogenic cell by blocking the entry and exit channels of the efflux pumps by the EPIs

4.12 Conclusion

Efflux pump inhibitors are therefore one of the easily accessible options via which the action of a wide range of antimicrobials can be revived. The application of EPIs along with these antimicrobials as accessory drugs might be the solution that the research world is waiting for in order to overcome hurdles such as MDR and antimicrobial resistance. Moreover, the use of botanicals as EPIs has several advantages. Being derived from medicinal plant sources, these phytotherapeutic molecules rarely show any adverse harmful side effects on hosts and do not have any long-term aftereffects as well (Prasch and Bucar 2015). These molecules have shown a very high percentage of efficiency in this field, and further investigation would definitely enlighten us further in this regard. To conclude, the inception of phytotherapeutics as EPIs might mark the end of the reign of antimicrobial resistance to pathogenic microorganisms.

Acknowledgments TB is supported by the Early Career Research (ECR) Award, Science & Engineering Research Board (SERB) JRF fellowship. MA is supported by the UGC junior research fellowship. The authors would like to thank ECR-SERB (File No. ECR/2018/001015), Government of India, for funding.

References

Abdi SN et al (2020) Acinetobacter baumannii efflux pumps and antibiotic resistance. Infect Drug Resist 13:423–434. https://doi.org/10.2147/IDR.S228089

Abushaheen MA et al (2020) Antimicrobial resistance, mechanisms and its clinical significance. Dis Mon 66:100971

Ahmad I et al (2018) Bacterial multidrug efflux proteins: a major mechanism of antimicrobial resistance. Curr Drug Targets 20:16–28

Annunziato G (2019) Strategies to overcome antimicrobial resistance (AMR) making use of non-essential target inhibitors: a review. Int J Mol Sci 20:5844

Bay DC, Rommens KL, Turner RJ (2008) Small multidrug resistance proteins: a multidrug transporter family that continues to grow. Biochim Biophys Acta Biomem 1778:1814–1838. https://doi.org/10.1016/j.bbamem.2007.08.015

Bay DC, Turner RJ (2009) Diversity and evolution of the small multidrug resistance protein family. BMC Evol Biol 9:140

Blair JMA, Richmond GE, Piddock LJV (2014) Multidrug efflux pumps in Gram-negative bacteria and their role in antibiotic resistance. Future Microbiol 9:1165–1177. https://doi.org/10.2217/FMB.14.66

Blair JMA, Webber MA, Baylay AJ, Ogbolu DO, Piddock LJV (2015) Molecular mechanisms of antibiotic resistance. Nat Rev Microbiol 13:42–51. https://doi.org/10.1038/nrmicro3380

Borges-Walmsley I, Mckeegan KS, Walmsley AR (2003) Structure and function of efflux pumps that confer resistance to drugs. Biochem J 376:313

Cole SPC, Deeley RG (1998) Multidrug resistance mediated by the ATP-binding cassette transporter protein MRP. BioEssays 20:931

Dashtbani-Roozbehani A, Brown MH (2021) Efflux pump mediated antimicrobial resistance by staphylococci in health-related environments: challenges and the quest for inhibition. Antibiotics 10:1502. https://doi.org/10.3390/antibiotics10121502

Davidson AL, Chen J (2004) ATP-binding cassette transporters in bacteria. Ann Rev Biochem 73: 241–268. https://doi.org/10.1146/annurev.biochem.73.011303.073626

Du D et al (2018) Multidrug efflux pumps: structure, function and regulation. Nat Rev Microbiol 16:523–539. https://doi.org/10.1038/s41579-018-0048-6

Dwivedi GR et al (2018) Antibiotics potentiating potential of catharanthine against superbug Pseudomonas aeruginosa. J Biomol Struct Dyn 36:4270–4284

Eswaran J, Koronakis E, Higgins MK, Hughes C, Koronakis V (2004) Three's company: component structures bring a closer view of tripartite drug efflux pumps. Curr Opin Struct Biol 14:741–747. https://doi.org/10.1016/j.sbi.2004.10.003

Fernando DM, Kumar A (2013) Resistance-nodulation-division multidrug efflux pumps in gram-negative bacteria: role in virulence. Antibiotics 2:163–181. https://doi.org/10.3390/antibiotics2010163

Fiamegos YC et al (2011) Antimicrobial and efflux pump inhibitory activity of caffeoylquinic acids from Artemisia absinthium against gram-positive pathogenic bacteria. PLoS One 6:e18127

Garvey MI, Rahman MM, Gibbons S, Piddock LJV (2011) Medicinal plant extracts with efflux inhibitory activity against gram-negative bacteria. Int J Antimicrob Agents 37:145–151

Golparian D, Shafer WM, Ohnishi M, Unemo M (2014) Importance of multidrug efflux pumps in the antimicrobial resistance property of clinical multidrug-resistant isolates of neisseria gonorrhoeae. Antimicrob Agents Chemother 58:3556–3559

Hassan KA et al (2019) Short-chain diamines are the physiological substrates of PACE family efflux pumps. Proc Natl Acad Sci U S A 116:18015–18020

Hollenstein K, Dawson RJ, Locher KP (2007) Structure and mechanism of ABC transporter proteins. Curr Opin Struct Biol 17:412–418. https://doi.org/10.1016/j.sbi.2007.07.003

Hughes D, Andersson DI (2017) Environmental and genetic modulation of the phenotypic expression of antibiotic resistance. FEMS Microbiol Rev 41:374–391. https://doi.org/10.1093/femsre/fux004

Jack DL, Storms ML, Tchieu JH, Paulsen IT, Saier MH (2000) A broad-specificity multidrug efflux pump requiring a pair of homologous SMR-Type proteins. J Bacteriol 182:2311. https://journals.asm.org/journal/jb

Joshi P et al (2014) Osthol and curcumin as inhibitors of human Pgp and multidrug efflux pumps of Staphylococcus aureus: reversing the resistance against frontline antibacterial drugs. Medchemcomm 5:1540–1547

Kabra R, Chauhan N, Kumar A, Ingale P, Singh S (2019) Efflux pumps and antimicrobial resistance: Paradoxical components in systems genomics. Prog Biophysics Mol Biol 141:15–24. https://doi.org/10.1016/j.pbiomolbio.2018.07.008

Kakarla P et al (2017) Inhibition of the multidrug efflux pump LmrS from Staphylococcus aureus by cumin spice Cuminum cyminum. Arch Microbiol 199:465–474

Kolbusz MA, ter Horst R, Slotboom DJ, Lolkema JS (2010) Orientation of small multidrug resistance transporter subunits in the membrane: correlation with the positive-inside rule. J Mol Biol 402:127–138

Kornelsen V and Kumar A (2021) Update on multidrug resistance efflux pumps in Acinetobacter spp. https://journals.asm.org/journal/aac

Kourtesi C et al (2013) Send orders of reprints at reprints@benthamscience.net microbial efflux systems and inhibitors: approaches to drug discovery and the challenge of clinical implementation. Open Microbiol J 7:34

Kumar S, Mukherjee MM, Varela MF (2013) Modulation of bacterial multidrug resistance efflux pumps of the major facilitator superfamily. Int J Bacteriol 2013:1–15

Kumar A, Schweizer HP (2005) Bacterial resistance to antibiotics: active efflux and reduced uptake. Adv Drug Deliv Rev 57:1486–1513. https://doi.org/10.1016/j.addr.2005.04.004

Kumar S et al (2013) Modulation of bacterial multidrug resistance efflux pumps of the major facilitator superfamily. Int J Bacteriol 2013:204141. https://doi.org/10.1155/2013/20414

Kumar S et al (2020) Functional and structural roles of the major facilitator superfamily bacterial multidrug efflux pumps. Microorganisms 8:266. https://doi.org/10.3390/microorganisms8020266

Kuroda T, Tsuchiya T (2009) Multidrug efflux transporters in the MATE family. Biochim Biophys Acta Proteins and Proteom 1794:763–768. https://doi.org/10.1016/j.bbapap.2008.11.012

Li XZ, Nikaido H (2009) Efflux-mediated drug resistance in bacteria: an update. Drugs 69:1555–1623. https://doi.org/10.2165/11317030-000000000-00000

Lomovskaya O, Watkins W (2001) Inhibition of efflux pumps as a novel approach to combat drug resistance in bacteria JMMB symposium. J Mol Microbiol Biotechnol 650:225. www.caister.com/bacteria-plant

Lomovskaya O et al (2001) Identification and characterization of inhibitors of multidrug resistance efflux pumps in Pseudomonas aeruginosa: novel agents for combination therapy. Antimicrob Agents Chemother 45:105–116

Mahamoud A, Chevalier J, Alibert-Franco S, Kern WV, Pagès JM (2007) Antibiotic efflux pumps in Gram-negative bacteria: the inhibitor response strategy. J Antimicrob Chemother 59:1223–1229. https://doi.org/10.1093/jac/dkl493

Marquez B (2005) Bacterial efflux systems and efflux pumps inhibitors. Biochimie 87:1137–1147. https://doi.org/10.1016/j.biochi.2005.04.012

Motta SS, Cluzel P, Aldana M (2015) Adaptive resistance in bacteria requires epigenetic inheritance, genetic noise, and cost of efflux pumps. PLoS One 10:e0118464

Murakami S, Nakashima R, Yamashita E, Matsumoto T, Yamaguchi A (2006) Crystal structures of a multidrug transporter reveal a functionally rotating mechanism. Nature 443:173–179

Negi N, Prakash P, Gupta ML, Mohapatra TM (2014) Possible role of curcumin as an efflux pump inhibitor in multi drug resistant clinical isolates of Pseudomonas aeruginosa. J Clin Diagn Res 8: DC04–DC07

Nikaido H (2011) Structure and mechanism of Rnd-type multidrug efflux pumps. Adv Enzymol Relat Areas Mol Biol 77:1–60

Oluwatuyi M, Kaatz GW, Gibbons S (2004) Antibacterial and resistance modifying activity of Rosmarinus officinalis. Phytochemistry 65:3249–3254

Pagès JM, Amaral L (2009) Mechanisms of drug efflux and strategies to combat them: challenging the efflux pump of gram-negative bacteria. Biochim Biophys Acta Proteins and Proteom 1794: 826–833. https://doi.org/10.1016/j.bbapap.2008.12.011

Pasqua M et al (2019) The varied role of efflux pumps of the mfs family in the interplay of bacteria with animal and plant cells. Microorganisms 7:285

Paulsen FT et al (1996) The SMR family: a novel family of multidrug efflux proteins involved with the efflux of lipophilic drugs. Mol Microbiol 19:1167

Piddock LJV (2006a) OPINION Multidrug-resistance efflux pumps-not just for resistance. www.nature.com/reviews/micro

Piddock LJV (2006b) Clinically relevant chromosomally encoded multidrug resistance efflux pumps in bacteria. Clin Microbiol Revi 19:382–402. https://doi.org/10.1128/CMR.19.2.382-402.2006

Poole K (2001) Multidrug efflux pumps and antimicrobial resistance in Pseudomonas aeruginosa and related organisms JMMB Symposium. J Mol Microbiol Biotechnol 613:www.caister.com/bacteria-plant

Poole K (2007) Efflux pumps as antimicrobial resistance mechanisms. Ann Med 39:162–176. https://doi.org/10.1080/07853890701195262

Poulsen BE, Rath A, Deber CM (2009) The assembly motif of a bacterial small multi drug resistance protein. J Biol Chem 284:9870–9875

Prasad R, Goffeau A (2012) Yeast ATP-binding cassette transporters conferring multidrug resistance. Annu Rev Microbiol 66:39–63

Prasch S, Bucar F (2015) Plant derived inhibitors of bacterial efflux pumps: an update. Phytochem Rev 14:961–974. https://doi.org/10.1007/s11101-015-9436-y

Prestinaci F, Pezzotti P, Pantosti A (2015) Antimicrobial resistance: a global multifaceted phenomenon. Pathog Glob Health 109:309–318. https://doi.org/10.1179/2047773215Y.0000000030

Pulingam T et al (2022) Antimicrobial resistance: prevalence, economic burden, mechanisms of resistance and strategies to overcome. Eur J Pharm Sci 170:106103. https://doi.org/10.1016/j.ejps.2021.106103

Radchenko M, Symersky J, Nie R, Lu M (2015) Structural basis for the blockade of MATE multidrug efflux pumps. Nat Commun 6:7995

Ranaweera I et al (2015) Structural comparison of bacterial multidrug efflux pumps of the major facilitator superfamily HHS public access. Trends Cell Mol Biol 10:131

Rath A, Melnyk RA, Deber CM (2006) Evidence for assembly of small multidrug resistance proteins by a ‘two-faced’ transmembrane helix. J Biol Chem 281:15546–15553

Reddy Bolla J, Howes AC, Fiorentino F, Robinson CV (2020) Assembly and regulation of the chlorhexidine-specific efflux pump AceI. Proc Natl Acad Sci U S A 117:17011. https://doi.org/10.1073/pnas.2003271117/-/DCSupplemental

Reygaert CW (2018) An overview of the antimicrobial resistance mechanisms of bacteria. AIMS Microbiol 4:482–501

Rouquette-Loughlin C, Dunham SA, Kuhn M, Balthazar JT, Shafer WM (2003) The NorM efflux pump of Neisseria gonorrhoeae and Neisseria meningitidis recognizes antimicrobial cationic compounds. J Bacteriol 185:1101–1106

Routh MD et al (2011) Efflux pumps of the resistance-nodulation-division family: a perspective of their structure, function and regulation in gram-negative bacteria. Adv Enzymol Relat Areas Mol Biol 77:109

Roy SK, Pahwa S, Nandanwar H, Jachak SM (2012) Phenylpropanoids of Alpinia galanga as efflux pump inhibitors in mycobacterium smegmatis mc2 155. Fitoterapia 83:1248–1255

Roy SK et al (2013) NorA efflux pump inhibitory activity of coumarins from Mesua ferrea. Fitoterapia 90:140–150

Sajjad Aghayan S et al (2017) The effects of Berberine and Palmatine on efflux pumps inhibition with different gene patterns in Pseudomonas aeruginosa isolated from burn infections. Avicenna J Med Biotechnol 9:2

Sana M, Jameel H, Rahman M (2015) Miracle remedy: inhibition of bacterial efflux pumps by natural products. J Biotechnology Biochem 4:1

Seukep AJ, Kuete V, Nahar L, Sarker SD, Guo M (2020) Plant-derived secondary metabolites as the main source of efflux pump inhibitors and methods for identification. J Pharm Anal 10:277–290. https://doi.org/10.1016/j.jpha.2019.11.002

Shafer WM, Waring AJ, Lehrer RI (1998) Modulation of Neisseria gonorrhoeae susceptibility to vertebrate antibacterial peptides due to a member of the resistance nodulationdivision efflux pump family. Proc Natl Acad Sci U S A 95:1829

Siriyong T et al (2017) Conessine as a novel inhibitor of multidrug efflux pump systems in Pseudomonas aeruginosa. BMC Complement Altern Med 17:405

Spengler G, Kincses A, Gajdács M, Amaral L (2017) New roads leading to old destinations: efflux pumps as targets to reverse multidrug resistance in bacteria. Molecules 22:468. https://doi.org/10.3390/molecules22030468

Stavri M, Piddock LJV, Gibbons S (2007) Bacterial efflux pump inhibitors from natural sources. J Antimicrob Chemother 59:1247–1260. https://doi.org/10.1093/jac/dkl460

Stermitz FR, Scriven LN, Tegos G and Lewis K (n.d.) Heruntergeladen von: UC Santa Barbara

Sun J, Deng Z, Yan A (2014) Bacterial multidrug efflux pumps: mechanisms, physiology and pharmacological exploitations. Biochem Biophys Res Commun 453:254–267. https://doi.org/10.1016/j.bbrc.2014.05.090

Thomas E, Renau R et al (1999) Inhibitors of efflux pumps in Pseudomonas aeruginosa potentiate the activity of the fluoroquinolone antibacterial levofloxacin. J Med Chem 49:5597

Verma P, Tiwari M, Tiwari V (2021) Efflux pumps in multidrug-resistant Acinetobacter baumannii: current status and challenges in the discovery of efflux pumps inhibitors. Microb Pathog 152: 104766

Watkins WJ et al (2003) The relationship between physicochemical properties, in vitro activity and pharmacokinetic profiles of analogues of diamine-containing efflux pump inhibitors. Bioorg Med Chem Lett 13:4241–4244

Yerushalmi H, Lebendiker M, Schuldiner S (1996) Negative dominance studies demonstrate the oligomeric structure of EmrE, a multidrug antiporter from Escherichia coli. J Biol Chem 271: 31044–31048

Zgurskaya H, Introduction I (2021) Transporters, porins, and efflux pumps. Chem Rev 121:5095–5097. https://doi.org/10.1021/acs.chemrev.1c00010

Chapter 5
Plant Endophytes: A Treasure House of Antimicrobial Compounds

Surbhi Agarwal, Garima Sharma, and Vartika Mathur

Abstract Recent decades have witnessed a significant progress in identifying novel and economically viable therapeutic compounds from natural resources. There is an increased interest in determining various plant-based compounds as an alternative to commercially available antibiotic drugs. However, endophytes, despite being a treasure house for a number of secondary metabolites, including antimicrobial compounds, are still less explored. These microbes are well known to produce hydrolytic enzymes, phenols, and terpenes which contribute to plants' defense against pathogens as well as predators. Endophytes using similar biosynthetic pathways to their host sometimes mimic the production of secondary metabolites because of which they become equally efficient for drug development. This chapter explores the endophytes known for producing antimicrobial compounds and describes various biosynthetic pathways used by them for their production. In addition, we have discussed the role of endophyte-derived antimicrobial metabolites in both their host plant and their benefits to mankind in combating the problem of antimicrobial resistance.

Keywords Coumarins · Bacteria · Fungi · Cyclic peptides · Secondary metabolites · Biosynthetic pathways · Multidrug resistance

5.1 Introduction

In recent decades, there has been a rapid shift from synthetic to natural bioactive compounds obtained mainly from plants and microorganisms (Okogun 2002; Rai et al. 2009). Investigation toward bioactive metabolites led to exploring the potential of endophytes, which are microorganisms harboring in the inter- and intracellular plant spaces within the plant tissues (Schulz and Boyle 2005; Joo et al. 2021). The mutualistic relationship between microbes and plants has been known for more than

S. Agarwal · G. Sharma · V. Mathur (✉)
Animal-Plant Interactions Lab, Department of Zoology, Sri Venkateswara College, New Delhi, India
e-mail: vmathur@svc.ac.in

V. Kumar et al. (eds.), *Medicinal Plants and Antimicrobial Therapies*,
https://doi.org/10.1007/978-981-99-7261-6_5

120 years (Hardoim et al. 2008). Based on the life strategies, bacterial endophytes have been classified as obligate and facultative (Correa 1994). Obligate endophytes are strictly dependent on the host for their growth and survival. However, facultative organisms do not reside inside the host for their complete life cycle but are associated with plants for most of their life (Hardoim et al. 2008; Abreu-Tarazi et al. 2010). In due course of evolution, these microbes have adapted to their host environment and play an important role in plants' growth, physiology, and metabolite production (Strobel and Daisy 2003; Nicoletti and Fiorentino 2015; Ludwig-Müller 2015). However, the structure of the endophyte community is complex and depends on various factors such as plant type, age, genetic makeup, season, and surroundings (Jasim et al. 2014; Sharma et al. 2020). Similar to their host plant, endophytes also synthesize secondary metabolites in response to environmental signals such as biotic and abiotic stress (Strobel et al. 1993; Lee et al. 1996; Mathur and Ulanova 2022). Studies also suggest that endophytes trigger induced systemic resistance (ISR) in their host plant which leads to the activation of its defense mechanism against pathogens (Van Loon et al. 1998; Kloepper and Ryu 2006).

Endophytes include both bacteria and fungi, and their diversity varies among the plant parts. Bacterial endophytes are more diversified in roots, whereas the fungi are distributed throughout the plant (Wang et al. 2016b; Furtado et al. 2019). Both bacterial and fungal endophytes are known for secondary metabolite production including antimicrobial compounds (Mousa and Raizada 2013; Ryan et al. 2008). Compounds produced by endophytes belong to different classes including xanthones, chinones, sterols, phenols, terpenoids, isocoumarins, tetralones, and benzopyranones (Krohn et al. 2002) and contribute to plant protection against pathogens and pests (Schulz et al. 1995, 2002). Accordingly, host–endophyte interaction sometimes also results in the production of antimicrobial compounds (Prado et al. 2012; Ludwig-Müller 2015). Antimicrobial compounds produced by microbes fall under the category of low molecular weight secondary metabolites (Demain 1981). The production of bioactive compounds may also depend on the cooperative interaction of the host plant and endophyte in a biosynthetic pathway (Schafhauser et al. 2019). Moreover, endophytes are known to produce antimicrobial cyclic peptides with higher potency through biosynthetic pathways that are unknown in plants and may serve as a potential candidate for novel drug development (Schwarzer et al. 2003; Abdalla and Matasyoh 2014; Sharma et al. 2023). The antimicrobial products obtained from both bacterial and fungal endophytes have applications in diverse fields such as agriculture, defense, and therapeutics. For example, endophyte-derived fungicides may find application in crop protection as a sustainable alternative to synthetic fungicides.

Another application of endophyte-derived antimicrobial compounds is in mitigating multidrug resistance (MDR). MDR is an emerging crisis prevalent due to the misuse of antibiotics, poor hygiene practices, and deferral in the diagnosis of disease (Costelloe et al. 2010; Mishra et al. 2017). It is estimated to be claiming approximately 7 lakh morbidities annually, and by 2050, this number is projected to increase to more than 1 crore annually (Pasrija et al. 2022). A recent study has shown that extracellular metabolites of endophytic fungus isolated from *Azadirachta indica* have potent antimicrobial activity against clinical strains multidrug-resistant bacteria and phytopathogenic fungi (Raihan et al. 2021).

Many endophytes are known to produce compounds with antimicrobial properties and might serve as a potential source of pharmaceutically important compounds. Hence, this chapter focuses on the role of bacterial and fungal endophyte-derived compounds in their host plant and their biosynthetic pathways with a special focus on the important antimicrobial compounds. These compounds may be developed for overcoming, placating, or reversing MDR or as alternative sources of drugs to the antimicrobial drugs currently available, many of which are now being rendered ineffective against MDR.

5.2 Endophyte-Mediated Pathways for Metabolite Production

Production of endophyte-derived bioactive compounds is facilitated through different biosynthetic pathways (Tanaka et al. 2005; Barry and Challis 2009; Miller et al. 2012). Sharma et al. (2023) have described that the production of bioactive compounds mainly takes place via four approaches: (1) modulation of host secondary metabolism, (2) elicitation, (3) endophyte exclusive pathways, and (4) *in-planta* biotransformation. Among these, the most significant pathways for metabolite production are the endophyte-exclusive pathways, which include the synthesis of non-ribosomal peptide synthetase (NRPS) and polyketide synthase. The biosynthesis of non-ribosomal peptides occurs via both proteinogenic and nonproteinogenic amino acids, through the production of polyketides by acetyl-CoA and malonyl-CoA precursors using respective synthases (Stierle et al. 1993). These pathways have been analyzed by targeting type I PKS genes and NRPS and in endophytes of medicinal plants (Miller et al. 2012). The synthesis of secondary metabolites takes place via various pathways such as isoprenoid, polyketide, and amino acid derivatives (Jalgaonwala 2013; Li et al. 2014). Bacteria and fungi are the major producers of polyketide decalin moiety containing bioactive compounds with a broad spectrum of bioactivities (Li et al. 2014). For example, in *Streptomyces cyaneofuscatus* (KY287599), nonribosomal peptide synthetases, aminodeoxyisochorismate synthase (phzE), and polyketide synthases type II have been detected, which are involved in the production of antimicrobial compounds (Zothanpuia Passari et al. 2017). The secondary metabolites produced via different pathways are classified as alkaloids, terpenes, flavonoids, coumarins, and other miscellaneous compounds (Schulz et al. 2002). The biosynthesis of alkaloids takes place via mevalonate and shikimic acid pathways from dimethylallyl pyrophosphates and aromatic amino acids, whereas terpenes are synthesized through the mevalonate pathway only by enzyme terpene cyclase. Similarly, benzylisoquinoline alkaloid (BIA) pathway is used by endophytes to enhance alkaloid production which helps in producing various therapeutic drugs (Pandey et al. 2016). Coumarins are produced via the

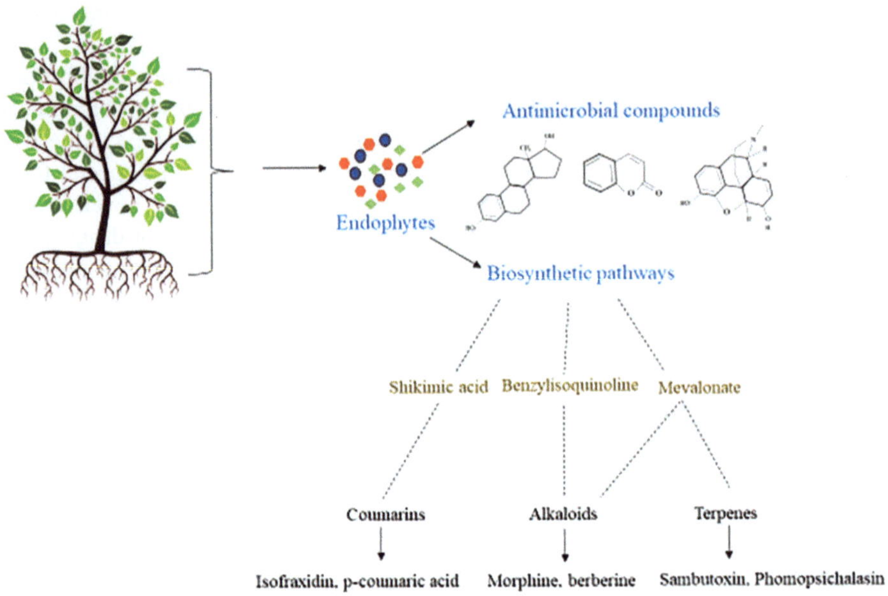

Fig. 5.1 Role of plant–endophyte interaction in the production of antimicrobial metabolites via different biosynthetic pathways

shikimic acid pathway through the metabolism of phenylalanine (Srinivasa et al. 2022). Figure 5.1 describes the role of plant and endophyte interaction in the production of antimicrobial compounds via various biosynthetic pathways.

5.3 Role of Endophyte-Derived Antimicrobial Compounds in Plants

Antimicrobial compounds produced by plant endophytes are eco-friendly and non-toxic to humans and have the ability to kill pathogens. The compounds produced due to the inoculation of endophytes have high medicinal properties (Singh et al. 2017). Besides this, endophytes also enhance plant growth in various ways including the production of growth hormone (Lee et al. 2004), siderophores such as enterobactin as well as essential vitamins and phosphate solubilization (Rodelas et al. 1993; Morales et al. 2011). It has been observed that solely plant–pathogen interaction is not able to produce secondary metabolites, as this relationship is one-sided as endophytes are also required for the production of metabolites (Ludwig-Müller 2015). Hence, plants benefit from their mutualistic endophytic partner in terms of secondary metabolite production which provides stress tolerance as well as defensive ability to the host (Clay and Schardl 2002; Strobel et al. 2004; Knop et al. 2007). Although endophytes produce a diverse group of compounds including phenols, terpenoids, and steroids, they sometimes constitute the derivatives of these compounds (Schulz et al. 2002). For example, the endophytic fungus *Chaetomium globosum* CGMCC 6882 isolated from *Gynostemma pentaphyllum* causes the biotransformation of xanthan to low molecular weight xanthan which possesses antibacterial activity against *Staphylococcus aureus* (Hu et al. 2019).

Furthermore, it has been found that bacterial endophytes chiefly induce allelopathic effects (Mishra et al. 2013), whereas fungal endophytes contribute to mitigating the damage caused by pests (Sánchez-Rodríguez et al. 2018). Table 5.1 provides a comprehensive list of the endophytes from different host plants producing antimicrobial compounds. Besides Table 5.1, below are examples of some noteworthy bacterial and fungal endophytes with the antimicrobial compounds produced by them:

5.3.1 Antimicrobial Products By Bacterial Endophytes

Obligate endophytes belonging to the genus *Burkholderia* are known to produce a wide range of antimicrobial compounds (Depoorter et al. 2016; Yao et al. 2017; Rodríguez-Cisneros et al. 2023). For example, *Burkholderia* sp. MS455 isolated from soybean shows potent antifungal activity against pathogens including *Aspergillus flavus* by the production of occidiofungin (Jia et al. 2022). Endophytic bacteria belonging to actinobacteria are also known for their potent antimicrobial and other therapeutic properties (Salam et al. 2017). Among actinobacteria, obligate endophytes belonging to actinomycetes are a promising source of secondary metabolites that can be pharmaceutically valuable. Actinomycetes are a group of Gram-positive bacteria known for their ability to produce a wide range of biologically active compounds, particularly antibiotics. *Streptomyces* is a diverse and abundant genus within the actinomycetes group, comprising filamentous bacteria commonly found as endophytes in majority of plant species. Plethora of studies have revealed that *Streptomyces* species are responsible for the majority of antibiotic production (~70%) among actinomycetes. The high percentage of antibiotic compounds produced by *Streptomyces* highlights the significance of this genus in the field of antibiotic discovery and development. There has been a heightened interest in identifying and harnessing the antibiotic-producing capabilities of *Streptomyces* species, leading to the discovery of many clinically important antibiotics such as streptomycin, tetracycline, erythromycin, and vancomycin (Berdy 2005; Subramani and Aalbersberg 2012). With upcoming advancements in functional library preparation of endophytes, these antimicrobial compounds are quickly gaining popularity in modern medicine. For example, *Streptomyces rochei* CH1 of *Cinnamomum* sp. produces several antibiotics effective against multidrug resistant bacteria (Roy and Banerjee 2015). Streptomyces KX852460 strain is known for its anti-fungal activity against plant pathogen *Rhizoctonia solani* AG-3 KX852461, which is a causative agent of target spot disease in tobacco leaf (Ahsan et al. 2017). Similarly, Alnumycin obtained from *Streptomyces* sp. isolated from *Alnus glutinosa* also

Table 5.1 Endophyte-derived antimicrobial compounds

Bioactive compounds	Activity	Endophyte	Host	References
Phenols and flavonoids				
Mycophenolic acid	Inhibit growth of phyto-pathogenic fungi	*Penicillium* sp.	*Urginea marituma*	Azar et al. (2023)
Phenolic compounds and flavonoids	Antibacterial	*Trichoderma harzianum*	*Rosmarinus officinalis*	Abdulhadi et al. (2023)
Phenolic compounds	Antimicrobial	*Streptomyces* sp. BO-07	*Boesenbergia rotunda*	Taechowisan et al. (2017)
4-(2,4,7-Trioxa bicycle [4.1.0] heptan-3-yl) phenol	Antibacterial	*Pestalotiopsis mangiferae*	*Mangifera indica*	Subban et al. (2013)
Isorhamnetin	Antibacterial activity against multidrug-resistant bacteria	*Penicillium sclerotigenum*	*Ageratina adenophora*	Wen et al. (2023)
Genistein	Antibacterial activity against multidrug-resistant bacteria	*Pestalotiopsis trachycarpicola*	*Ageratina adenophora*	Wen et al. (2023)
Primary and fatty alcohols				
2-Butyl-1-dodecanol	Antibacterial	*Aspergillus fumigatus*	*Dillenia indica*	Bora and Devi (2023)
1-Octanol	Antibacterial	*Aspergillus fumigatus*	*D. indica*	Bora and Devi (2023)
Dicerandrol A	Antibacterial	*Phomopsis longicolla*	Plant	Lim et al. (2010)
2-Isopropyl-5-methyl-1-heptanol	Antibacterial	*Aspergillus fumigatus*	*D. indica*	Bora and Devi (2023)
Fatty acids and carboxylic acids				
1-Fluoro-pentanoic acid	Antibacterial	*Aspergillus fumigatus*	*D. indica*	Bora and Devi (2023)
Phenyllactic acid and citric acid		*Diaporthe kochmanii* and *Pestalotiopsis trachycarpicola*	*Ageratina adenophora*	Wen et al. (2023)

	Antibacterial activity against multidrug-resistant bacteria			
1,2-Benzene dicarboxylic acid	Antibacterial activity against multidrug-resistant pathogens	*Daldinia eschscholtzii*	*Polianthes tuberosa*	Sundar and Arunachalam (2023)
Palmitic acid, carbonochloridic acid, and 3-[(1-*carboxyvinyl*)*oxy*] benzoic acid	Antimicrobial	*Actinobacteria*	*Harpagophytum procumbens*	Lang et al. (2023)
Alkanes				
Dodecane	Antibacterial	*Aspergillus fumigatus*	*Dillenia indica*	Bora and Devi (2023)
Esters				
Diheptyl ester	Antibacterial activity against multidrug-resistant pathogens	*Daldinia eschscholtzii*	*Polianthes tuberosa*	Sundar and Arunachalam (2023)
Chlorogenic acid	Antibacterial and antifungal	*Sordariomycete* sp.	*Eucommia ulmoides*	Chen et al. (2010)
Diethyl phthalate and benzyl benzoate	Antimicrobial	*Nigrospora sphaerica, Fusarium oxysporum, Gibberella moniliformis,* and *Beauveria bassiana*	*Crescentia cujete*	Prabukumar et al. (2015)
2-Chloroethyl ester	Antimicrobial	*Actinobacteria*	*Harpagophytum procumbens*	Lang et al. (2023)
2-Ethylhexyl ester	Antibacterial	*Aspergillus fumigatus*	*Dillenia indica*	Bora and Devi (2023)
Ether				
Alternariol methyl ether	Antifungal	*Ginkgo biloba*	*Paraphaeosphaeria* sp.	Su et al. (2023)
Peptides and polyketides				
Surfactins and iturins	Antimicrobial	*Bacillus subtilis*	*Triticum aestivum*	Taheri et al. (2023)

(continued)

Table 5.1 (continued)

Bioactive compounds	Activity	Endophyte	Host	References
Beauvericin	Antibacterial and antifungal	*Fusarium oxysporum*	*Cinnamomum kanehirae*	Wang et al. (2011)
Lijiquinone 1	Antifungal	*Ascomycete* sp. F53	*Taxus yunnanensis*	Cain et al. (2020)
Azaphilones	Antibacterial	*Colletotrichum* sp. BS4	*Buxus sinica*	Wang et al. (2016a)
Helvolic acid	Antibacterial	*Fusarium solani*	*Ficus carica*	Zhang et al. (2012)
Bikaverin	Antibacterial and antifungal	*Fusarium oxysporum*	*Cinnamomum kanehirae*	Wang et al. (2011)
Terpenes				
Paraphaone A, paraphaterpenes A, C, and D	Antifungal	*Ginkgo biloba*	*Paraphaeosphaeria* sp.	Su et al. (2023)
Sambutoxin	Antibacterial and antifungal	*Fusarium oxysporum*	*Cinnamomum kanehirae*	Wang et al. (2011)
Phomopsichalasin	Antimicrobial	*Phomopsis* sp.	*Triticum* sp.	Horn et al. (1995)
Xanthones				
Phomoxanthone A and phomoxanthone B	Broad-range antimicrobial activity	*Phomopsis* sp.	*Tectona grandis*	Isaka et al. (2001)
Alkaloids				
Bisdethiobis (methylthio) gliotoxin	Antibacterial	*Fusarium solani*	*Ficus carica*	Zhang et al. (2012)
Diketopiperazines				
Gliotoxin	Antibacterial	*Fusarium solani*	*Ficus carica*	Zhang et al. (2012)
Amines				
5-(1H-Indol-3-yl)-4,5-dihydro-[1,2,4]triazin-3-ylamine	Antibacterial	*Aspergillus cejpii*	*Nelumbo nucifera*	Techaoei et al. (2020)
Miscellaneous compounds				
Oxysporidinone, 6-*epi*-oxysporidinone	Antibacterial and antifungal	*Fusarium oxysporum*	*Cinnamomum kanehirae*	Wang et al. (2011)
Fumonisin		*Fusarium oxysporum*		Wang et al. (2011)

	Antibacterial and antifungal		*Cinnamomum kanehirae*	
NA	Antimicrobial	*Trichoderma harzianum*, *Guignardia* sp., and *Phomopsis* sp.	*Coffea arabica* and *C. robusta*	Sette et al. (2006)
Silver nanoparticles, squalene, 1,2-dibutyl phthalate	Antibacterial and antimicrobial	*Aspergillus fumigatus* and *Alternaria tenuissima*	*Cannabis sativa* and *Artemisia judaica*	Bala and Arya (2013), Al Mousa et al. (2021)
Pyrrocidines A and B	Antimicrobial	*Acremonium zeae*	*Zea mays*	Wicklow and Pol-ing (2009)
Asperlone A and asperlone B	Antimicrobial	*Aspergillus* sp.	*Sonneratia apetala*	Xiao et al. (2015)

possesses narrow-spectrum antibiotic activity against *Bacillus subtilis*, *Arthrobacter crystallopoites*, *Rhodococcus* sp., and *Micrococcus luteus* (Bieber et al. 1998). Other strains of *Streptomyces* have also been reported to produce a wide variety of antimicrobial compounds such as alkanes, benzoic acid esters, decenes, amines, and so on, which possess broad spectrum antibiotic potential (Ahsan et al. 2017; Zothanpuia Passari et al. 2017).

Another study showed that endophytic *Actinobacteria* associated with *Dracaena cochinchinensis* Lour. can produce antibiotic compounds that can be an alternative medicinal source for bioactive compounds (Salam et al. 2017). Some bacterial NRPS sequences are highly similar to pathways of pharmaceutically important antibiotics. One such example is of endophytic *Paenibacillus* sp. The B3 strain of *Bletilla striata* containing bacterial NRPS1 is 93% identical to the PKS of *Paenibacillus polymyxa* which produces antibiotic polymyxin. Similarly, bacterial NRPS of endophyte *Bacillus* sp. B61 isolated from *Belamcanda chinensis* has 93% identity to PKS involved in the biosynthesis of lichenysin antibiotic (Miller et al. 2012). Though it is evident that endophytes play a pivotal role in host plant metabolism and metabolite production, however, the precise mechanism is still to be explored.

5.3.2 Antimicrobial Products By Fungal Endophytes

Fungal endophytes have been the most common, rich, and diverse source of secondary metabolite production, imparting stress tolerance in plants (Sirikantaramas et al. 2007; Yan et al. 2019). Production of antimicrobial metabolites is one of the mechanisms of endophytic fungi through which it increases the capacity of its host plant to resist pathogens (Mousa and Raizada 2015; Yan et al. 2019). Fungal endophytes are known as a rich source of coumarins. For example, isofraxidin, a hydroxy coumarin compound, produced by fungal endophytes of *Schleichera oleosa* has potential antibacterial activity (Gagana et al. 2020; Yamazaki and Tokiwa 2010). Similarly, p-coumaric acid obtained from *Alternaria alternata* isolated from the leaves of *Catharanthus roseus* possesses potent antimicrobial activity against a wide range of fungi and bacteria (Sudharshana et al. 2019). Moreover, 7-amino-4-methylcoumarin extracted from *Xylaria* sp. YX-28 of *Ginkgo biloba* is well known for its antibacterial and antifungal activities (Liu et al. 2008). Among decalins, two derivatives are obtained from endophytic *Penicillium* sp. that are known for their antifungal activity against plant pathogenic *Sclerotinia sclerotiorum* (Stierle et al. 1999).

Another class of antibiotics includes proteinaceous bioactive compounds. For example, trtesin, a defensin-like peptide, from *Fusarium tricinctum*, a fungal endophyte of the medicinal herb *Rhododendron tomentosum*, possesses antimicrobial activity against various human pathogens (Tejesvi et al. 2013). Another peptide antibiotic, munumbicins A-D40, obtained from *Streptomyces* sp. NRRL30562, an endophyte of the snake vine plant, has broad-spectrum antibiotic activity against both human and plant pathogenic bacteria as well as fungi (Singh et al. 2017).

Another class of compounds is quinols, endophytic fungus, *Chloridium* sp. isolated from neem is known to produce a napthaquinone named Javanicin which displays potential antimicrobial activity against *Pseudomonas* sp. (Nisa et al. 2015). As *Pseudomonas* sp. is pathogenic to both plants and humans, Javanicin can be used as a potential antibiotic compound commercially.

Endophytic *Penicillium purpurogenum* ED76 possesses high antimicrobial activity against *Acintobacter anitratus*, *Candida albicans,* and *S. aureus* through the production of stigmasterol. The activity of this compound has been observed via scanning electron micrographs to understand the morphology of treated bacterial cells (Yenn et al. 2017).

Another fungal endophyte *Muscodor tigerii*, which inhabits the stem of *Cinnamomum camphora*, is a source of many antibacterial and antifungal volatile organic compounds (VOCs). These VOCs include 1-tetradecanamine, mono (2-ethylhexyl) ester, 4-octadecylmorpholine, squalene, *N*, *N*-dimethyl, 1,2-benzenedicarboxylic acid, and phytol (Saxena et al. 2015). These studies suggest that the isolation of these endophyte-derived antimicrobial compounds will provide a platform for the production of antibiotics from natural sources.

Some studies have shown the potential of endophytes against various fungal and bacterial pathogens. However, the information about the chemical nature of antimicrobial compounds is yet to be known. For example, *Bauhinia forficata* (Brazilian folk medicine plant) is a repository of beneficial endophytes such as *Khuskia oryzae* which exhibits strong antibiotic activity against *Salmonella typhi*. Another endophyte of the same plant, *Penicillium glabrum*, has shown the potential to kill *S. aureus*, *Streptococcus pyogenes*, and *Bacillus subtilis* (Bezerra et al. 2015). Similarly, fungal endophytes isolated from *Nyctanthes arbor-tristis* show potent broad-spectrum antimicrobial activity against a number of bacterial and fungal pathogens (Gond et al. 2012). There are a number of studies that suggest the role of endophytes obtained from different hosts in the production of antimicrobial secondary metabolites. As a diverse range of bacterial and fungal endophytes exist in nature, extraction of metabolites and further research on them can lead to the manufacture of cost-effective drugs.

5.4 Conclusion

Endophytes have high species diversity and are adapted to various environments which makes them a rich source of novel secondary metabolites for both agricultural, biotechnological, and pharmaceutical industries. Similar to their host, the antimicrobial metabolites produced by these microbes play an important role in host plant defense and are also effective against phytopathogens. However, the mechanism due to which the endophytes mimic its host plant in terms of metabolite production is less understood. Due to the co-evolution and inter-dependency of host plants on endophytes and vice versa, production of these antimicrobial compounds and/or their derivations takes place which are limited in either quantity or potency. The use of transgenic technologies can be considered one of the effective methods to produce cost-effective endophytes with better efficacy. Another way to enhance the production of desired antimicrobial compounds is to use a complete plant-endophyte system for large-scale production. However, the key to unlocking the potential of such a system is to understand the cross-talk between the plant and its endophyte (viz. biosynthetic pathway). Further research on endophyte-derived natural antimicrobial compounds will also be beneficial in the production of cost-effective drugs with lesser or no side effects. Moreover, this approach will also mitigate the global issue of increasing antimicrobial resistance.

References

Abdalla MA, Matasyoh JC (2014) Endophytes as producers of peptides: an overview about the recently discovered peptides from endophytic microbes. Nat Prod Bioprospect 4:257–270

Abdulhadi SY, Hasan GQ, Gergees RN (2023) Molecular detection and antimicrobial activity of endophytic fungi isolated from a medical plant Rosmarinus officinalis. arXiv preprint arXiv:2303.05242

Abreu-Tarazi MF, Navarrete AA, Andreote FD, Almeida CV, Tsai SM, Almeida M (2010) Endophytic bacteria in long-term in vitro cultivated “axenic” pineapple microplants revealed by PCR–DGGE. World J Microbiol Biotechnol 26:555–560

Ahsan T, Chen J, Zhao X, Irfan M, Wu Y (2017) Extraction and identification of bioactive compounds (eicosane and dibutyl phthalate) produced by Streptomyces strain KX852460 for the biological control of Rhizoctonia solani AG-3 strain KX852461 to control target spot disease in tobacco leaf. AMB Express 7(1):1–9

Al Mousa AA, Mohamed H, Hassane AM, Abo-Dahab NF (2021) Antimicrobial and cytotoxic potential of an endophytic fungus Alternaria tenuissima AUMC14342 isolated from Artemisia judaica L. growing in Saudi Arabia. J King Saud Univ Sci 33(5):101462

Azar N, Liarzi O, Zavitan M, Samara M, Nasser A, Ezra D (2023) Endophytic Penicillium species secretes mycophenolic acid that inhibits the growth of Phytopathogenic fungi. Microbial Biotechnol 16:1629

Bala M, Arya V (2013) Biological synthesis of silver nanoparticles from aqueous extract of endophytic fungus Aspergillus fumigatus and its antibacterial action. Int J Nanomat Biostruct 3(2):37–41

Barry SM, Challis GL (2009) Recent advances in siderophore biosynthesis. Curr Opin Chem Biol 13(2):205–215

Berdy J (2005) Bioactive microbial metabolites. J Antibiot 58(1):1–26

Bezerra JD, Nascimento CC, Barbosa RDN, da Silva DC, Svedese VM, Silva-Nogueira EB, Souza-Motta CM (2015) Endophytic fungi from medicinal plant Bauhinia forficata: diversity and biotechnological potential. Braz J Microbiol 46:49–57

Bieber B, Nüske J, Ritzau M, Gräfe U (1998) Alnumycin a new naphthoquinone antibiotic produced by an endophytic Streptomyces sp. J Antibiot 51(3):381–382

Bora P, Devi NN (2023) Exploration of the chemical constituents and its antioxidant, antibacterial activities of endophytic fungi isolated from the medicinal plant Dillenia indica. Arch Microbiol 205(2):67

Cain JW, Miller KI, Kalaitzis JA, Chau R, Neilan BA (2020) Genome mining of a fungal endophyte of Taxus yunnanensis (Chinese yew) leads to the discovery of a novel Azaphilone polyketide, Lijiquinone. Microb Biotechnol 13(5):1415–1427

Chen X, Sang X, Li S, Zhang S, Bai L (2010) Studies on a chlorogenic acid-producing endophytic fungi isolated from Eucommia ulmoides Oliver. J Ind Microbiol Biotechnol 37(5):447–454

Clay K, Schardl C (2002) Evolutionary origins and ecological consequences of endophyte symbiosis with grasses. Am Nat 160(S4):S99–S127

Correa JA (1994) Infections by pigmented algal endophytes: misuse of concepts and terminology

Costelloe C, Metcalfe C, Lovering A, Mant D, Hay AD (2010) Effect of antibiotic prescribing in primary care on antimicrobial resistance in individual patients: systematic review and meta-analysis. BMJ 340:c2096

Demain AL (1981) Industrial microbiology. Science 214(4524):987–995

Depoorter E, Bull MJ, Peeters C, Coenye T, Vandamme P, Mahenthiralingam E (2016) Burkholderia: an update on taxonomy and biotechnological potential as antibiotic producers. Appl Microbiol Biotechnol 100:5215–5229

Furtado BU, Gołębiewski M, Skorupa M, Hulisz P, Hrynkiewicz K (2019) Bacterial and fungal endophytic microbiomes of Salicornia europaea. Appl Environ Microbiol 85(13):e00305–e00319

Gagana SL, Kumaraswamy BE, Shivanna MB (2020) Diversity, antibacterial and antioxidant activities of the fungal endophytes associated with Schleichera oleosa (Lour.) Merr. S Afr J Bot 134:369–381

Gond SK, Mishra A, Sharma VK, Verma SK, Kumar J, Kharwar RN, Kumar A (2012) Diversity and antimicrobial activity of endophytic fungi isolated from Nyctanthes arbor-tristis, a well-known medicinal plant of India. Mycoscience 53(2):113–121

Hardoim PR, van Overbeek LS, van Elsas JD (2008) Properties of bacterial endophytes and their proposed role in plant growth. Trends Microbiol 16(10):463–471

Horn WS, Simmonds MSJ, Schwartz RE, Blaney WM (1995) Phomopsichalasin, a novel antimicrobial agent from an endophytic Phomopsis sp. Tetrahedron 51(14):3969–3978

Hu X, Wang K, Yu M, He P, Qiao H, Zhang H, Wang Z (2019) Characterization and antioxidant activity of a low-molecular-weight Xanthan gum. Biomol Ther 9(11):730

Isaka M, Jaturapat A, Rukseree K, Danwisetkanjana K, Tanticharoen M, Thebtaranonth Y (2001) Phomoxanthones A and B, novel Xanthone dimers from the endophytic fungus Phomopsis species. J Nat Prod 64(8):1015–1018

Jalgaonwala RE (2013) Bioprospecting for microbial endophytes and their natural products (Ph. D Thesis). North Maharastra University, Jalgaon

Jasim B, Joseph AA, John CJ, Mathew J, Radhakrishnan EK (2014) Isolation and characterization of plant growth promoting endophytic bacteria from the rhizome of Zingiber officinale. 3 Biotech 4:197–204

Jia J, Ford E, Hobbs SM, Baird SM, Lu SE (2022) Occidiofungin is the key metabolite for antifungal activity of the endophytic bacterium Burkholderia sp. MS455 against Aspergillus flavus. Phytopathology 112(3):481–491

Joo HS, Deyrup ST, Shim SH (2021) Endophyte-produced antimicrobials: a review of potential lead compounds with a focus on Quorum-sensing disruptors. Phytochem Rev 20:543–568

Kloepper JW, Ryu CM (2006) Bacterial endophytes as elicitors of induced systemic resistance. Microbial Root Endophytes:33–52

Knop M, Pacyna S, Voloshchuk N, Kant S, Müllenborn C, Steiner U, Schulz M (2007) Zea mays: Benzoxazolinone detoxification under sulfur deficiency conditions—a complex allelopathic alliance including endophytic Fusarium verticillioides. J Chem Ecol 33:225–237

Krohn K, Biele C, Drogies KH, Steingröver K, Aust HJ, Draeger S, Schulz B (2002) Fusidilactones, a new group of Polycyclic lactones from an endophyte, Fusidium sp. Eur J Org Chem 2002(14): 2331–2336

Lang ME, Sibanda T, Louw S, Uzabakiriho JD (2023) Antimicrobial potential of the endophytic actinobacteria isolated from Harpagophytum procumbens: a Southern African medicinal plant. S Afr J Bot 156:268–277

Lee JC, Strobel GA, Lobkovsky E, Clardy J (1996) Torreyanic acid: a selectively cytotoxic quinone dimer from the endophytic fungus Pestalotiopsis microspora. J Org Chem 61(10):3232–3233

Lee S, Flores-Encarnacion M, Contreras-Zentella M, Garcia-Flores L, Escamilla JE, Kennedy C (2004) Indole-3-acetic acid biosynthesis is deficient in Gluconacetobacter diazotrophicus strains with mutations in cytochrome c biogenesis genes. J Bacteriol 186(16):5384–5391
Li G, Kusari S, Spiteller M (2014) Natural products containing 'decalin'motif in microorganisms. Nat Prod Rep 31(9):1175–1201
Lim CS, Kim JY, Choi JN, Ponnusamy K, Jeon YT, Kim SU, Lee CH (2010) Identification, fermentation, and bioactivity against Xanthomonas oryzae of antimicrobial metabolites isolated from Phomopsis longicolla S1B4. J Microbiol Biotechnol 20(3):494–500
Liu X, Dong M, Chen X, Jiang M, Lv X, Zhou J (2008) Antimicrobial activity of an endophytic Xylaria sp. YX-28 and identification of its antimicrobial compound 7-amino-4-methylcoumarin. Appl Microbiol Biotechnol 78:241–247
Ludwig-Müller J (2015) Plants and endophytes: equal partners in secondary metabolite production? Biotechnol Lett 37:1325–1334
Mathur V, Ulanova D (2022) Microbial metabolites beneficial to plant hosts across ecosystems. Microb Ecol 86:1–24
Miller KI, Qing C, Sze DMY, Roufogalis BD, Neilan BA (2012) Culturable endophytes of medicinal plants and the genetic basis for their bioactivity. Microb Ecol 64:431–449
Mishra S, Upadhyay RS, Nautiyal CS (2013) Unravelling the beneficial role of microbial contributors in reducing the allelopathic effects of weeds. Appl Microbiol Biotechnol 97:5659–5668
Mishra VK, Passari AK, Chandra P, Leo VV, Kumar B, Uthandi S, Singh BP (2017) Determination and production of antimicrobial compounds by Aspergillus clavatonanicus strain MJ31, an endophytic fungus from Mirabilis jalapa L. using UPLC-ESI-MS/MS and TD-GC-MS analysis. PLoS One 12(10):e0186234
Morales A, Alvear M, Valenzuela E, Castillo CE, Borie F (2011) Screening, evaluation and selection of phosphate-solubilising fungi as potential biofertiliser. J Soil Sci Plant Nutr 11(4): 89–103
Mousa WK, Raizada MN (2013) The diversity of anti-microbial secondary metabolites produced by fungal endophytes: an interdisciplinary perspective. Front Microbiol 4:65
Mousa WK, Raizada MN (2015) Biodiversity of genes encoding anti-microbial traits within plant associated microbes. Front Plant Sci 6:231
Nicoletti R, Fiorentino A (2015) Plant bioactive metabolites and drugs produced by endophytic fungi of Spermatophyta. Agriculture 5(4):918–970
Nisa H, Kamili AN, Nawchoo IA, Shafi S, Shameem N, Bandh SA (2015) Fungal endophytes as prolific source of phytochemicals and other bioactive natural products: a review. Microb Pathog 82:50–59
Okogun JI (2002) Drug discovery through ethnobotany in Nigeria: some results. In: Advances in phytomedicine, vol 1. Elsevier, Amsterdam, pp 145–154
Pandey SS, Singh S, Babu CV, Shanker K, Srivastava NK, Kalra A (2016) Endophytes of opium poppy differentially modulate host plant productivity and genes for the biosynthetic pathway of benzylisoquinoline alkaloids. Planta 243:1097–1114
Pasrija P, Girdhar M, Kumar M, Arora S, Katyal A (2022) Endophytes: an unexplored treasure to combat multidrug resistance. Phytomed Plus 100249:100249
Prabukumar S, Rajkuberan C, Ravindran K, Sivaramakrishnan S (2015) Isolation and characterization of endophytic fungi from medicinal plant Crescentia cujete L. and their antibacterial, antioxidant and anticancer properties. Int J Pharm Pharm Sci 7(11):316–321. Nat Prod Res, 32(11), 1353–1356
Prado S, Li Y, Nay B (2012) Diversity and ecological significance of fungal endophyte natural products. Stud Nat Prod Chem 36:249–296
Rai M, Yadav A, Gade A (2009) Silver nanoparticles as a new generation of antimicrobials. Biotechnol Adv 27(1):76–83
Raihan T, Azad AK, Ahmed J, Shepon MR, Dey P, Chowdhury N, Suhani S (2021) Extracellular metabolites of endophytic fungi from Azadirachta indica inhibit multidrug-resistant bacteria and phytopathogens. Future Microbiol 16(8):557–576

Rodelas B, Salmeron V, Martinez-Toledo MV, González-López J (1993) Production of vitamins by Azospirillum brasilense in chemically-defined media. Plant Soil 153:97–101

Rodríguez-Cisneros M, Morales-Ruíz LM, Salazar-Gómez A, Rojas-Rojas FU, Santos E-d L, P. (2023) Compilation of the antimicrobial compounds produced by Burkholderia sensu stricto. Molecules 28(4):1646

Roy S, Banerjee D (2015) Bioactive endophytic actinomycetes of Cinnamomum sp.; isolation, identification, activity guided purification and process optimization of active metabolite. Am J Microbiol 6(1):4

Ryan RP, Germaine K, Franks A, Ryan DJ, Dowling DN (2008) Bacterial endophytes: recent developments and applications. FEMS Microbiol Lett 278(1):1–9

Salam N, Khieu TN, Liu MJ, Vu TT, Chu-Ky S, Quach NT, Li WJ (2017) Endophytic actinobacteria associated with Dracaena cochinchinensis Lour.: isolation, diversity, and their cytotoxic activities. BioMed Res Int 2017:1

Sánchez-Rodríguez AR, Raya-Díaz S, Zamarreño ÁM, García-Mina JM, del Campillo MC, Quesada-Moraga E (2018) An endophytic Beauveria bassiana strain increases spike production in bread and durum wheat plants and effectively controls cotton leafworm (Spodoptera littoralis) larvae. Biol Control 116:90–102

Saxena S, Meshram V, Kapoor N (2015) Muscodor tigerii sp. nov.-volatile antibiotic producing endophytic fungus from the Northeastern Himalayas. Ann Microbiol 65(1):47–57

Schafhauser T, Jahn L, Kirchner N, Kulik A, Flor L, Lang A, Ludwig-Müller J (2019) Antitumor astins originate from the fungal endophyte Cyanodermella asteris living within the medicinal plant Aster tataricus. Proc Natl Acad Sci 116(52):26909–26917

Schulz B, Boyle C (2005) The endophytic continuum. Mycol Res 109(6):661–686

Schulz B, Sucker J, Aust HJ, Krohn K, Ludewig K, Jones PG, Döring D (1995) Biologically active secondary metabolites of endophytic Pezicula species. Mycol Res 99(8):1007–1015

Schulz B, Boyle C, Draeger S, Römmert AK, Krohn K (2002) Endophytic fungi: a source of novel biologically active secondary metabolites. Mycol Res 106(9):996–1004

Schwarzer D, Finking R, Marahiel MA (2003) Nonribosomal peptides: from genes to products. Nat Prod Rep 20(3):275–287

Sette LD, Passarini MRZ, Delarmelina C, Salati F, Duarte MCT (2006) Molecular characterization and antimicrobial activity of endophytic fungi from coffee plants. World J Microbiol Biotechnol 22:1185–1195

Sharma G, Guleria R, Mathur V (2020) Differences in plant metabolites and microbes associated with Azadirachta indica with variation in air pollution. Environ Pollut 257:113595

Sharma G, Agarwal S, Verma K, Bhardwaj R, Mathur V (2023) Therapeutic compounds from medicinal plant endophytes: molecular and metabolic adaptations. J Appl Microbiol:lxad074

Singh M, Kumar A, Singh R, Pandey KD (2017) Endophytic bacteria: a new source of bioactive compounds. 3 Biotech 7:1–14

Sirikantaramas S, Asano T, Sudo H, Yamazaki M, Saito K (2007) Camptothecin: therapeutic potential and biotechnology. Curr Pharm Biotechnol 8(4):196–202

Srinivasa C, Mellappa G, Patil SM, Ramu R, Shreevatsa B, Dharmashekar C, Shivamallu C (2022) Plants and endophytes–a partnership for the coumarin production through the microbial systems. Mycology 13(4):243–256

Stierle A, Strobel G, Stierle D (1993) Taxol and taxane production by Taxomyces andreanae, an endophytic fungus of Pacific yew. Science 260(5105):214–216

Stierle DB, Stierle AA, Ganser BK (1999) Isolation of two highly methylated polyketide derivatives from a yew-associated Penicillium species. J Nat Prod 62(8):1147–1150

Strobel G, Daisy B (2003) Bioprospecting for microbial endophytes and their natural products. Microbiol Mol Biol Rev 67(4):491–502

Strobel G, Stierle A, Stierle D, Hess WM (1993) Taxomyces andreanae, a proposed new taxon for a bulbilliferous hyphomycete associated with Pacific yew (Taxus brevifolia). Mycotaxon 47:71–80

Strobel G, Daisy B, Castillo U, Harper J (2004) Natural products from endophytic microorganisms. J Nat Prod 67(2):257–268

Su S, Yang XQ, Yang YB, Ding ZT (2023) Antifungal and antifeedant terpenoids from Paraphaeosphaeria sp. cultured by extract of host Ginkgo biloba. Phytochemistry 210:113651

Subban K, Subramani R, Johnpaul M (2013) A novel antibacterial and antifungal phenolic compound from the endophytic fungus Pestalotiopsis mangiferae. Nat Prod Res 27(16): 1445–1449

Subramani R, Aalbersberg W (2012) Marine actinomycetes: an ongoing source of novel bioactive metabolites. Microbiol Res 167(10):571–580

Sudharshana TN, Venkatesh HN, Nayana B, Manjunath K, Mohana DC (2019) Anti-microbial and anti-mycotoxigenic activities of endophytic Alternaria alternata isolated from Catharanthus roseus (L.) G. Don.: molecular characterisation and bioactive compound isolation. Mycology 10(1):40–48

Sundar RDV, Arunachalam S (2023) Effect of endophytic fungi Daldinia eschscholtzii against multidrug resistant pathogens. Bangladesh J Pharmacol 18(1):17–23

Taechowisan T, Chaisaeng S, Phutdhawong WS (2017) Antibacterial, antioxidant and anticancer activities of biphenyls from Streptomyces sp. BO-07: an endophyte in Boesenbergia rotunda (L.) Mansf A. Food Agric Immunol 28(6):1330–1346

Taheri E, Tarighi S, Taheri P (2023) An endophytic bacterium with biocontrol activity against important wheat pathogens. Biol Control 183:105243

Tanaka A, Tapper BA, Popay A, Parker EJ, Scott B (2005) A symbiosis expressed non-ribosomal peptide synthetase from a mutualistic fungal endophyte of perennial ryegrass confers protection to the symbiotum from insect herbivory. Mol Microbiol 57(4):1036–1050

Techaoei S, Jirayuthcharoenkul C, Jarmkom K, Dumrongphuttidecha T, Khobjai W (2020) Chemical evaluation and antibacterial activity of novel bioactive compounds from endophytic fungi in Nelumbo nucifera. Saudi J Biol Sci 27(11):2883–2889

Tejesvi MV, Segura DR, Schnorr KM, Sandvang D, Mattila S, Olsen PB, Pirttilä AM (2013) An antimicrobial peptide from endophytic Fusarium tricinctum of Rhododendron tomentosum Harmaja. Fungal Divers 60:153–159

Van Loon LC, Bakker PAHM, Pieterse CMJ (1998) Systemic resistance induced by Rhizosphere bacteria. Annu Rev Phytopathol 36(1):453–483

Wang QX, Li SF, Zhao F, Dai HQ, Bao L, Ding R, Liu HW (2011) Chemical constituents from endophytic fungus Fusarium oxysporum. Fitoterapia 82(5):777–781

Wang WX, Kusari S, Laatsch H, Golz C, Kusari P, Strohmann C, Spiteller M (2016a) Antibacterial azaphilones from an endophytic fungus, Colletotrichum sp. BS4. J Nat Prod 79(4):704–710

Wang W, Zhai Y, Cao L, Tan H, Zhang R (2016b) Endophytic bacterial and fungal microbiota in sprouts, roots and stems of rice (Oryza sativa L.). Microbiol Res 188:1–8

Wen J, Okyere SK, Wang J, Huang R, Wang Y, Liu L, Hu Y (2023) Endophytic fungi isolated from Ageratina adenophora exhibits potential antimicrobial activity against multidrug-resistant Staphylococcus aureus. Plan Theory 12(3):650

Wicklow DT, Poling SM (2009) Antimicrobial activity of pyrrocidines from Acremonium zeae against endophytes and pathogens of maize. Phytopathology 99(1):109–115

Xiao ZE, Lin SE, Tan C, Lu Y, He L, Huang X, She Z (2015) Asperlones A and B, dinaphthalenone derivatives from a mangrove endophytic fungus Aspergillus sp. 16-5C. Mar Drugs 13(1): 366–378

Yamazaki T, Tokiwa T (2010) Isofraxidin, a coumarin component from Acanthopanax senticosus, inhibits matrix metalloproteinase-7 expression and cell invasion of human hepatoma cells. Biol Pharm Bull 33(10):1716–1722

Yan L, Zhu J, Zhao X, Shi J, Jiang C, Shao D (2019) Beneficial effects of endophytic fungi colonization on plants. Appl Microbiol Biotechnol 103:3327–3340

Yao GW, Duarte I, Le TT, Carmody L, LiPuma JJ, Young R, Gonzalez CF (2017) A broad-host-range tailocin from Burkholderia cenocepacia. Appl Environ Microbiol 83(10):e03414–e03416

Yenn TW, Ibrahim D, Chang LK, Ab Rashid S, Ring LC, Nee TW, bin Muhamad Noor MI (2017) Antimicrobial efficacy of endophytic Penicillium purpurogenum ED76 against clinical pathogens and its possible mode of action. 미생물학회지 제 53(3):1

Zhang HC, Ma YM, Liu R (2012) Antimicrobial additives from endophytic fungus Fusarium solani of Ficus carica. In: Applied mechanics and materials, vol 178. Trans Tech Publications Ltd., pp 783–786

Zothanpuia Passari AK, Chandra P, Leo VV, Mishra VK, Kumar B, Singh BP (2017) Production of potent antimicrobial compounds from Streptomyces cyaneofuscatus associated with fresh water sediment. Front Microbiol 8:68

Chapter 6
Exploring Medicinal Plant Resources for Combating Viral Diseases, Including COVID-19

Anirban Goutam Mukherjee, Pragya Bradu, Antara Biswas, Uddesh Ramesh Wanjari, Kaviyarasi Renu, Sandra Kannampuzha, Balachandar Vellingiri, and Abilash Valsala Gopalakrishnan

Abstract This book chapter provides insight into the medicinal plant extracts and phytochemicals essential to combat several viral diseases, including coronavirus disease (COVID-19). First, this work focuses on the current antiviral treatments and the medicinal plant extracts that can be utilized for the treatment or combined with other antiviral drugs to prevent viral diseases and reduce toxic side effects. Next, this chapter discusses polyphenols, flavonoids, proanthocyanidins (PACs), and terpenes—monoterpenes, triterpenes, sesquiterpenes, and glucosides—and how they are effective in treating viral diseases like influenza, hepatitis B and C viruses, and COVID-19. The next part specifically talks about the medicinal plants like amla, giloy, and extracts like Lianhua Qingwen, polyphenols, flavonoids, and more, which can help mitigate the severe acute respiratory syndrome-related coronavirus-2 (SARS-CoV-2) virus. Lastly, the book chapter explains the mechanisms by which the medicinal plants activate to prevent the replication and proliferation activity of enzymes and inhibit the binding of the virus with the angiotensin-converting enzyme-2 (ACE2) receptors. Therefore, this book chapter gives brief and concise

A. G. Mukherjee · P. Bradu · A. Biswas · U. R. Wanjari · S. Kannampuzha ·
A. V. Gopalakrishnan (✉)
Department of Biomedical Sciences, School of Bio-Sciences and Technology, Vellore Institute of Technology, Vellore, Tamil Nadu, India
e-mail: sandrajacobkannampuzha@gmail.com; abilash.vg@vit.ac.in

K. Renu
Centre of Molecular Medicine and Diagnostics (COMManD), Department of Biochemistry, Saveetha Dental College and Hospitals, Saveetha Institute of Medical and Technical Sciences, Saveetha University, Chennai, Tamil Nadu, India

B. Vellingiri
Department of Zoology, School of Basic Sciences, Central University of Punjab (CUPB), Bathinda, Punjab, India

V. Kumar et al. (eds.), *Medicinal Plants and Antimicrobial Therapies*,
https://doi.org/10.1007/978-981-99-7261-6_6

information regarding the different viral diseases and medicinal plants and their extracts for treating and preventing such viral diseases.

Keywords COVID-19 · Medicinal plant · Viral

6.1 Introduction

The recent COVID-19 outbreak and the pandemic have created a great concern among the doctors and scientists in the medical and health sectors. There were a series of epidemics that took place in 2003 in China with the severe acute respiratory syndrome coronavirus (SARS-CoV), Swine flu in 2009 due to the H1N1 virus, Middle East respiratory syndrome coronavirus (MERS-CoV) in Saudi Arabia in 2012, Nipah virus, avian influenza, etc. (Jansi et al. 2021; Mehta et al. 2021). Despite these epidemics, there was not much consideration ever given that, one day, one of these viral diseases would end up making people sit in their homes and develop into a worldwide pandemic. According to Jansi et al. (2021), viral diseases have now become a colossal threat to human beings and animals and, in recent years, have resulted in the loss of life and big economic losses (Jansi et al. 2021). Hence, looking for feasible, safe, and effective patient treatments is essential. Some studies suggest using small interfering ribonucleic acid (siRNA) silencing antiviral drugs (Mehta et al. 2021), while many scientists are also focusing on the benefits of medicinal plants.

Medicinal plants, their extracts, and phytochemicals such as polyphenols and flavonoids prove to be promising antiviral drugs since they provide safe and therapeutic results on the side effects caused by the infection. Phytochemicals possess excellent antiviral, anticancer, and antimicrobial properties and exhibit direct and indirect mechanisms to prevent the virus's entry, replication, and proliferation and express immunomodulatory properties (Anand et al. 2021). Several types of traditional medicines originated from India, China, Japan, Algeria, and many more countries, where these indigenous treatments were used earlier in treating viral diseases, fever, cancer, etc. Now, scientists are looking back into these traditional medicines and using various compounds extracted from them to treat viral diseases, including COVID-19.

This book chapter will make us understand the current treatments for viral diseases, including medicinal plants for influenza, respiratory syncytial virus (RSV), SARS-CoV, MERS-CoV, and the recent SARS-CoV-2, which resulted in the pandemic. It will also explore the various phytochemicals like flavonoids, polyphenols, types of terpenes, and glucosides in the treatment of viral diseases. A special focus is also on the medicinal plant for treating COVID-19, and finally, we explain the different mechanisms involved in inhibiting the replication and side effects of the infection caused by the viral disease, including the SARS-CoV-2.

6.2 Current Treatment Including Medicinal Plant Resources for Viral Diseases

Since the COVID-19 pandemic, there has been a huge concern in the health sector regarding viral diseases and how they should be prevented. Viral diseases can easily spread from one organism to another—from plant to animal, animal to human, plant to human, etc. These microorganisms require a host—a plant, an animal, or a human to survive and mutate itself to become stronger and spread faster. Hence, the requirement for medications has increased due to more interaction between humans and animals to protect oneself from deadly viral diseases and their harmful and sometimes fatal consequences. Currently, many countries use antiviral drugs that are useful in controlling disease transmission but may not cure the effects of the virus on the host body. However, when combined with antimicrobials, they produce a very effective action against resistant strains, and direct-acting antiviral (DAA) agents can also treat severe infections like hepatitis C virus (Jansi et al. 2021). There has been a growing interest in many health sectors of the world to focus on natural resources like plants, to produce effective and safe medicines. These medicinal plants can act on specific viruses, degrade their deoxyribonucleic acid (DNA) or RNA to treat viral diseases and provide additional nutritional benefits to the host. In this chapter, we will be studying how medicinal plants can act against several types of viruses. These include influenza–parainfluenza, respiratory syncytial virus, severe acute respiratory syndrome (SARS), Middle East respiratory syndrome, and COVID-19.

6.2.1 Medicinal Plants Acting Against Influenza–Parainfluenza Viruses

Influenza is a negative-sense single-stranded RNA virus that belongs to the Orthomyxoviridae family and has a size of 14 kb. It has three major types that can infect humans. It includes influenza A, influenza B, and influenza C. Influenza C is less severe and has lower mortality rates than influenza A and influenza B. Their gene structure consists of surface glycoproteins, neuraminidase (NA), and hemagglutinin (HA) (Jansi et al. 2021). The most common forms of influenza virus are H1N1 swine flu and H3N2. Several antiviral drugs have been proven effective in treating the influenza virus, along with HA and NA inhibitors combined with monoclonal antibodies (Jansi et al. 2021). When it comes to natural sources like plants, medicinal plants contain bioactive agents that can help enhance antiviral functions. Certain medicinal plants including *Canavalia ensiformis* (canavanine), *Theobroma cacao* (caffeine), *Chelidonium majus* (chelidonine), *Aspergillus nidulans* (cordycepin), *Cabucula erythrocarpa* (ochropamine and epi-16-ochropamine), and *Catharanthus roseus* (leurocristine and vincaleukoblastine) are effective for the treatment of the influenza virus. *Arachnoiditis aureus* (serotonin and gliotoxin) is effective for the treatment of both influenza and parainfluenza, and garlic, also known as *Allium sativa* (allyl methyl thiosulfinate, allicin, etc.), is effective for the treatment of parainfluenza virus (Jansi et al. 2021). The Indian medicinal plants and herbal extracts from Ayurveda and Unani medicines have antioxidant properties, which alleviate the effects of viral diseases, including influenza, and strengthen the immune system. Certain Chinese medicinal herbal extracts from dried ginger, cassia twig, etc. exhibited antiviral activity and prevented the H1N1 virus. A Japanese herbal medicine, maoto, exhibited antiviral activity against the influenza virus (Jansi et al. 2021). According to Sen et al. (2021), plant polysaccharides can also help in mediating the antiviral and immunomodulating activity to ensure the inhibition of viral diseases. Some Indian medicines, such as seer, zanjabeel, kalonji, behi dana, etc., were found to be virucidal toward parainfluenza virus type 3 and inhibited its replication (Sen et al. 2021). Plant extracts from polyphenols and flavonoids, such as quercetin, catechins, saponins, luteolin, and apigenin, play a significant role in preventing influenza, SARS, and other viral diseases. Flavonoids, apigenin, baicalin, and kaempferol all inhibit the NA activity, and saponins inhibit the HA activity in the influenza virus. Certain lignans like diphyllin prevent downstream replication, and clemastanin B exhibits antiviral properties against the influenza virus (Anand et al. 2021). Indian plants such as yashtimadhu or liquorice, green chirata, tulsi, and garlic have a strong antiviral effect on the influenza virus. The seed extract from the *Caryophyllaceae* plants can also exhibit antiviral properties against human immunodeficiency virus (HIV) and parainfluenza-3 virus (Anand et al. 2021).

6.2.2 Medicinal Plants Protecting from Respiratory Syncytial Virus

Respiratory syncytial virus (RSV) is a droplet infection that can cause lower respiratory tract infections in young children and adults and can also cause pneumonia (Chen et al. 2019; Venu and Austin 2020). It is a single-stranded RNA virus that is part of the Paramyxoviridae family. The rationale treatments given to patients mostly bring about side effects for the patients. Ribavirin is one of the antiviral drugs given to children to treat RSV suffering from bronchiolitis, but it has also been found to be highly toxic and it is a major concern (Chen et al. 2019; Venu and Austin 2020). Palivizumab is another drug given to premature babies suffering from immunoprophylaxis, but it is costly, and many patients cannot afford it (Chen et al. 2019). Hence, a natural solution from medicinal plants can be a safe and an effective method for treating this viral disease (Venu and Austin 2020). Certain plants that exhibit excellent antiviral properties against RSV include *Amaryllis belladonna*, *Blumea laciniata*, *Elephantopus scaber*, *Mussaenda pubescens*, *Narcissus tazetta*, *Scutellaria indica*, and *Selaginella sinensis*. The bioactive compounds extracted from these plants, such as lycorine, apigenin, naringenin, glycyrrhizin, scopoletin, carnosic acid, and many more, can help to develop novel antiviral drugs that will be safer and more effective for treating RSV (Venu and Austin 2020). Another experimental study has shown that ethanol extracted from the medicinal plant, *Lophatherum gracile*, is highly effective and shows high antiviral activity against RSV, under both in vitro and in vivo conditions. It successfully reduced inflammation and expressed excellent antitumor, antibacterial, and antipyretic properties (Chen et al. 2019). Another study by Chathuranga et al. (2019) found that the extracts from the medicinal plants *Plantago asiatica* and *Clerodendrum trichotomum* have been proven to have antiviral activity against RSV. The extract, the phenolic glycoside, was the active chemical compound found in both herbs to effectively prevent RSV and exhibit antiviral activity under both in vitro and in vivo conditions (Chathuranga et al. 2019).

6.2.3 Severe Acute Respiratory Syndrome Protected By Medicinal Plants

Severe acute respiratory syndrome coronavirus (SARS-CoV) is a positive-sense single-stranded RNA virus with a size of 30 kb and is responsible for causing respiratory and intestinal infections in humans and animals. It belongs to the virus family of Coronaviride and order of Nidovirales (Jansi et al. 2021). No antiviral drugs can target the SARS-CoV (Jansi et al. 2021), but we can use certain medicinal plants and extract to develop antiviral drugs. The SARS-CoV mostly identifies the ACE2 receptors in the lungs and activates the coronavirus spike (S) protein that enhances the entry of the virus into the host (Remali and Aizat 2021). According to a study conducted in 2003, it was found that when patients suffering from SARS-CoV in China 2003 were treated with traditional Chinese medicine, the patients presented fewer side effects and exhibited an improved condition of the viral symptoms (Remali and Aizat 2021). Several herbs effectively treat SARS-CoV, including *Glycyrrhiza glabra*, *Lycoris radiata*, and *Veronica linariifolia*, and their extracts like ethanol, methanol, chloroform, and luteolin have antiviral properties and prevent the viral disease (Remali and Aizat 2021).

6.2.4 Medicinal Plants Alleviating the Cause of Middle East Respiratory Syndrome Coronavirus (MERS-CoV)

The Middle East respiratory syndrome coronavirus (MERS-CoV) was a very lethal form of coronavirus that resulted in respiratory and intestinal infections in many individuals, and the most effects of this virus were observed in Saudi Arabia, Japan, and China. The first case was found in Saudi Arabia in a 60-year-old man who died from the virus (Mahrosh et al. 2021). The receptor involved in the MERS-CoV activation is dipeptidyl peptidase-4 (DPP4) (Anand et al. 2021). As a treatment, combination therapy with ribavirin and β-interferon, which was given to early-stage patients affected by the MERS-CoV, does possess antiviral potential. Through molecular docking techniques, it was found that phytochemicals such as quercetin, cyclocalamin, citromitin, rosavin, ichangin, pluviation, betaxanthin, digitogenin, kobusinol A, and methyl deacetylnomilinate could inhibit the replication of the MERS-CoV and target the viral protein, thereby treating the viral disease (Mahrosh et al. 2021). The Asian and Middle Eastern populations, around 85%, rely on medicinal plants to treat viral diseases. Medicinal plants like *Isatis indigotica* exhibited antiviral properties where clemastatin B, β-sitosterol, and sinigrin were the bioactive agents produced (Adeleye et al. 2022). Another plant called *Saposhnikovia divaricata* inhibits the S protein, enhancing the viral activity and preventing MERS-CoV. The main bioactive agents are coumarins, lignans, sterols, and more, which exhibit antioxidant, antimicrobial, and immunomodulating properties (Adeleye et al. 2022). *Psychotria ipecacuanha* is another bioactive agent that has been found to inhibit MERS. *Vaccinium* species found in blueberry and other plants contain bioactive compounds like anthocyanins, gallic acid, and procyanidins prevent the production of tumor necrosis factor (TNF)-α, inhibit binding of the virus to the ACE2 receptor, and prevent nitric oxide (NO) production, which in turn prevents MERS-CoV (Llivisaca-Contreras et al. 2021). *Morus alba* has aliphatic, aromatic, and heterocyclic bioactive compounds, and *Echinacea purpurea* has caftaric acid and echinacoside as bioactive compounds and exhibited antiviral mechanisms, and the latter blocked the infectivity caused by the virus (Llivisaca-Contreras et al. 2021).

6.2.5 Medicinal Plants Mitigating the Cause of Severe Acute Respiratory Syndrome-Related Coronavirus 2 (SARS-CoV-2) or Novel Coronavirus Disease (COVID-19)

The severe acute respiratory syndrome-related coronavirus-2 (SARS-CoV-2), also known as the novel coronavirus disease (COVID-19), was identified in Wuhan, China, in late 2019. It became a pandemic declared on March 11, 2020, and caused millions of deaths (Yepes-Perez et al. 2021). SARS-CoV-2 is an RNA virus and has a positive strand. Its symptoms include fever, shortness of breath, and cough, but most patients do not show the symptoms but are often carriers. There are several proteins responsible for the high virulence, which includes—spike proteins, the 3-chymotrypsin-like cysteine protease (3CLpro), papain-like protease (PLpro), and the RNA-dependent RNA polymerase (RdRp) (Adeleye et al. 2022; Gani et al. 2021). The spike protein of COVID-19 recognizes and interacts with the angiotensin-converting enzyme-2 (ACE2) receptor, which facilitates viral entry and infection (Adhikari et al. 2021). An experiment found that the medicinal plant, *Uncaria tomentosa* or cat's claw plant, could inhibit the effect of SARS-CoV-2 through its hydroalcoholic extract. It was conducted as an in vitro study, exhibiting anti-inflammatory and antiviral effects on the novel virus (Yepes-Perez et al. 2021). *Saffron* can be a very effective plant extract that can be utilized to prevent the harmful consequences of SARS-CoV-2. It acts as an antioxidant, prevents inflammation, and helps to maintain respiratory, cardiovascular, and renal functions (Husaini et al. 2021). Certain Indonesian herbal plants like *Justicia procumbens*, *Curcuma longa*, *Calophyllum inophyllum*, and *Zingiber officinale*; extracts like quercetin, lycorine, and luteolin; and compounds like carabine and justicidin D exhibit antiviral activity (Gani et al. 2021). According to an experiment, when molecular docking was conducted for medicinal plant extract to treat COVID-19, compounds like quercetin, capsaicin, naringin, gallic acid, and psychotrine exhibited antiviral properties against the virus (Alrasheid et al. 2021). Lim et al. (2021) concluded that four medicinal plants—*Nigella sativa*, *Eurycoma longifolia*, *Azadirachta indica*, and *Vernonia amygdalina*—can function in an antiviral, anti-inflammatory, and multimodal manner of managing COVID-19 by restoring the immunomodulatory functions. Nevertheless, more studies are required to understand the process of these mechanisms (Lim et al. 2021). Curcumin and phytochemicals from the *Citrus* plant species exhibited excellent antiviral properties by preventing the replication and progression of the COVID-19. In order to prevent the severe symptoms of COVID-19, another herbal bioactive compound derived from the sweet root called diammonium glycyrrhizinate was utilized and can be a potential drug against COVID-19 (Adhikari et al. 2021).

6.3 Various Compounds from the Medicinal Plants Acting Against Viral Diseases

Many studies have been conducted to understand the various compounds that can be extracted from medicinal plants, and they provide excellent antimicrobial, antiviral, and anticancer properties. They can be used on their own in nutraceuticals or in combination with other medicinal drugs to prevent several types of viral diseases. This topic will study polyphenols, flavonoids, proanthocyanidins, monoterpenes and triterpenes, glucosides, and sesquiterpenes. We will understand how they are effec tive and help alleviate the harmful consequences of viral disease and treat them.

6.3.1 Polyphenol Against Viral Diseases

Polyphenols are derived from fruits, vegetables, and herbs that express antioxidant, anti-inflammatory, anticancer, anti-allergic, and antiviral properties (Montenegro-Landívar et al. 2021). As natural antiviral agents, polyphenols can directly interact with the virus, prevent its replication, and bind to the target receptor, which prevents the DNA or RNA virus from releasing its particles into the host body. For example, antiviral polyphenols such as morin, coumarin, and quercetin directly affect viral replication and can help block viral replication. They are feasible and have very low to no side effects on consumption (Montenegro-Landívar et al. 2021). Polyphenols can help provide antioxidant properties, form stable free radicals, and prevent cellular damage caused due to oxidative stress (Montenegro-Landívar et al. 2021). Using polyphenols as nutraceuticals can effectively treat viral diseases, including COVID-19 (Besednova et al. 2021). The molecular docking studies found that polyphenols extracted from *Curcuma* and *Citrus* species, like curcumin, hesperetin, hesperidin, and tangeretin, can inhibit COVID-19 infections via the interaction with the S protein in the receptor-binding domain of the spike protein. Hesperidin was found to bind in the target receptor, which facilitated the interaction between the S protein domain and the ACE2 receptor (Paraiso et al. 2020). The polyphenol resveratrol extracted from the skin of *Vitis vinifera*, a red grape wine, regulated the ACE2 receptor function and reduced the damaging effects of SARS-CoV-2 (Paraiso et al. 2020).

6.3.2 Flavonoids Protecting from Viral Diseases

Flavonoids are phytochemicals with antiviral properties and can effectively treat several viral diseases like SARS, hepatitis, and COVID-19. It can prevent viral entry, replication, and development of the infection and has exhibited antiviral activity in in vitro, in vivo, and in silico conditions (Badshah et al. 2021). They can also provide benefits in terms of antiviral chemotherapy and can prevent the toxicity caused by antiviral drugs in patients. The epigallocatechin gallate (ECCG) flavonoid is very effective in treating viruses such as influenza A and influenza B, Zika, and hepatitis B viruses. Quercetin flavonoid was found to be effective against dengue, chikungunya, rhinovirus, Zika virus, poliovirus, and hepatitis C virus. The baicalin flavonoid is effective against chikungunya, enterovirus A71, and the HIV-1 virus (Badshah et al. 2021). A recent study found that flavonoids such as ECCG and thymoquinone can be highly effective in inducing the Nrf-2 transcription factor, which acts against COVID-19 infection but also exhibits high antiviral activity (Mendonca and Soliman 2020). ECCG, an extract from green tea, has good antioxidant properties, which helps express the phase II antioxidant genes and leads to the Nrf-2 electrophile response element (EpRE) signaling. When green tea is consumed daily, the ECCG increases the number of Nrf-2-induced proteins and antiviral response genes such as RIG-1, IFN-β, and MxA. These genes get activated when infection occurs, directly preventing the virus's entry and reducing its replication (Mendonca and Soliman 2020). Another flavonoid called thymoquinone also helps in inducing the Nrf-2 activity. This flavonoid is derived from black cumin or the *Nigella sativa* plant that exhibits excellent antiviral activity against many types of viral diseases. Thymoquinone will help inhibit the proinflammatory cytokines released due to the harmful effects of the coronavirus infection and reduce the severity of the disease (Mendonca and Soliman 2020). One issue with flavonoids is their poor bioavailability, which can be sorted out using nanotechnology tools and systems and can help to deliver the flavonoid at the exact location to express its antiviral activity and increase its bioavailability (Jannat et al. 2021). Nevertheless, more experimental research is required to understand nano-formulations and excellent combinations of nano-delivery systems with flavonoids to provide a therapeutic effect on old and novel forms of viruses (Jannat et al. 2021).

6.3.3 Proanthocyanidins Protecting from Viral Diseases

Proanthocyanidins (PACs) are polyphenolic bioflavonoids extracted from many fruits, vegetables, and nuts, especially flowers, bark, and grape seed (Zhang et al. 2018). According to the experiment conducted by Sugamoto et al. (2022), it was found that the PACs extracted from the blueberry leaf and stem are successfully able to inhibit the COVID-19/SARS-CoV-2 infection by blocking the ACE2 receptors and the 3CLpro enzyme. The PAC from blueberry was already found to have a strong antiviral response to the hepatitis C virus. It possesses antioxidant, anti-inflammatory, and anti-obesity effects, and since it is a flavonoid, it will have a safe and therapeutic effect on patients suffering from COVID-19 (Sugamoto et al. 2022). Another experiment was conducted in which the virucidal activity of PAC against the Mayaro virus, an arbovirus released by mosquitoes, is sublethal (Ferraz et al. 2019). PAC is extracted from the roots of Maytenus imbricata, a dimer of epigallocatechin, which possesses antiviral activity against the Mayaro virus. PAC eliminated the Mayaro virus proliferation and inactivated the virus, thereby treating the infection (Ferraz et al. 2019). One experimental study showed that PAC A2, a dimer of PAC formed after the condensation of catechins, successfully inhibited the infection caused by the porcine reproductive and RSV (PRRSV) under in vitro conditions (Zhang et al. 2018). It exhibited the antiviral condition under in vivo conditions and targeted multiple genes simultaneously, and proinflammation cytokines were produced due to the infection (Zhang et al. 2018). PAC can also be derived from grapeseed as the grapeseed PAC extract (GSPE), and a study proved that it could downregulate the proinflammatory cytokine production in the epithelial cells due to RSV (Kim et al. 2019). GSPE possesses antioxidant, antiviral, anticarcinogenic, and anti-allergic properties. GSPE was able to suppress interleukin (IL)-1β, IL-6, and IL-8 cytokine expression and reduce the inflammation of the airway epithelial cells and prevent RSV-induced airway disease (Kim et al. 2019).

6.3.4 *Monoterpenes and Triterpenes Acting Effectively Against Viral Diseases*

Terpenes are secondary metabolites in many plants and comprise five carbon isoprene units. Terpenes are found in the essential oils of many flowers and herbs and are modified with different functional groups to form terpenoids. These terpenoids are classified into monoterpenes, diterpenes, triterpenes, and sesquiterpenes (Aydin and Gürü 2022). Terpenes show strong effects on the immune system, exhibit anti-inflammatory activity toward several viral diseases, reduce cytokines, and enhance anti-inflammatory cytokines, that is, IL-4, IL-10, and transforming growth factor (TGF)-β1. Glycyrrhizin, a triterpenoid, inhibited SARS-CoV-2 infection by reducing the proinflammatory cytokines and reactive oxygen species (Aydin and Gürü 2022). Saikosaponin B2, another triterpenoid, inhibits hepatitis C virus and human coronavirus 229E by neutralizing the viral particles and exhibiting antiviral activity (Aydin and Gürü 2022). Hence, triterpenes showcase their antiviral activity against several types of viruses (Darshani et al. 2022). The triterpene, 3β-friedelanol, extracted from *Euphorbia neriifolia*, showed high antiviral activity against human coronavirus 229E (Darshani et al. 2022). 1,8-Cineole, a monoterpene oxide, prevented infectious bronchitis virus (IBV) N-protein binding and the RNA (Aydin and Gürü 2022).

6.3.5 *Glucosides and Sesquiterpenes Alleviating Causes of Viral Diseases*

Sesquiterpenes are phytochemicals belonging to terpenes, but these are extracted from the plant *Ferula vesceritensis*. It is found in Algeria and Libya as traditional medicine for treating cancer, fever, and throat infections (Mohamed et al. 2020). Previous studies on this phytochemical reveal its accumulation in the body as sesquiterpenes and sesquiterpenes coumarins. However, recent molecular docking studies found that these can block the main protease and RdRp, which is responsible for enhancing the SAR-CoV-2 infection (Mohamed et al. 2020). According to Hafez Ghoran et al. (2021), sesquiterpenes such as cedran-3β-12, diol, and α-cardinol found in the ethyl extract derived from the *Juniperus formosana* inhibit the SARS-CoV activity. It prevents its proliferation and replication of the virus (Hafez Ghoran et al. 2021). The sesquiterpenes extracted from the leaf's essential oils, called *Pogostemon cablin*, exhibit antiviral activity against the influenza A virus under in vitro conditions (Zrig 2022). Glucosides such as luteolin-7-glucoside and apigenin-7-glucoside can be utilized to inhibit the COVID-19 infection. Also, glucosides such as 3-β-d-glucoside can block the 3CLpro in MERS-CoV, which will block the enzymatic activity (Zrig 2022).

6.4 Various Medicinal Plants Mitigating a Viral Disease, SARS-COV-2

Crucial structural and nonstructural SARS-CoV-2 proteins interact with several phytocompounds and medicinal plants. Herbal substances interact strongly with the SARS-CoV-2 protease active site and undergo significant morphological changes. These phytochemicals may antagonize the spike protein, 3CLpro, and PLpro (Nallusamy et al. 2021). Angiotensin-converting enzyme 2 receptor, control of inflammatory mediators, transmembrane protease, serine 2 blockage, toll-like receptors, suppression of the activation of endothelium, and nuclear factor erythroid-derived 2-related factor 2 activation are some of the key anti-SARS-CoV-2 responses of medicinal plants and their bioactive components (Malekmohammad and Rafieian-Kopaei 2021).

Lianhua Qingwen was a part of clinical trials (phase II) in the United States and was conventionally used to treat cough, fever, influenza, pneumonia, tiredness, early stages of measles, and bronchitis. Lianhua Qingwen is a concoction of 11 different medicinal species, known as gypsum and menthol (a mineral medicine). The Chinese National Health Commission advised using this herbal cocktail to cure or control COVID-19. Using tests for the diminution of plaque and the suppression of cytopathic impact, the anti-SARS-CoV-2 activity was evaluated in Vero E6 cells. The herbal concoction caused dose-dependent inhibition of SARS-CoV-2 replication (Benarba and Pandiella 2020). Different plant bioactive compounds, including phenolics, flavonoids, saponins, polyphenols, alkaloids, tannins, and alkaloids, hold great potential against SARS-CoV-2 and may serve as a model for the emergence of new anti-COVID-19 preventatives (Majnooni et al. 2021). *Emblica officinalis*, locally known as amla, *Phyllanthus niruri Linn.* (Bhumi amla), and *Tinospora cordifolia* (Giloy) are significant medicinal herbs native to India and are frequently utilized in several ayurvedic preparations for healing various ailments. These medicinal plants' bioactive substances have immunomodulatory properties and help lessen cytokine outbursts in response to viral infections (Murugesan et al. 2021).

Natural component curcumin is employed because it delivers a unique flavor and color profile (Sharifi-Rad et al. 2020). Since ancient times, curcumin has been widely used in ayurvedic herbal medicines. Numerous in vitro and in vivo investigations have demonstrated that curcumin can treat and prevent oral, skin, intestinal, and colon cancers. If such foods are advised for oral intake, their immunomodulatory qualities can enhance humans' immunity and help people battle against the COVID-19 (Tripathy et al. 2021).

6.5 Different Mechanistic Actions of Medicinal Plants and Their Compounds on Viral Diseases, Including SARS-COV-2

Numerous antiviral medications are useless when treating resistant or mutant virus variants. As a result, there is a continuing need for nontoxic antiviral medications that are effective at entirely curing viral infections. The COVID-19 pandemic has become a growing global health emergency (Wu 2021). Thirty-six bioactive metabolites have been found to have significant antiviral activity, even though the phytochemical profiles of these plants have not yet been fully uncovered. The possible mechanisms of these activities are yet unknown. Bioactive metabolites from medicinal plants can suppress and interact with coronavirus's molecular and cellular mechanisms (Bachar et al. 2021). Except for vaccines, no medications can treat SARS-CoV-2. As a result, finding medicines derived from plants is necessary to treat COVID-19. Therefore, evaluating the medicinal plants used to treat various viral diseases is imperative (Tegen et al. 2021).

6.5.1 Medicinal Plants and Their Compounds Blocking ACE2 Receptor

The various herbal remedies and purified compounds may prevent the reproduction or entrance of the SARS-CoV-2 directly. It is interesting to note that some substances may block the ACE2 receptor or the serine protease TMPRRS2, which are important for SARS-CoV-2 to taint human cells (Samavati and Uhal 2020). It has also been demonstrated that natural substances, such as papain and chymotrypsin-like proteases, block proteins associated with the SARS-CoV-2 life cycle. Using natural items alone or in combinations can be an alternative to pharmaceuticals for treating and preventing COVID-19 infection. Their structural details may also guide the creation of anti-SARS-CoV-2 medications (Benarba and Pandiella 2020). The Polygonaceae family of plants includes the herb *Rheum palmatum L.* Due to its possible effectiveness in blocking the ACE2 receptor, which prevents virus entry into cells and CoV replication, it was used to treat SARS-CoV-1 infection. The main active ingredient in this plant is emodin, which prevents SARS-CoV-1 S protein from attaching to the ACE2 receptor. As a result, the potential therapeutic administration of COVID-19 can be evaluated using emodin derived from Radix et Rhizoma Rhei (Siddiqui 2020).

6.5.2 Medicinal Plants and Their Compounds Targeting TMPRSS2

By blocking TMPRSS2 from priming the spike (S) protein, COVID-19-causing SARS-CoV-2 is prevented from infecting human cells. To treat SARS-CoV-2 infection, phytochemicals that can reduce TMPRSS2 activity are promising therapeutic candidates. The Moringa oleifera plants' quercetin, niazirin, and moringyne all demonstrated promising abilities as possible TMPRSS2 inhibitors (Oyedara et al. 2021). Along with the coronavirus, this protein interacts with the influenza hemagglutinin protein, making it a possible therapeutic target for the flu virus. Variants and TMPRSS2 expression acted as COVID-19 modulators (Rahman et al. 2020).

6.5.3 Medicinal Plants and Their Compounds Targeting Papain-Like Proteinase (PLpro)

The proteolytic cleavage delivers nonunderlying proteins of replicase polyprotein 1a and replicase polyprotein 1ab by the activity of viral PLpro on the N-ends, bringing about three products and 3-chymotrypsin-like protease (3CLpro, otherwise called fundamental protease) at the focal region and c-termini. Papain-like protease is a proteolytic compound that catalyzes the breakdown the N-end of viral replicase polyprotein at three distinct areas, delivering mature nonprimary proteins 1, 2, and 3. This strategy is fundamental for forestalling viral replication. Furthermore, papain-like protease plays a significant capability in contradicting the host's natural resistance. Because of its critical role in COVID multiplication and host disease, papain-like protease has arisen as a typical objective for COVID inhibitors (Shawky et al. 2020).

6.5.4 Medicinal Plants Targeting Chymotrypsin-Like Protease (3CLpro)

A prospective therapeutic target for fighting coronavirus infection is the preserved 3-chymotrypsin-like protease, which directs the multiplication of the COVID. The investigation of ligand–protein connection found that the greater part of the main 20 terpenoids and alkaloids interfered decidedly with the viruses and 3CLpro, restricting intensities surpassing those of ritonavir and lopinavir (Gyebi et al. 2021).

3CLpro, the main protease, is fundamental for viral replication. The proton–proton (PP) chain is separated into 16 nonstructural proteins (NSPs) by 3CLpro and PLpro, with 3CLpro producing 11 of the 16 NSPs, delivering this protease one of the critical goals for planning hostile to SARS-CoV drugs. The rationed

3-chymotrypsin-like protease, which oversees COVID development, is a potential therapeutic in the battle against COVID contamination (Mody et al. 2021).

6.6 Conclusion

In this book chapter, we understand how important these medicinal plants are in treating viral diseases. They possess excellent antiviral properties that can counter the infection caused by these viruses and protect the body from other damaging side effects. The common point between SARS-CoV and SARS-CoV-2 is their binding to the ACE2 receptor, which can damage the lungs. Flavonoids like quercetin and plant extract from cat's claw and other plants mentioned in the above points block the binding of the virus onto the ACE2 receptor and exhibit antiviral activity. Medicinal plants, phytochemicals, and herb extracts provide excellent potential for treating viral diseases. There might be issues of bioavailability of certain flavonoids to be utilized as an antiviral drug, but it can be solved by using nano-delivery vehicles, which can help enhance the therapeutic effect of flavonoids. The different mechanisms by which medicinal plants and their compounds prevent viral diseases show that the bioactive compounds act as metabolites to suppress the replication of the SARS-CoV-2. These metabolites can be incorporated into the antiviral drugs or can be used as antiviral drugs by themselves for treating viral diseases. These not only exhibit antiviral activity but also maintain the renal, cardiovascular, and other organ system processes to completely remove the toxins released by the virus and heal the patient. To conclude, we require more experimental studies to be conducted to convert these medicinal plants and their extracts into effective antiviral drugs that are safe for use, affordable, and effective.

References

Adeleye OA, Bamiro OA, Bakre LG, Odeleye FO, Adebowale MN, Okunye OL, Menaa F (2022) Medicinal plants with potential inhibitory bioactive compounds against coronaviruses. Adv Pharm Bull 12(1):7. https://doi.org/10.34172/apb.2022.003

Adhikari B, Marasini BP, Rayamajhee B, Bhattarai BR, Lamichhane G, Khadayat K, Parajuli N (2021) Potential roles of medicinal plants for the treatment of viral diseases focusing on COVID-19: a review. Phytother Res 35(3):1298–1312. https://doi.org/10.1002/ptr.6893

Alrasheid AA, Babiker MY, Awad TA (2021) Evaluation of certain medicinal plants compounds as new potential inhibitors of novel corona virus (COVID-19) using molecular docking analysis. In Silico Pharm 9(1):1–7. https://doi.org/10.1007/s40203-020-00073-8

Anand AV, Balamuralikrishnan B, Kaviya M, Bharathi K, Parithathvi A, Arun M, Al-Dhabi NA (2021) Medicinal plants, phytochemicals, and herbs to combat viral pathogens including SARS-CoV-2. Molecules 26(6):1775. https://doi.org/10.3390/molecules26061775

Aydin DY, Gürü S (2022) Potential of natural therapeutics against SARS-CoV-2: phenolic compounds and terpenes. Namik Kemal Tip Dergisi 10(2). https://doi.org/10.4274/nkmj.galenos.2021.28290

Bachar SC, Mazumder K, Bachar R, Aktar A, Al Mahtab M (2021) A review of medicinal plants with antiviral activity available in Bangladesh and mechanistic insight into their bioactive metabolites on SARS-CoV-2, HIV and HBV. Front Pharm 12:732891

Badshah SL, Faisal S, Muhammad A, Poulson BG, Emwas AH, Jaremko M (2021) Antiviral activities of flavonoids. Biomed Pharmacother 140:111596. https://doi.org/10.1016/j.biopha.2021.111596

Benarba B, Pandiella A (2020) Medicinal plants as sources of active molecules against COVID-19. Front Pharmacol 11:1189

Besednova NN, Andryukov BG, Zaporozhets TS, Kryzhanovsky SP, Fedyanina LN, Kuznetsova TA, Shchelkanov MY (2021) Antiviral effects of polyphenols from marine algae. Biomedicine 9(2):200. https://doi.org/10.3390/biomedicines9020200

Chathuranga K, Kim MS, Lee H-C, Kim T-H, Kim J-H, Gayan Chathuranga W, Kim HI (2019) Anti-respiratory syncytial virus activity of Plantago asiatica and Clerodendrum trichotomum extracts in vitro and in vivo. Viruses 11(7):604. https://doi.org/10.3390/v11070604

Chen L-F, Zhong Y-L, Luo D, Liu Z, Tang W, Cheng W, Li M-M (2019) Antiviral activity of ethanol extract of Lophatherum gracile against respiratory syncytial virus infection. J Ethnopharmacol 242:111575. https://doi.org/10.1016/j.jep.2018.10.036

Darshani P, Sen Sarma S, Srivastava AK, Baishya R, Kumar D (2022) Anti-viral triterpenes: a review. Phytochem Rev 21:1761. https://doi.org/10.1007/s11101-022-09808-1

Ferraz AC, Moraes TDFS, da Cruz Nizer WS, Dos Santos M, Totola AH, Ferreira JMS, de Brito Magalhaes CL (2019) Virucidal activity of proanthocyanidin against Mayaro virus. Antivir Res 168:76–81. https://doi.org/10.1016/j.antiviral.2019.05.008

Gani MA, Nurhan AD, Maulana S, Siswodihardjo S, Shinta DW, Khotib J (2021) Structure-based virtual screening of bioactive compounds from Indonesian medical plants against severe acute respiratory syndrome coronavirus-2. J Adv Pharm Technol Res 12(2):120. https://doi.org/10.4103/japtr.JAPTR_88_21

Gyebi GA, Ogunro OB, Adegunloye AP, Ogunyemi OM, Afolabi SO (2021) Potential inhibitors of coronavirus 3-chymotrypsin-like protease (3CLpro): an in silico screening of alkaloids and terpenoids from African medicinal plants. J Biomol Struct Dyn 39(9):3396–3408

Hafez Ghoran S, El-Shazly M, Sekeroglu N, Kijjoa A (2021) Natural products from medicinal plants with anti-human coronavirus activities. Molecules 26(6):1754. https://doi.org/10.3390/molecules26061754

Husaini AM, Jan KN, Wani GA (2021) Saffron: a potential drug-supplement for severe acute respiratory syndrome coronavirus (COVID) management. Heliyon 7(5):e07068. https://doi.org/10.1016/j.heliyon.2021.e07068

Jannat K, Paul AK, Bondhon TA, Hasan A, Nawaz M, Jahan R, Pereira M (2021) Nanotechnology applications of flavonoids for viral diseases. Pharmaceutics 13(11):1895. https://doi.org/10.3390/pharmaceutics13111895

Jansi RS, Khusro A, Agastian P, Alfarhan A, Al-Dhabi NA, Arasu MV, Al-Tamimi A (2021) Emerging paradigms of viral diseases and paramount role of natural resources as antiviral agents. Sci Total Environ 759:143539. https://doi.org/10.1016/j.scitotenv.2020.143539

Kim SJ, Lee JW, Eun YG, Lee KH, Yeo SG, Kim SW (2019) Pretreatment with a grape seed proanthocyanidin extract downregulates proinflammatory cytokine expression in airway epithelial cells infected with respiratory syncytial virus. Mol Med Rep 19(4):3330–3336. https://doi.org/10.3892/mmr.2019.9967

Lim XY, Teh BP, Tan TYC (2021) Medicinal plants in COVID-19: potential and limitations. Front Pharmacol 12:611408. https://doi.org/10.3389/fphar.2021.611408

Llivisaca-Contreras SA, Naranjo-Morán J, Pino-Acosta A, Pieters L, Vanden Berghe W, Manzano P, Cevallos-Cevallos JM (2021) Plants and natural products with activity against various types of coronaviruses: a review with focus on SARS-CoV-2. Molecules 26(13):4099. https://doi.org/10.3390/molecules26134099

Mahrosh HS, Tanveer M, Arif R, Mustafa G (2021) Computer-aided prediction and identification of phytochemicals as potential drug candidates against MERS-CoV. Bio Med Res Int 2021:1. https://doi.org/10.1155/2021/5578689

Majnooni MB, Fakhri S, Bahrami G, Naseri M, Farzaei MH, Echeverría J (2021) Alkaloids as potential phytochemicals against SARS-CoV-2: approaches to the associated pivotal mechanisms. Evid Based Complement Alternat Med 6632623:1

Malekmohammad K, Rafieian-Kopaei M (2021) Mechanistic aspects of medicinal plants and secondary metabolites against severe acute respiratory syndrome coronavirus 2 (SARS-CoV-2). Curr Pharm Des 27(38):3996–4007

Mehta A, Michler T, Merkel OM (2021) siRNA therapeutics against respiratory viral infections—what have we learned for potential COVID-19 therapies? Adv Healthc Mater 10(7):2001650. https://doi.org/10.1002/adhm.202001650

Mendonca P, Soliman KF (2020) Flavonoids activation of the transcription factor Nrf2 as a hypothesis approach for the prevention and modulation of SARS-CoV-2 infection severity. Antioxidants 9(8):659. https://doi.org/10.3390/antiox9080659

Mody V, Ho J, Wills S, Mawri A, Lawson L, Ebert M, Taval S (2021) Identification of 3-chymotrypsin like protease (3CLPro) inhibitors as potential anti-SARS-CoV-2 agents. Commun Biol 4(1):93

Mohamed TA, Elshamy AI, Ibrahim MA, Zellagui A, Moustafa MF, Abdelrahman AH, Hegazy M-EF (2020) Carotane sesquiterpenes from ferula vesceritensis: in silico analysis as SARS-CoV-2 binding inhibitors. RSC Adv 10(57):34541–34548. https://doi.org/10.1039/d0ra06901arsc.li/rsc-advances

Montenegro-Landívar MF, Tapia-Quirós P, Vecino X, Reig M, Valderrama C, Granados M, Saurina J (2021) Polyphenols and their potential role to fight viral diseases: an overview. Sci Total Environ 801:149719. https://doi.org/10.1016/j.scitotenv.2021.149719

Murugesan S, Kottekad S, Crasta I, Sreevathsan S, Usharani D, Perumal MK, Mudliar SN (2021) Targeting COVID-19 (SARS-CoV-2) main protease through active phytocompounds of ayurvedic medicinal plants—Emblica officinalis (Amla), Phyllanthus niruri Linn. (Bhumi Amla) and Tinospora cordifolia (Giloy)—a molecular docking and simulation study. Comput Biol Med 136:104683

Nallusamy S, Mannu J, Ravikumar C, Angamuthu K, Nathan B, Nachimuthu K, Ramasamy G, Muthurajan R, Subbarayalu M, Neelakandan K (2021) Exploring phytochemicals of traditional medicinal plants exhibiting inhibitory activity against main protease, spike glycoprotein, RNA-dependent RNA polymerase and non-structural proteins of SARS-CoV-2 through virtual screening. Front Pharmacol 12:667704

Oyedara OO, Agbedahunsi JM, Adeyemi FM, Juárez-Saldivar A, Fadare OA, Adetunji CO, Rivera G (2021) Computational screening of phytochemicals from three medicinal plants as inhibitors of transmembrane protease serine 2 implicated in SARS-CoV-2 infection. Phytomed Plus 1(4): 100135

Paraiso IL, Revel JS, Stevens JF (2020) Potential use of polyphenols in the battle against COVID-19. Curr Opin Food Sci 32:149–155. https://doi.org/10.1016/j.cofs.2020.08.004

Rahman N, Basharat Z, Yousuf M, Castaldo G, Rastrelli L, Khan H (2020) Virtual screening of natural products against type II transmembrane serine protease (TMPRSS2), the priming agent of coronavirus 2 (SARS-CoV-2). Molecules 25(10):2271

Remali J, Aizat WM (2021) A review on plant bioactive compounds and their modes of action against coronavirus infection. Front Pharmacol 11:589044. https://doi.org/10.3389/fphar.2020.589044

Samavati L, Uhal BD (2020) ACE2, much more than just a receptor for SARS-COV-2. Front Cell Infect Microbiol 10:317

Sen IK, Chakraborty I, Mandal AK, Bhanja SK, Patra S, Maity P (2021) A review on antiviral and immunomodulatory polysaccharides from Indian medicinal plants, which may be beneficial to COVID-19 infected patients. Int J Biol Macromol 181:462–470. https://doi.org/10.1016/j.ijbiomac.2021.03.162

Sharifi-Rad J, Rayess YE, Rizk AA, Sadaka C, Zgheib R, Zam W (2020) Turmeric and its major compound curcumin on health: bioactive effects and safety profiles for food, pharmaceutical, biotechnological and medicinal applications. Front Pharmacol 11:01021

Shawky E, Nada AA, Ibrahim RS (2020) Potential role of medicinal plants and their constituents in the mitigation of SARS-CoV-2: identifying related therapeutic targets using network pharmacology and molecular docking analyses. RSC Adv 10(47):27961–27983

Siddiqui A (2020) Medicinal plants against SARS-CoV. In: Encyclopedia, p 377

Sugamoto K, Tanaka YL, Saito A, Goto Y, Nakayama T, Okabayashi T, Morishita K (2022) Highly polymerized proanthocyanidins (PAC) components from blueberry leaf and stem significantly inhibit SARS-CoV-2 infection via inhibition of ACE2 and viral 3CLpro enzymes. Biochem Biophys Res Commun 615:56–62. https://doi.org/10.1016/j.bbrc.2022.04.072

Wu T (2021) The socioeconomic and environmental drivers of the COVID-19 pandemic: a review. Ambio 50(4):822–833

Tegen D, Dessie K, Damtie D (2021) Candidate anti-COVID-19 medicinal plants from Ethiopia: a review of plants traditionally used to treat viral diseases. Evid Based Complement Alternat Med 2021:6622410

Tripathy S, Verma DK, Thakur M, Patel AR, Srivastav PP, Singh S, Utama GL (2021) Curcumin extraction, isolation, quantification and its application in functional foods: a review with a focus on immune enhancement activities and COVID-19. Front Nutr 8:747956

Venu LN, Austin A (2020) Antiviral efficacy of medicinal plants against respiratory viruses: respiratory syncytial virus (RSV) and coronavirus (CoV)/COVID 19. J Phytopharm 9:281–290. https://doi.org/10.31254/phyto.2020.9412

Yepes-Perez AF, Herrera-Calderón O, Oliveros CA, Flórez-Álvarez L, Zapata-Cardona MI, Yepes L, Zapata W (2021) The Hydroalcoholic extract of Uncaria tomentosa (Cat's claw) inhibits the infection of severe acute respiratory syndrome coronavirus 2 (SARS-CoV-2) in vitro. Evid Based Complement Alternat Med 2021:1. https://doi.org/10.1155/2021/6679761

Zhang M, Wu Q, Chen Y, Duan M, Tian G, Deng X, Chen W (2018) Inhibition of proanthocyanidin A2 on porcine reproductive and respiratory syndrome virus replication in vitro. PLoS One 13(2):e0193309. https://doi.org/10.1371/journal.pone.0193309

Zrig A (2022) The effect of Phytocompounds of medicinal plants on coronavirus (2019-NCOV) infection. Pharm Chem J 55(10):1080–1084. https://doi.org/10.1007/s11094-021-02540-8

Chapter 7
Cultivation of Corn Silk: Remunerative Venture for Medicinal Boon and Antimicrobial Therapies

Priyanka Devi, Prasann Kumar, and Joginder Singh

Abstract Corn silk (CS; *Stigma maydis*) has been used traditionally by Chinese and Native American medicine for the treatment of a variety of illnesses for thousands of years. In many countries worldwide, including Turkey, the United States, and France, traditional medicine practitioners use it for various purposes. Several publications have asserted that it has antioxidant properties and potential medical applications, such as a diuretic, a diabetic aid, an antidepressant, and an anti-fatigue agent. Furthermore, corn silk can be used for a variety of medical purposes, including teas and dietary supplements to treat voiding dysfunctions. In spite of the fact that corn silk has been viewed as a waste product for a long time, it has recently gained a lot of attention in Asia and Africa because of its many health benefits. There have been many studies conducted on extracts and bioactive components derived from maize silk that have shown to have antibacterial, anticancer, anti-hepatotoxic, anti-obesity, and anti-hyperlipidemic properties. Researchers have shown that maize silk contains a wide variety of bioactive substances, some of which may account for the purported health benefits associated with this product. It is believed that these chemicals include proteins, carbohydrates, vitamins, minerals, volatile and fixed oils, steroids, flavonoids, and phenolic acids, among others. It has been shown that corn silk contains a variety of bioactive chemicals, many of which contain antioxidant properties, which have helped to protect against a variety of age-related illnesses, including diabetes, hypertension, cancer, heart disease, and hepatic problems. Flavonoids and terpenoids are only two examples of bioactive plant elements that may be responsible for the plant's potential usefulness. Here, we will examine the available literature on the phytochemical and pharmacological activity of maize silk, as well as its prospective medical uses. The purpose of this chapter is to describe both the botanical and toxicological aspects of this plant.

P. Devi · P. Kumar
Department of Agronomy, School of Agriculture, Lovely Professional University, Phagwara, Punjab, India

J. Singh (✉)
Department of Microbiology, School of Bioengineering and Biosciences, Lovely Professional University, Phagwara, Punjab, India
e-mail: joginder.15005@lpu.co.in

V. Kumar et al. (eds.), *Medicinal Plants and Antimicrobial Therapies*,
https://doi.org/10.1007/978-981-99-7261-6_7

Keywords Antioxidant · Anti-hyperlipidemic · Antidiabetic · Corn silk · Health care · Pharmacology · Phytochemical · Therapeutic agent · Zero hunger · No poverty

Abbreviation

AChE	Inhibition of acetylcholinesterase
AF	Acetic ether fraction
AGEs	Advanced glycation end products
AS	Atherosclerosis
BChE	Butyrylcholinesterase
BF	n-butanol fraction
BHA	Butylated hydroxyanisole
BHT	Butylated hydroxytoluene
CECS	Chloroform
Cg	Carrageenan
CHD	Coronary heart disease
COX-2	Cyclooxygenase-2
CS	Corn silk
CVDs	Cardiovascular disorders
DNA	Deoxyribonucleic acid
DPPH	2,2-diphenyl-1-picryl-hydrazyl-hydrate
EDTA	Ethylenediaminetetraacetic acid
EF	Ethanol extract
EtOAc	Ethyl acetate
FRAP	Ferric reducing antioxidant power
FST	Forced-switch test
GPX	Glutathione peroxidase
GR	Glutathione reductase
GSDS	Glutathione disulfide
GSH	Glutathione
HbAl	Glycohemoglobin
iNOS	Inducible nitric oxide synthase
LDL-c	Low-density lipoprotein concentrations
LPS	Lipopolysaccharide
MDA	Malondialdehyde
MIC	Minimum inhibitory concentration
NRF2	Natural redox factor 2
OH	Hydroxyl radical
PECS	Petroleum ether
PF	Petroleum ether fraction
ROS	Reactive oxygen species
SO	Superoxide
SOD	Superoxide dismutase

TC	Triglyceride
TG	Triglyceride
TNF	Tumor necrosis factor
TST	The tail-flick test
VEGF	Vascular endothelial growth factor
VEGF-	Vascular endothelial growth factor-alpha
WF	Water fraction

7.1 Introduction

Traditional herbal remedies have indeed been used for millennia to treat many medical conditions, and they continue to be the foundation of contemporary medicine today (Wan Rosli et al. 2010). There is no doubt that natural antioxidants, notably phenolic chemicals, are primarily responsible for the medicinal effects of many traditional plants (Liu et al. 2011a, b). By eliminating reactive oxygen species (ROS), these chemicals can reduce the risk of cancer, high blood pressure, cognitive decline, and other ailments linked to oxidative stress. In order to protect humans from the effects of oxidative stress, various herbs and plants have been used for their potential benefits in preventing illnesses associated with oxidative stress and in maintaining health. There are many plants in this category, but one of them is corn silk (*Stigma maydis*). The stigmas on the female maize flower, which are found on the ovaries, are responsible for creating corn silk (CS). It can be found in large quantities in the world today as a by-product of maize farming (Maksimović et al. 2005). For centuries, people have been using it as a natural remedy for a variety of ailments, and it is becoming more and more popular as a result of its many benefits. The use of this plant medicinally has been practiced for centuries by many different cultures, such as those in China, Turkey, the United States, and France. There are quite a few conditions that it can be used to treat, including cystitis, kidney stones, diuretics, prostate problems, urinary tract infection, bedwetting, and obesity (Hu et al. 2010; Bastien 1982). It has the property of calming the bladder and urinary tract, which in turn increases the amount of urine produced by decreasing discomfort as a consequence (Steenkamp df2003). In addition to being a useful remedy for fatigue, antidepression, and kaliuretic, CS is also an anti-fatigue agent, an antidepressive agent, and a kaliuretic agent (Ebrahimzadeh et al. 2009; Velazquez et al. 2005). In addition, it has been shown to be protective against radiation and nephrotoxicity in animal studies (Bai et al. 2010; Sepehri et al. 2011). However, a recent study found that CS did not appear to have any antibacterial effect against a panel of bacteria that included *Pseudomonas aeruginosa*, *Klebsiella pneumonia, Staphylococcus aureus, Streptococcus pneumonia, Escherichia coli*, and *Streptococcus pyogenes* (Alam 2011). There is no doubt that this herb is regarded as one of the most vital medicinal herbs in China when it comes to treating prostate issues. Meanwhile, CS was also used by the Native Americans as a treatment for malaria, heart conditions, and diseases of the urinary tract in the past. CS tea has been touted for its supposed health benefits, some of which include the reduction of high blood pressure, the reduction of inflammation in the prostate, the reduction of diabetes, the reduction of urinary tract infections, the prevention of edema and obesity, as well as the reduction of stress levels associated with the consumption of CS tea. There are currently a number of commercially available CS goods online (Wan Rosli et al. 2008) that have been proven to be medically effective (Wan Rosli et al. 2008). There has been evidence that CS is a rich source of flavonoids, other phenolic chemicals (Liu et al. 2011a, b), and a mixture of them. In addition to this, there are also volatile oils and steroid compounds such as stigmasterol and sitosterol, as well as alkaloids, saponins, as well as a variety of vitamins and carbohydrates (Ebrahimzadeh et al. 2008). Because of this potential usefulness, a number of studies have reported CS's pharmacological actions as a result of its potential usefulness. The fresh corn silk (*Stigma maydis*), produced by the corn plant, looks like silky strands of fabric of varying lengths that measure between 10 and 20 cm in length. In general, it is available in shades ranging from light green to yellow-brown color (Guo et al. 2009). Among the foodstuffs found in corn, silk is maizeric acid, resin, sugar, mucilage, fibers, and other compounds of food-grade quality (Nurhanan and Rosli 2014). Besides water pills (diuretics), blood sugar regulators (corn silk) and inflammation-fighting compounds (corn silk) have also been found to be present in the plant (Gwendin et al. 2015). Further, maize silk is also rich in a wide range of bioactive chemicals, such as carbohydrates, fixed and volatile oils, steroids, as well as other natural antioxidants such as flavonoids (Hu and Deng 2011; Ren et al. 2013; Bhuvaneshwari and Sridevi 2015; Zilic et al. 2016; Vijitha and Saranya 2017). Corn silk is used in traditional Asian cultures, specifically in China, to make tea, as it is known to have medicinal benefits when consumed (Cuina et al. 2011). A number of phytochemicals in maize silk are known to have antioxidant properties, which would suggest that they could be of medical benefit to humans (Hasanudin et al. 2012). Therefore, this means that it has the potential to act both as a form of dietary fiber and as a food additive that could reduce the risk of several different types of chronic illnesses (Hasanudin et al. 2012). The traditional Chinese medicine industry has been using corn silk for centuries to treat kidney-related diseases, including edema, cystitis, gout, rheumatism, rheumatoid arthritis, and infections of all kinds, (Zhao et al. 2012; Amreen et al. 2012; Chen et al. 2013). The incorporation of maize silk into beef patties at a dosage of 2%, 4%, and 6% has been shown to increase the protein and mineral content as well as reduce the fat content of the meat products without altering the patties' sensory quality (Wanrosli et al. 2011). Several studies have been performed utilizing male and female rats as subjects to prove that corn silk is safe and nontoxic (Wang et al. 2011a, c). There were no deaths or abnormal symptoms found in the acute and subacute corn silk toxicity studies, nor did any of the treatment groups show any detrimental effects on body weight, water intake, food consumption, urine parameters, clinical chemistry, or organ weight in any of the treatment groups (Ha et al. 2018). A corn silk extract of 100, 200, and 400 mg/kg body weight was given to rats for 28 days at doses of 100, 200, and 400 mg/kg body weight without causing any adverse effects on the hematological parameters tested (Saheed et al. 2012). As part of the review, the present work focuses on the results of experimental studies conducted in vitro and in vivo using PubMed and reputable indexed journals to demonstrate that corn silk has the potential to provide health benefits against chronic diseases and age-linked diseases, particularly liver and kidney diseases, diabetes, and cancer.

7.2 Botanical Description

Corn, or *Zea mays* L., is a grass of Poaceae or Gramineae. Originating in Mesoamerica, it was first cultivated in Mexico around the year 9000 AD before making its way to the rest of the Americas. It is now widely farmed in many parts of the globe. The grass family includes wheat, oats, barley, and rice, along with more than 10,000 varieties of native corn (Leon and Coors 2008) that are also part of the grass family (Leon and Coors 2008). As with the rest of the maize, the silks are also used to make different products. The flowers of corn are monoecious, which means they are divided into male and female inflorescences on a single stalk (Inglett 1970). The male flowers (tassel) at the top of the plant's pollen of yellow color are produced and released into the air. The female flowers that make CS are tucked up in the leaf axils. A silk tuft is an elongated stigma that looks like a tuft of hair that is elongated. It is common for the CS to start off as a pale green but quickly change to red, yellow, or pale brown. The purpose of CS is to ensnare the pollen in order to facilitate pollination. In the case of silk, pollination can result in one kernel of maize being produced. With a length of up to 30 cm, the CS has a sweet flavor that is somewhat similar to that of chocolate. It is possible to use either fresh or dried CS, and the harvesting occurs just before the pollination occurs.

7.3 Phytochemical Composition of Corn Silk Extraction

CS extracts are known for their biological effects as a result of the flavonoids that take part in their synthesis. In plants, flavonoids are a class of phenolic chemicals that are found in their leaves. There is no doubt that these molecules are powerful antioxidants (Pietta 2000). Researchers have found a strong correlation between the total flavonoid content (TFC) and the total phenolic content (TPC) of the butanol fraction of the CS extract. There is a much larger butanol proportion of CS in TPC than in TFC (164 ng gallic acid equivalent [GAE]/g dry corn silk (DCS) versus 69 ng retinol equivalent (RE)/g DCS) and vice versa. Total phenolics (180 g GAE/g fresh weight (F.W.)), total anthraquinones (17.22 g/g F.W.), and total flavonoids (119.47 g/g F.W.) were all greater in the top (dark brown) portions of CS than they were in the lower portions (151.33 g GAE/g F.W., 8.61 g/g F.W., and 101.66 g/g F.W.). From the CS of several corn inbreeds, the flavonoid 3′-methoxymaysin and its reduced counterparts have been extracted and identified. Among the identified compounds are 2″,2″-O-L-rhamnosyl-6-fucosylluteolin and 2″-O-L-rhamnosyl-6-C-methoxyluteolin. From the 80% ethanol extract of CS, five more flavonoid derivatives were recovered and named: 2″-O-L-rhamnosyl-6-C-3″-deoxyglucosyl-3′-methoxyluteolin; 6,4′-dihydroxy-3′-methoxyflavone-7-O-glucosides; ax-5″-methane-3′-methoxymaysin; ax-4″-OH-3′-″-O-alpha-L-rhamnosyl; and 3′-O-methylquercetin-7-O-glucoside 6-C-fucoside (Ren et al. 2009). Other flavonoid molecules than maysin were recovered from CS methanol extract (100%), and they were c-glycosylflavones (Elliger et al. 1980). Two flavone glycosides, namely isoorientin-2-2″-O-L-rhamnoside and 3′-methoxymaysin, were found in CS 95% ethanol extract. Maysin has been linked to CS's biological activities, particularly its ability to deter maize earworm infestation, according to a number of studies (Snook et al. 1989). Gas chromatography–mass spectrometry (GC-MS) analysis of a dichloromethane extract of Egyptian CS yielded the identification of 36 different components. Major components of the extract were cis-terpinol (24.22%), citronellol (16.18%), 6,11-oxidoacor-4-ene (18.06%), trans-pinocamphone (5.74%), eugenol (4.37%), neo-iso-3-thujanol (2.59%), and cis-sabinene hydrate (2.28%). Many different types of goods, including soaps, cleaning supplies, and cosmetics, utilize these chemicals for their fragrant and flavorful properties. Moreover, the CS possessed a high level of cinnamic acid as well as the cinnamic derivatives, glucose, and rhamnose as well as natural minerals, which included sodium (0.05%), potassium (15%), iron (0.0082%), zinc (0.016%), and chloride (0.25%) in addition to the cinnamic derivatives. According to Wang et al. (2011a, c), corn silk has an approximate composition of 9.65% water, 3.91% ash, 0.29% crude fat, 17.6% crude protein, and 40% crude fiber.

7.4 Potential Health Care of Pharmacological Studies

There is no doubt that the health-care industry is recognized for its ability to prevent human sickness. This compound and its bioactive components provide several advantages and have a wide range of therapeutic applications. The pharmacological research of CS takes into account both in vitro and in vivo experimental paradigms (Tables 7.1 and 7.2).

7.4.1 Corn Silk Extracts' Antimicrobial Activity

Several investigations have shown that solvent extracts of corn silk have antibacterial properties that are beneficial to human health. Corn silk extracts, such as those derived from petroleum ether (PECS), chloroform (CECS), and methanol (MECS), as well as two isolated flavonoid glycosides (2 mg/mL), have been studied for their antibacterial properties. It was found that PECS, MECS, and flavonoids were effective against both Gram-positive and Gram-negative bacteria (Nessa et al. 2012) Corn silk extracts, both ethanolic and aqueous, have been shown to suppress

Table 7.1 Different pharmacological activities of corn silk extract in vivo conditions

In vivo condition			
Pharmacological activity	Procedure	Result	References
Activities of antioxidant	γ-Radiation caused oxidative damage in mice after a 10-day treatment	Activities of antioxidant against γ-radiation	Bai et al. (2010)
	Exercise caused oxidative damage in mice given a 28-day treatment	Acute exercise at the time of oxidative stress against the activities of antioxidant	Hu et al. (2011)
Kaliuresis and diuresis	Wistar rats were given CS extract via orogastric catheter, and urine was collected continuously for 3 and 5 h	Exhibition of kaliuresis and diuresis effects	Velazquez et al. (2005)
	Wistar rats were given CS extract intragastrically for 90 min, and urine collection and urinary flow were monitored using a cannula inserted into the urinary bladder	It has a diuretic effect	Pinheiro et al. (2011)
Reduction of hyperglycemia	Hyperglycemic included by adrenaline in mice animals were given CS extract orally for 45 and 14 days	Blood glucose levels are reduced	Guo et al. (2009)
Reduction of nephrotoxicity	Mice with GM-induced nephrotoxicity were given CS extract for 8 days	Improved nephropathy	Sepehri et al. (2011)
Activities of anti-fatigue	Swimming activity was performed by 10 mice after 14 days of treatment of flavonoid CS and loading with 5% of their body weight. Made with galvanized wire	Anti-fatigue action is quite strong	Hu et al. (2010)
Activities of antidepressant	FST and TST were performed on 10 male Swiss mice 1 h after being treated with CS extract for 6 and 5 min, respectively	Significant antidepressant activity	Ebrahimzadeh et al. (2009)
	In a black box, activity periods of CS-treated mice (normal and diabetic animals) were recorded	Excellent antidepressant action	Zhao et al. (2012)
Effect of anti-hyperlipidemic	For 20 days, rats with hyperlipidemia were given CS extract	It has an anti-hyperlipidemic action	Kaup et al. (2011)
Effects of antidiabetic	Streptozotocin-induced diabetic rats were given polysaccharides from CS intragastrically for 4 weeks	It has an antidiabetic effect	Zhao et al. (2012)

(continued)

Table 7.1 (continued)

In vivo condition			
Pharmacological activity	Procedure	Result	References
Effects of anti-inflammatory	CS was given orally to rats with carrageenan-induced pleurisy for 6 h	Reduce the inflammatory reaction	Wang et al. (2011a, c)

the growth of bacteria, including *E. coli*, in prior research (Surjee and Zwain 2015). Similarly, both *S. aureus* and *Bacillus subtilis* were inhibited by a minimum inhibitory concentration (MIC) of 500 mg/mL and 62.5 mg/mL, respectively, when exposed to an aqueous extract of maize silk (Xing et al. 2012). This investigation reveals that maize silk extracts are more effective against Gram-positive than Gram-negative bacteria. So, in addition to its stated antibacterial action, corn silk aqueous extract has also been shown to have an antifungal impact against *Candida albicans* (Xing et al. 2012). It indicates that maize silk has an antibacterial effect on microorganisms that cause diseases.

7.4.2 Antioxidant Activity

To counteract the oxidation that occurs during oxygen consumption, which can cause cell damage, aerobic organisms require antioxidants in order to reduce the risk of oxidative stress (Hashim 2011). In addition to atherosclerosis (AS), neurological disorders, cancer, diabetes, inflammation, and senescence, multiple diseases can be brought on by oxidation (Vladimir-Knežević et al. 2011). As they are more efficient than synthetic antioxidants like butylated hydroxyanisole (BHA) and butylated hydroxytoluene (BHT) at neutralizing free radicals, researchers have looked at natural antioxidants extracted from fruits, vegetables, grains, and medicinal plants as a means of neutralizing free radicals (Zhu et al. 2011). Many types of antioxidants are found in nature, including vitamins such as tocopherols and vitamin C, as well as plant-based compounds such as flavonoids and phenolic acids (Hassas-Roudsari et al. 2009). In recent research, it has been discovered that extracts from the CS plant have the potential to function as a bioactive source of natural antioxidants, according to the results of many recent scientific studies (Fig. 7.1). An in vitro model was developed for studying the antioxidant properties of five fractions of CS, namely ethanol extract (EF), petroleum ether fraction (PF), acetic ether fraction (AF), n-butanol fraction (BF), and water fraction (WF). The study found that the total phenolic content (164.1 g GAE/g DCS) and total flavonoid content (69.4 g RE/g DCS) were both the greatest in the BF fraction (100 g/mL). Also, BF demonstrated the greatest antioxidant activity compared to the other CS fractions, with a radical scavenging activity of 72.93% and an iron-chelating activity of 62.06% at 120 g/mL for BF. These levels of antioxidant protection were on par with those seen in vitamin

Table 7.2 Different pharmacological activities of corn silk extract under in vitro conditions

In vitro condition			
Pharmacological activity	Procedure	Result	References
Activities of antioxidant	In ethanol extract fraction (EF), petroleum ether fraction (PF), acetic ether fraction (AF), n-butanol fraction (BF), and water fraction, total antioxidant capacity, DPPH radical scavenging activity, reducing power, and iron-chelating capacity were measured (WF)	BF had the highest antioxidant activity	Liu et al. (2011a, b)
	The total antioxidant capacity of CS ethanolic extract was determined using DPPH radical scavenging activity	The antioxidant activity of upper regions of CS was greater than that of lower parts of CS	Alam (2011)
	The DPPH radical scavenging activity, metal-chelating activity, nitric oxide-scavenging activity, reducing power determination, and ferric thiocyanate (FTC) technique were all examined with a 50% ethanolic extract	The antioxidant activity of ethanol extract was equivalent to that of the reference compounds (BHA, BHT, vitamin C, quercetin, and EDTA)	Ebrahimzadeh et al. (2009)
	The antioxidant activity of dichloromethane extract, petroleum ether extract, 95% ethanol extract, and water extract was tested using the DPPH and carotene bleaching assays	The highest antioxidant activity was found in ethanol extract	El-Ghorab et al. (2007)
	Using various hybrids, 70% aqueous acetone extract was evaluated for ferric reducing antioxidant power (FRAP)	The antioxidant activity of the NS 640 hybrid acetone extract was the greatest	Maksimović et al. (2005)
	The antioxidant potential of CS methanolic extract was determined by inhibiting lipid peroxidation in liposomes generated by the Fe2+/ascorbate system	Antioxidant activity of developed CS is greater than that of juvenile CS	Maksimovic and Kovačević (2003)
	In ethyl acetate and ethanol extracts, DPPH radical scavenging activity, superoxide (SO) scavenging activity, iron-chelating	All of the extracts had the modest DPPH radical scavenging activity The maximum iron-chelating capability was	Kan et al. (2011)

(continued)

Table 7.2 (continued)

In vitro condition			
Pharmacological activity	Procedure	Result	References
	capacity, and ferric reducing antioxidant power (FRAP) assays were performed	found in an ethanol extract of *Z. mays* var. indurate FRAP test revealed higher antioxidant activity in ethyl acetate extract	
Effects of anti-glycation	AGE generation test inhibition in 80% methanolic extract	Nonenzymatic glycation is prevented	Farsi et al. (2008)
Effect of anti-inflammatory	Endothelial-monocyte adhesion assay, TNF-mediated cytotoxicity molecule expression therapy, and LPS-induced TNF release were tested in chloroform, ethyl acetate, butanol, and water extract	Endothelial cell adhesion and ICAM-1 expression are inhibited by ethanol extract	Habtemariam (1998)
	COX-2 levels were evaluated in macrophages treated with CS, and PGE2 levels were assessed using a PGE2 enzyme immunoassay kit	COX-2 and PGE2 secretion were enhanced by CS	Kim et al. (2005)
Effect of neuroprotective	Acetylcholinesterase (AChE) and butyrylcholinesterase (BChE) inhibition assays were performed in ethyl acetate and ethanol extracts	*Z. mays* var. indentata ethyl acetate extract highly inhibits AChE, while *Z. mays* var. everta ethyl acetate extract greatly inhibits BChE	Kan et al. (2011)

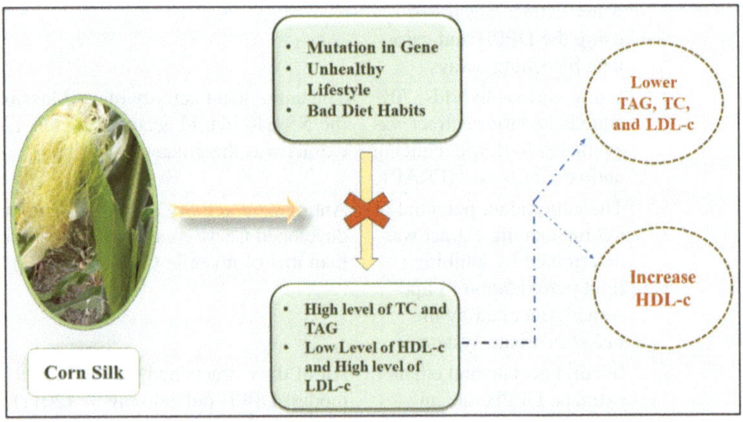

Fig. 7.1 Effects of corn silk extraction on hyperglycemia and its antioxidative and anticancer properties

C (the gold standard) and ethylenediaminetetraacetic acid (EDTA). At doses of 0.8 and 1.6 mg/mL, CS ethanol extract showed strong reducing power equivalent to vitamin C ($p > 0.05$) in another antioxidant trial. Therefore, the extract has the potential to operate as an electron donor, thereby bringing an end to the free radical chain reactions. Inhibiting free radical scavenging activity (84%) and beta-carotene bleaching (75%), further investigations revealed that CS ethanol extract (400 g/mL) offered substantial antioxidant activity (El-Ghorab et al. 2007). Ferric reducing antioxidant power (FRAP) values are linearly related to the total level of polyphenols, tannins, proanthocyanidins, and flavonoids in CS, suggesting that polyphenols may play an important role in the antioxidant properties of the substance. In addition to the antioxidant properties of CS, it can also prevent liposomes from becoming peroxided by Fe2+ or ascorbate. In order to prepare the herbal medicine, the young CS plant is used as the source material. The methanol extract (4 mg/sample) of mature CS can inhibit lipid peroxidation with a greater degree of intensity (45.9% inhibition compared with 22.8% inhibition in the immature CS) compared with the methanol extract of immature CS. Two different methods were used to measure the amount of antioxidant activity present in CS ethanolic extract both from the upper parts of the plant (dark brown) and from the lower parts of the plant (light yellow), which are not exposed to air. It was found that the upper sections had the highest overall antioxidant capacity (2735 mg/g gallic acid [GA] equivalents) and the highest 2,2-diphenyl-1-picryl-hydrazyl-hydrate (DPPH) scavenging activity (half-maximal inhibitory concentration [IC50] = 0.704 mg/mL) of the two sections. As a result of the increased exposure to sunlight experienced by the top regions of the CS, an abundance of flavonoids and other phenolic compounds has accumulated in these regions as a protection against the induction of ultraviolet (UV) damage to maize deoxyribonucleic acid (DNA). A study was conducted on silks of four different *Zea mays* varieties (var. indentata, indurata, everta, and saccharata) to determine the antioxidant power of the silks using various methods (DPPH, superoxide [SO] scavenging activity, iron-chelating capacity, and ferric reducing antioxidant power, etc.) (Kan et al. 2011). When evaluated at 500, 1000, and 2000 g/mL, all ethanol (EtOH) and ethyl acetate (EtOAc) extracts showed poor DPPH radical scavenging activity (10%–23%), whereas only the EtOH extract of *Z. mays var. indentata* showed SO scavenging activity (25%). EtOH extract of *Z. mays var. indurata* (63%) had the maximum iron-chelating ability at 500, 1000, and 2000 g/mL. It was found that both varieties performed similarly on the FRAP test, but for the *Z. mays* intent data, the maximum activity was observed for both the extracts when treated with EtOAc. The damage to biological systems caused by ionizing radiation, such as the release of free radicals and reactive oxygen species (ROS), causes the release of free radicals. A number of reactive oxygen species (ROS), including the hydroxyl radical (OH) and superoxide anion (O_2), can damage the body's lipid membranes and other vital biomolecules within the body. Consequently, lipid peroxidation will occur, which will cause polyunsaturated fatty acids to be oxidized into peroxides and lipid hydroperoxides, which will then degrade into several harmful compounds, such as malondialdehyde (MDA). There are several functions of glutathione (GSH) in the body, such as protecting by scavenging free radicals, repairing damaged molecules, and reducing peroxide molecule formation. In mice with radiation-induced oxidative stress, 75, 150, or 300 mg/kg/day of CS ethanol extract supplementation was administered intragastrically for 10 days. As a result of the extract, the liver was protected from GSH depletion, and the levels of MDA decreased in a dose-dependent manner (from 0.04 to 0.025 M/g protein), and red blood cells and hemoglobin levels were also increased (Bai et al. 2010). According to another study (Hu et al. 2011), flavonoids found in CS extract can protect mice against the oxidative damage caused by strenuous exercise, which was shown to be caused by strenuous exercise. During the study, mouth-to-mouth gavage of flavonoids from corn silk (FCS) (100 and 400 mg/kg body weight) was given for 28 days before treadmill activity in order to determine its protective effects. As a part of this study, mice were forced to exhaustion and then sedated to death to obtain skeletal muscle tissue, which was used to test the MDA content, as well as antioxidant enzymes superoxide dismutase (SOD) and glutathione peroxidase (GPX). When compared with the control group, the FCS-treated groups had significantly longer running times than the control group ($p < 0.05$): 100 mg/kg of FCS increased running time by 41.52%, and 400 mg/kg of FCS increased running time by 72.32%. A dose of 100 mg/kg of FCS was found to reduce MDA by 26.46%, while a dose of 400 mg/kg of FCS was found to lower MDA by 37.91%.

As a result of the study, we were able to demonstrate that flavonoids taken from corn silk mitigate the effects of oxidative stress caused by short-term exercise on skeletal muscle. As compared to the control group, the exercisers had significantly higher levels of superoxide dismutase (SOD) and glutathione peroxidase (GPX), two antioxidative enzymes that play a crucial role in protecting cells from reactive oxygen species (ROS). The lipid membranes of bone marrow and peripheral blood cells are susceptible to degradation by radiation-induced reactive oxygen species. This results in a decrease in new blood cell production and an increase in their destruction. This ultimately results in a decrease in the number and mass of blood cells.

7.5 Natural Redox Factor 2 (Nrf2) Expression

It has also been reported that when cells were treated with an extract of CS, there was an increase in the protein expression of Nrf2 within the cells (depending on the dosage) and an increase in the activity of antioxidant enzymes such as glutathione reductase (GR) and superoxide dismutase (SOD), which destroy free radicals in the body. It is known that many genes encoding enzymes that are involved in detoxification processes, such as glutathione reductase (GR) and superoxide dismutase (SOD), contain the antioxidant response element, and Nrf2 is a transcription factor that binds to this element (Nguyen et al. 2003). This defense against oxidative stress is provided by the proteins encoded by these genes. Among other things, SOD catalyzes the decomposition of superoxide anion and hydrogen peroxide, which makes it responsible for the oxidative damage caused by ROS. As for GR, it is responsible for maintaining a stable ratio between glutathione (GSH) and glutathione disulfide (GSDS) in the body. This antioxidase activity can be controlled so that damage from oxidative stress can be prevented if they are allowed to function properly. As a result of these results, CS has been given the green light to be used as a potent antioxidant and in cutting-edge treatments for diseases caused by oxidative damage.

7.5.1 Reduction of Hyperglycemia

There is a condition known as hyperglycemia that occurs when blood glucose levels rise above the normal range (Preedy and Watson 2010). The ability of CS aqueous extract to lower blood sugar levels means that diabetics can take it as a hypoglycemic meal replacement as a result of its ability to lower blood sugar levels (Guo et al. 2009). Researchers used alloxan-induced hyperglycemia in mice as a model to test the effects of CS aqueous extract (0.5, 1.0, 2.0, and 4.0 g/kg body weight) on glycemic metabolism. As a positive control, Xiaoke tablets (an alternative treatment for diabetics in China) were used, as well as saline as a negative control. It is known that animals are given a substance called alloxan, which is itself diabetogenic, in order to study diabetes. Insulin release in response to glucose is reduced because it blocks glucokinase selectively in response to glucose. It has been shown that selective necrosis of cells contributes to the development of type 1 and type 2 diabetes mellitus, which in turn lowers the function of the cell's glucose sensor and makes the cell dependent on insulin. There was a progressive increase in body weight in both the Xiaoke tablet and the CS-treated groups after 20 days of treatment. As a result of these results, it seems that CS might act as a source of additional nourishment for the mice. When hyperglycemic mice were given CS extract at doses of 2.0 and 4.0 g/kg, respectively, their blood glucose levels dropped from the control level (21.2 mmol/L) to 15.6 and 11.5 mmol/L, respectively. A 4.0 g/kg dosage resulted in a level of blood insulin equal to 9.8 U/mL, whereas the level of insulin in the control group was 3.8 U/mL. Animals fed 4.0 g/kg of CS extract had some of their injured pancreatic cells heal by Day 15, and no cell damage was seen in CS-treated mice. In the meanwhile, diabetic mice exhibited cell apoptosis and alterations to islet cells, such as loss of plasma membrane, condensed nuclei, dissolved cytoplasm, and broad intercellular gaps. After 45 days, the extract lowered glycohemoglobin (HbAl) content in the plasma of diabetic mice by 6.9% (4 g/kg) compared with the control (11.8%). Hepatic glycogen was higher in CS-treated mice (17.0 mg/g) than in control mice (14.2 mg/g), but the difference was not statistically significant ($p > 0.05$). Adrenaline-induced hyperglycemic rats have also had their blood glucose levels examined. Adrenaline stimulates glyconeogenesis, leading to elevated serum glucose levels; however, 15 days of CS extract treatment had no effect on blood sugar. These findings indicated that the effect of CS extract on glycemic metabolism is not due to an increase in glycogen and an inhibition of gluconeogenesis but rather due to an increase in insulin level and the recovery of damaged cells. Fructose and glucose, two reducing sugars, can react nonenzymatically with proteins to make Schiff bases and Amadori products, which in turn produce advanced glycation end products (AGEs) in biological phenomena. Nonenzymatic glycation and advanced glycation end products (AGEs) have been linked to the onset of a wide range of illnesses, including diabetes and its related consequences, neurodegenerative diseases, and cardiovascular problems (Ramasamy et al. 2005). The CS extract (40 g/mL) showed inhibitory action for AGEs and nonenzymatic anti-glycation in a research study including 14 maize genotypes (Farsi et al. 2008). Glycation inhibition was shown to be the most efficient with the most active maize genotype (CO441; IC50 = 9.5 g/mL; known as a glycation inhibitor). Because of its total phenol content and resistance to some fungal diseases like Fusarium graminearum (Gibberella ear rot), CS extracts have shown promise as a source for the creation of natural AGE inhibitors in the prevention of diabetes and aging-related problems (Farsi et al. 2008).

7.5.2 The Effect of Diuresis and Kaliuresis

"Kaliuresis" refers to an increase in the quantity of potassium secreted in urine, whereas "diuresis" refers to an increase in the volume of pee discharged. The role of CS aqueous extract on glomerular function and potassium excretion in the urine was investigated by Velazquez et al. (2005). At 350 mg/kg body weight (100.42 equivalent (EQ)/5 h/100 g body weight) and 500 mg/kg body weight (94.97 EQ/5 h/100 g body weight), the CS aqueous extract caused kaliuresis (K+ urine excretion). Diuresis (an increase in urinary volume) was observed at a dosage of 500 mg/kg body weight (1.98 mL/5 h/100 g body weight) compared to its water control (p 0.05). At a dosage of 500 mg/kg body weight, the effects of CS extract on urinary volume, salt, potassium, uric acid excretions, and glomerular and proximal tubular function were evaluated using creatinine and lithium clearance. Although there is no change in urine volume, sodium, lithium, or uric acid excretion, there is a marked rise in potassium excretion (0.2289 EQ/min/100 g body weight). Creatinine clearance decreased from 295 to 241 L/min/100 g body weight, and the sodium-filtered load dropped from 41.9 to 34.3, both indicators of decreased glomerular function; however, proximal tubular performance, as assessed by lithium and sodium excretion, remained unchanged. Cannulation of the urinary bladder for urine collection and measurement of urinary flow was used to quantify diuresis in another experiment using anaesthetized Wistar rats fed water ad libitum (Pinheiro et al. 2011). After 90 min of intragastric injection of 1 mL of 20% aqueous CS extract, the findings demonstrated its diuretic effects, with urine flow increasing by 135% compared with baseline (p 0.05). Sodium (Na+) and potassium (K+) excretion were also considerably enhanced in the CS-treated group over the course of 60 and 90 min, with Na + excretion increasing by 127.5 and 86%, respectively. Diuretics can cause water and solute loss in the blood, which can have an influence on blood pressure regulation. Since blood volume decreases in response to a fall in blood pressure, this is the case. The infusion of CS extract exhibits hypotensive effects, as shown in the same experiment when CS aqueous extract considerably (*p* 0.05) reduced blood pressure. CS aqueous extract has traditionally been used as a diuretic, and scientific evidence backs up this practice. Researchers looked into the diuretic properties of CS extract and found that it increased urine flow by 159%. Results from a 1-week study with CS (600 mL water extract) in 38 healthy volunteers (Dat et al. 1992) indicated no significant changes in urinary sodium or potassium excretion. Since it is possible that the current dose of CS is too low to be useful as a diuretic, more clinical studies are needed to confirm this hypothesis.

7.5.3 Extracts from Corn Silk Having Anti-Hyperlipidemic Properties

Raised levels of lipids in the blood, including triglycerides, total cholesterol, cholesterol esters, and phospholipids, are called hyperlipidemia (Yadav et al. 2019). This illness is a key contributor to atherosclerosis and a major contributor to the start of cardiovascular disorders (CVDs) (Fig. 7.2). It has been shown in a previous study that giving hyperlipidemic rats flavonoids from corn silk extract at three different dosages (200, 400, and 800 mg/kg) over the course of 20 days decreases their total cholesterol, triglyceride, and low-density lipoprotein levels but has no effect on high-density lipoprotein concentrations (HDL-c; Wu et al. 2017). On the other hand, Zhang et al. (2015) found that corn silk flavonoids reduced serum triglyceride (TC), triglyceride (TG), and low-density lipoprotein concentrations (LDL-c) and increased serum HDL-c in a mouse model. This suggests that corn silk flavonoids may have

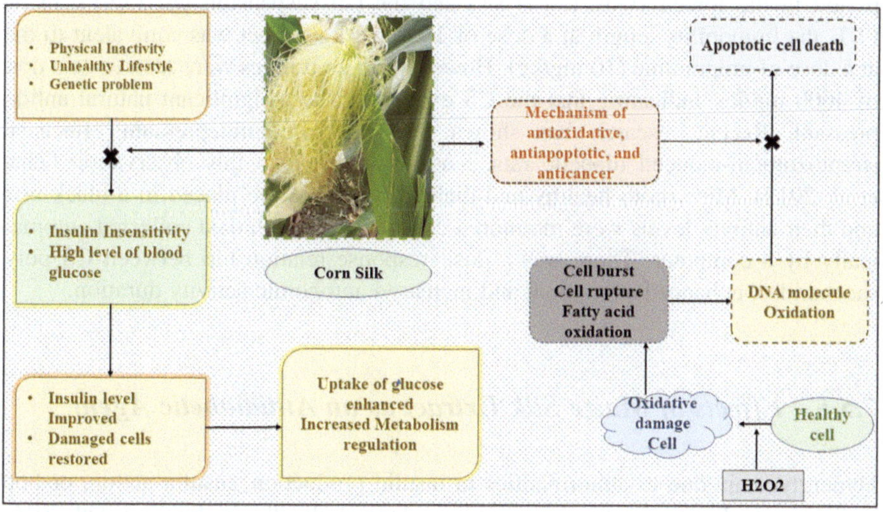

Fig. 7.2 Effect of corn silk extract on cardiovascular disease and hyperlipidemia

anti-hyperlipidemic effects and protect against atherosclerosis (Yan et al. 2011). The primary strategies for preventing atherosclerosis (AS) and coronary heart disease (CHD) involve modulating and counteracting the harmful effect of hyperlipoproteinemia, which is a major risk factor connected to both of these diseases (Yan et al. 2011). Similar decreases in atherogenic markers (TC, TG, and LDL-c) and improvements in nonatherogenic lipoprotein (HDL-c) levels were also observed in corn silk total flavonoid (CSTF)-treated hyperlipidemic rats (Zhao et al. 2012). Corn silk aqueous extract has long been known to have an anti-hyperlipidemic effect, but recently it has also been found to have an anti-hypertensive effect by bringing down blood pressure (George and Idu 2015; Yulina et al. 2013). Maysin, a bioactive ingredient from maize silk, significantly decreased body weight, kidney weight, and epididymal fat weight in high-fat diet-induced obese rats after 8 weeks of therapy, suggesting that Maysin had a weight-reducing impact via lowering fat accumulation in the body (Lee et al. 2016). In addition, Min et al. (2011) found that feeding mice maize silk extract at doses of 100 and 400 mg/kg body weight for 2 weeks resulted in a substantial reduction in body weight. This information added support to the idea that maize silk has a weight-reducing impact (Min et al. 2011).

7.5.4 *Effects on Depression*

Mice were given 125, 250, 500, 1000, and 1500 mg/kg of CS ethanol extract 1 h before undergoing forced swim tests (FSTs) and tail suspension tests (TSTs) for 6 and 5 min, respectively, to assess the extract's antidepressant effects (Ebrahimzadeh et al. 2009). The extract significantly shortened the immobility period in the forced swim test (FST) and the tail suspension test (TST). In the FST, the immobility length at a dose of 1500 mg/kg extract was equivalent to that at a dose of imipramine (10 mg/kg). However, no mortalities were seen up to a dose of 4000 mg/kg, indicating that the CS extract can be a significant natural antidepressant. Recent research has shown that CS has antidepressant effects in streptozotocin-induced diabetic rats, lending credence to this observation (Zhao et al. 2012). Mice (both healthy and diabetic strains) were placed in a black box, and their activity levels were monitored continuously, with data collected automatically by a computer. There was a dose–response relationship between CS polysaccharides, enhanced excitation, and increased autonomic activity duration.

7.5.5 *Effects of Maize Silk Extract as an Antidiabetic Agent*

Hyperglycemia due to abnormalities in insulin production, insulin action, or both characterizes diabetes mellitus, a complicated metabolic illness (Ibrahim et al. 2016). Polyuria, polydipsia, polyphagia, glycosuria, impaired vision, nausea, and systemic weight loss are the most common symptoms of diabetes (Vijitha and Saranya 2017). Insulin production, glucose uptake by target organs, and food absorption are a few of the glucose metabolic routes that traditional antidiabetic drugs might affect (Ghada et al. 2014). Corn silk has been shown in scientific studies to have a favorable effect on glycemic metabolism by increasing insulin secretion, with the increase in insulin level and the regeneration of cells being proposed as potential methods by which corn silk regulates hyperglycemia (Chen and Guo 2018). Corn silk polysaccharide, administered daily at doses ranging from 100 to 500 mg/kg of body weight, was associated with an antidiabetic effect in streptozotocin-induced diabetic rats, as evidenced by an improvement in both blood glucose control and serum lipid profile (Chen and Guo 2018). Polysaccharides from corn silk also enhanced glucose tolerance in diabetic rats, as shown by an oral glucose tolerance test (OGTT) (Chen and Guo 2018). Weight loss, blood sugar, insulin secretion, and glucose intolerance are all improved in mice with type 2 diabetes when maize silk polysaccharides are administered (Pan et al. 2019). Also, another study found that diabetic mice whose blood glucose levels were raised had a dose-dependent reduction in the amount of glucose in their systems after receiving either 250, 500, or 750 mg/kg of corn silk methanol extract (Umar 2016).

Furthermore, in vitro studies using isolated rat hemidiaphragms showed that corn silk increased glucose absorption by direct peripheral action, suggesting that it may be more efficient than insulin (Chen et al. 2013). In addition to its influence on glucose absorption, corn silk also reduces the pace of starch digestion and prevents a spike in blood sugar after eating (Chen et al. 2013). Like glucosidase activity can be inhibited, and glucose metabolism can be regulated by targeting signal pathways, corn silk can boost insulin action and enhance glucose metabolism (Chen and Guo 2018). This suggests that maize silk may be a useful bioactive agent for diabetes control and therapy (Vijitha and Saranya 2017).

7.5.6 Corn Silk Extract Inhibiting Tumor Growth Via an Antioxidant Mechanism

Maysin, a kind of flavonoid unique to maize, is abundant in corn silk extract (Fossen et al. 2001; Kim and Jung 2001). Lutein, found in maize silk extract as maysin, is a physiologically active molecule with antioxidant and anticancer properties (Lee et al. 1998). Maysin has been shown to have antioxidative, antiallergenic, and anticancer properties in previous investigations (Lee et al. 2016; Bai et al. 2010). In addition, maysin, which is extracted from maize silk, protects nerve cells by inhibiting oxidative damage and cell death (Choi et al. 2014; Lee et al. 2014; Kan et al. 2011). Antitumor activity was also shown to be improved by maize silk polysaccharide, perhaps as a result of its capacity to boost immune function and reduce inflammation (Yang et al. 2014; Wang et al. 2012).

7.5.7 The Prevention of Nephrotoxicity

Negative effects on the structure or function of the kidneys are collectively referred to as nephrotoxicity. The production of toxic metabolites causes these alterations from injected, swallowed, breathed, or absorbed chemical or biological products (Schreiner 1965), which then have a detrimental effect on the kidneys. In rats treated with gentamicin (GM), consuming CS together with the drug reduces the drug's nephrotoxicity. However, it does not completely reverse GM-induced parameter alterations such as urea levels, acute tubular necrosis (ATN), hyperemia, hyaline cast formation, and glomerular abnormalities (Sepehri et al. 2011). Serum creatinine levels decreased from 0.55 to 0.58 mg/dL after consuming CS methanol extract (80%) at doses of 200 and 300 mg/kg in GM-induced nephrotoxicity. This suggests a lessening of GM-induced nephrotoxicity. Compared with the GM group, CS administration with GM therapy significantly reduced GM-induced interstitial nephritis at doses up to 500 mg/kg, demonstrating a protective effect. However, higher doses of CS (400 and 500 mg/kg) caused nephrotoxicity, which included hyaline cast formation, apoptosis, congestion, and cell edema.

7.5.8 Neuroprotective Effects

By analyzing the inhibition of acetylcholinesterase (AChE) and butyrylcholinesterase (BChE), researchers were able to determine whether or not ethyl acetate (EtOAc) or ethanol extract (EtOH) of CS from four maize types (var. indentata, indurata, everta, and saccharata) had neuroprotective benefits (Hassas-Roudsari et al. 2009). Hydrolysis of the neurotransmitter acetylcholine by the enzymes acetylcholinesterase (AChE) and butyrylcholinesterase (BChE) leads to Alzheimer's disease; hence, blocking these enzymes may be useful for preventing this condition (Senol et al. 2009). The maximum AChE inhibition (96.69%) was found in var. indentata EtOAc extracts (200 g/mL), whereas the highest BChE inhibition (41.46%) was found in var. everta EtOAc extract (200 g/mL). The fact that CS EtOAc extracts had a strong effect on AChE suggested that CS extracts could be used to protect nerve cells.

7.5.9 Inhibition of Inflammation

Inflammatory processes are triggered by various stimuli, including antigen–antibody interactions, heat or physical harm, pathogens, and ischemia. Analgesic mediators are released in response to inflammation, causing pain (Nguemfo et al. 2007). It is widely accepted that inflammatory mediators like tumor necrosis factor (TNF) or *E. coli* lipopolysaccharide (LPS) play a significant role in triggering a wide range of physiologic responses (Stoecklin et al. 2003). TNF and LPS can cause inflammation by increasing leukocyte adherence to endothelial cells (EAHY 926) to the production of adhesion molecules such as intercellular adhesion molecule-1 (ICAM-1), endothelial leukocyte adhesion molecule-2 (ELAM-2), and vascular cell adhesion molecule-1 (VCAM-1). Treating inflammatory illnesses sometimes requires blocking adhesion or the expression of adhesion molecules on leukocytes. The TNF antagonistic activity of the crude ethanolic extract of CS was much higher than that of other herbal medicines used for inflammation (Habtemariam 1998). The ability of TNF and LPS to cause EAHY 926 endothelial cells to adhere to monocytic U937 cells was successfully blocked by a 9–250 g/mL ethanolic extract. TNF, LPS, and phorbol myristate acetate (PMA)-mediated ICAM-1 expression were studied, as well as the effect of CS extract. At 4 and 18 h, CS extract suppressed ICAM-1 expression by 65%. At 18 h after treatment, it also suppressed the LPS (1 and 10 g/mL)-induced production of ICAM-1. The capacity of CS ethanolic extracts to block TNF- and LPS-induced production of ICAM-1 and endothelial cell adhesion provided scientific support for the use of CS as conventional therapy for inflammatory illnesses. In a rat model of carrageenan (Cg)-induced pleurisy, Wang et al. (2011a, c) looked at how CS extract affected things like cellular infiltration, exudate formation, TNF, interleukin-1 beta (IL-1β), IL-17, vascular endothelial growth factor alpha (VEGF-α), C3 and C4 complement protein levels, ICAM-1, and inducible nitric oxide synthase (iNOS). When CG was injected into the pleural cavity of rats, it triggered inflammatory responses including the production of exudate and the movement of cells. Extract from CS substantially decreased cell migration by 60.8% and 82.4% at dosages of 2 and 4 g/kg body weight, respectively, and exudate formation by 28.6% and 54% at these same concentrations. TNF levels were reduced by 33.1% and 54.8% after treatment with CS extract (2 and 4 g/kg body weight), IL-1 levels were reduced by 25.7% and 42.9%, IL-17 levels were reduced by 79.8% and 91.3%, and VEGF-α levels were reduced by 41.1% and 51.4% after treatment. During inflammation and tissue injury, C3 and C4 protein levels shift. As the immune system matures, the C3 level rises in response to inflammation. The inflammation process can be slowed or stopped altogether by making this adjustment. It was shown that CS extracts (2 and 4 g/kg body weight) dramatically reduced C3 protein, which in turn inhibited inflammatory processes. While C3 protein is significantly different between the two dosages of CS extract, C4 protein is not. Many illnesses, particularly inflammatory disorders, can be avoided if antioxidants are taken into account (Miguel 2010). Superoxide anion (O_2) production causes inflammatory disorders by increasing vascular permeability and proinflammatory cytokine levels. Furthermore, it stimulates the transcription factor nuclear factor-kappa B (NF-κB), which controls inflammatory cytokines. The expression of tumor necrosis factor (TNF), interleukin (IL)-1 beta, vascular endothelial growth factor (VEGF), and interleukin (IL)-17 is suppressed by CS extracts (2 and 4 g/kg body weight). In contrast to the control group, ICAM-1 expression was dramatically elevated after CG injection. The expression of ICAM-1 was considerably lower in the CS extract treatment group than in the CG-saline group (p 0.05 and p 0.01, respectively). The protein expression of iNOS was downregulated dose-dependently by CS extracts. Superoxide anion (O_2) can cause tissue injury; hence, reducing iNOS is crucial. O_2 is produced when nitric oxide (NO) production drops. The findings of this research backed the idea that CS extract might be effective in the treatment of inflammatory illnesses caused by oxidative stress. Murine macrophages were used in an in vitro investigation, and cyclooxygenase-2 (COX-2) was induced by NF-κB activation at concentrations as low as 2.5 g/mL corn silk. After 24 h of incubation, the maximum concentrations were 12–25 g/mL (Kim et al. 2005). COX-2 overexpression induced prostaglandin E2 receptor 2 (EP2) and EP4 receptor-mediated upregulation of iNOS expression and prostaglandin E2 (PGE2) generation in macrophages. These results support the idea that iNOS and COX-2 help each other reach their therapeutic goals of reducing inflammation and making blood vessels more permeable.

7.5.10 Toxicity

There has been a rise in the popularity of herbal remedies in recent years. Herbal medications have a good reputation for being risk-free because of their traditional use and their all-natural origins. As a result, it is crucial to test the toxicity of herbs and establish their safety. Male and female Wistar rats were used in a recent investigation to validate CS's lack of toxicity (Wang et al. 2011a, c). When 8.0% (w/w) CS was eaten for 90 days, there were no histopathological or deleterious consequences. This amount of CS is equivalent to around 9.354 and 10.308 g/day/kg body weight for males and females, respectively. So, there are no negative reactions to CS ingestion, and it seems safe for human use.

7.6 Conclusion

This analysis emphasizes corn silk's possible medical uses as herbal medicine. Its antioxidant, hyperglycemia-lowering, antidepressant, anti-fatigue, and diuretic bio-activities have been the subject of extensive pharmacological research (in vitro and in vivo). CS's safety and lack of toxicity have been validated by more investigations, and recent discoveries have shown that CS poses no health risks whatsoever. Clinical trials are necessary to back up the health claims and boost faith in the substance's purported positive therapeutic benefits when taken by humans. Corn silk extract, long considered a useless by-product, has been shown to have therapeutic effects against a variety of debilitating conditions associated with aging and chronic illness. Corn silk extracts have been shown to be effective in preventing a variety of diseases caused by pathogens. Therefore, corn silk may be useful for human health.

However, more research needs to be done to figure out the molecular pathways that allow maize silk extract to be good for human health.

Acknowledgment It is with great appreciation that the authors would like to acknowledge the Department of Agronomy at Lovely Professional University for their constant support and encouragement throughout the research process.

Author Contributions In addition to contributing to the outline, the author is responsible for leading the draft and editing of the manuscript. It was PD, PK, and JS who conducted an extensive literature search and contributed to writing sections and constructing figures and tables. In addition, they contributed to the construction of figures and tables. In addition to providing professional advice, PK and JS helped revise the final version of the document and participated in its revision. In the end, all authors read and approved the final version of the manuscript.

Conflicts of Interest There is no conflict of interest.

References

Alam EA (2011) Evaluation of anti-oxidant and antibacterial activities of Egyptian Maydis stigma (*Zea mays* hairs) rich in some bioactive constituents. J Am Sci 7:726–729

Amreen F, Agrawal P, Singh PP (2012) Herbal option for diabetes: an overview. Asian Pac J Trop Dis 2:S536–S544

Bai H, Hai C, Xi M, Liang X, Liu R (2010) Protective effect of maize silks (Stigma maydis) ethanol extract on radiation-induced oxidative stress in mice. Plant Foods Hum Nutr 65:271–276

Bastien JW (1982) Pharmacopeia of Qollahuayaandeans. J Ethnopharmacol 8:97–111

Bhuvaneshwari K, Sridevi S (2015) Analysis of nutrients and photochemical contents in corn silk (*Zea mays*). Int J Sci Res 78(96):2319–7064

Chen H, Guo Q (2018) Corn silk (*Zea mays* L.) as a traditional herb in diabetes mellitus. EC Pharmacol Toxicol 6(6):477

Chen S, Chen H, Tian J, Wang Y, Xing L, Wang J (2013) Chemical modification, anti-oxidant and α-amylase inhibitory activities of corn silk polysaccharides. Carbohydr Polym 98(1):428–437

Choi SY, Lee Y, Kim SS, Ju HM, Baek JH, Park CS, Lee DH (2014) Inhibitory effect of corn silk on skin pigmentation. Molecules 19:2808–2818

Cuina W, Tiehua Z, Jun L, Shuang L, Cheng Z, Erlei W, Zuozhao W, Yan Z, Jingbo L (2011) Subchronic toxicity study of corn silk with rats. J Ethnopharmacol 137(1):36–43

Dat DD, Ham NN, Khac DH, Lam NT, Son PT, Dau NV, Grabe M, Johansson R, Lindgren G, Stjernstrom NE (1992) Studies on the individual and combined diuretics effects of four Vietnamese traditional herbal remedies (*Zea mays*, Imperata cylindrica, Plantago major and Orthosiphon stamineus). J Ethnopharmacol 36:225–231

Ebrahimzadeh MA, Pourmorad F, Hafe S (2008) Antioxidant activities of Iranian corn silk. Turk J Biol 32:43–49

Ebrahimzadeh MA, Mahmoudi M, Ahangar N, Ehteshami S, Ansaroudi F, Nabavi SF, Nabavi SM (2009) Anti-depressant activity of corn silk. Pharmacologyonline 3:647–652

El-Ghorab A, El-Massry KF, Shibamoto T (2007) Chemical composition of the volatile extract and anti-oxidant activities of the volatile and nonvolatile extracts of Egyptian corn silk (*Zea mays* L.). J Agric Food Chem 55:9124–9127

Elliger CA, Chan BG, Waiss AC Jr, Lundin RE, Haddon WF (1980) C-Glycosylflavones from *Zea mays* that inhibit insect development. Phytochemistry 19:293–297

Farsi DA, Harris CS, Reid L, Bennett SAL, Haddad PS, Martineau LC, Arnason JT (2008) Inhibition of non-enzymatic glycation by silk extracts from a Mexican land race and modern inbred lines of maize (*Zea mays*). Phytother Res 22:108–112

Fossen T, Slimestad R, Andersen OM (2001) Anthocyanins from maize (*Zea mays*) and reed canarygrass (Phalaris arundinacea). J Agric Food Chem 49:2318–2321

George GO, Idu FK (2015) Corn silk aqueous extracts and intraocular pressure of systemic and non-systemic hypertensive subjects. Clin Exp Optom 98:138–149

Ghada M, Eltohami MS, Adurahman HN, Mahmoud ME, Mudawi (2014) In vitro study of the effect of corn silk on glucose uptake by isolated rat hemi-diaphragm. World J Pharm Res 3:1–2191

Guo J, Liu T, Han L, Liu Y (2009) The effects of corn silk on glycaemic metabolism. Nutr Metab (Lond) 6:47. https://doi.org/10.1186/1743-7075-6-47

Gwendin JM, Induja TA, Manoj JB, Shivasamy MS (2015) Recent trends in effective utilization of by-product corn. Indian J Sci 22(76):18–26

Ha AW, Kang HJ, Kim SL, Kim MH, Kim WK (2018) Acute and subacute toxicity evaluation of corn silk extract. Nutr Food Sci 23(1):70–76

Habtemariam S (1998) Extract of corn silk (stigma of *Zea mays*) inhibit the tumour necrosis factor-α and bacterial lipopolysaccharide-induced cell adhesion and ICAM-1 expression. Planta Med 64: 314–318

Hasanudin K, Hashim P, Mustafa S (2012) Corn silk (Stigma maydis) in healthcare: a phytochemical and pharmacological review. Molecules 17:9697–9715

Hashim P (2011) *Centella asiatica* in food and beverage applications and its potential anti-oxidant and neuroprotective effect. J Int Food Res 18:2217–2222

Hassas-Roudsari M, Chang PR, Pegg RB, Tyler RT (2009) Anti-oxidant capacity of bioactive extracted from canola meal by subcritical water, ethanolic and hot water extraction. Food Chem 114:717–726

Hu Q, Deng Z (2011) Protective effects of flavonoids from corn silk on oxidative stress induced by exhaustive exercise in mice. Afr J Biotechnol 10:3163–3167

Hu, X., Yang, T., Qi, X., Guo, X., Hu, J. (2011). Effects of different drying methods on phenolic composition and antioxidant activity in corn silk (Stigma maydis). Journal of Food Processing and Preservation, 46(11), e17101

Hu QL, Zhang LJ, Li YN, Ding YJ, Li FL (2010) Purification and anti-fatigue activity of flavonoids from corn silk. Int J Phys Sci 5:321–326

Ibrahim MA, Lawal S, Onojah A, Abdukadir A, Adamu A (2016) Modulation of sialic acid levels among some organs during insulin resistance or hyperglycemic states. Mol Cell Biochem 411(1–2):235–239

Inglett GE (1970) Corn: culture, processing, products. The AVI Publishing Company, Westport, CT

Kan A, Orhan I, Coksari G, Sener B (2011) In-vitro neuroprotective properties of the Stigma maydis extracts from four corn varieties. Int J Food Sci Nutr 63:1–4

Kaup SR, Arunkumar N, Bernhardt LK, Vasari RG, Shetty SS, Pai SR, Arunkumar B (2011) Antihyperlipedemic activity of Cynodondactylon extract in high-cholesterol diet fed Wistar rats. Genom Med Biomark Health Sci 3:98–102

Kim SL, Jung TW (2001) Maysin and other flavonoid contents in corn silk. Korean J Breed 33:338–343

Kim K, Shin HH, Choi SK, Choi HS (2005) Corn silk induced Cyclooxygenase-2 in murine macrophages. Biosci Biotechnol Biochem 69:1848–1853

Lee EA, Byrne PF, McMullen MD, Snook ME, Wiseman BR, Widstrom NW, Coe EH (1998) Genetic mechanisms underlying apimaysin and maysin synthesis and corn earworm antibiosis in maize (*Zea mays* L.). Genetics 149:1997–2006

Lee J, Kim SL, Lee S, Chung MJ, Park YI (2014) Immunostimulating activity of maysin isolated from corn silk in murine RAW 264.7 macrophages. BMB Rep 47:382–387

Lee EY, Kim SL, Kang HJ, Kim MH, Ha AW, Kim WK (2016) High maysin corn silk extract reduces body weight and fat deposition in C57BL/6J mice fed high-fat diets. Nutr Res Pract 10: 575–582

Leon ND, Coors JG (2008) Genetic improvement of corn for lingo-cellulosic biofeedstock production. In: Vermerris W (ed) Genetic improvement of bioenergy crops. New York, NY, Springer Science, pp 185–210

Liu J, Lin S, Wang Z, Wang C, Wang E, Zhang Y, Liu J (2011a) Supercritical fluid extraction of flavonoids from Maydis stigma and its nitrite-scavenging ability. Food Bioprod Process 89: 333–339

Liu J, Wang C, Wang Z, Zhang C, Lu S, Liu J (2011b) The anti-oxidant and free-radical scavenging activities of extract and fractions from corn silk (*Zea mays* L.) and related flavone glycosides. Food Chem 126:261–269

Maksimovic ZA, Kovačević N (2003) Preliminary assay on the antioxidative activity of Maydis stigma extracts. Fitoterapia 74:144–147

Maksimović Z, Malenčić Đ, Kovačević N (2005) Polyphenol contents and anti-oxidant activity of Maydis stigma extracts. Bioresour Technol 96:873–877

Miguel MG (2010) Antioxidant and anti-inflammatory activities of essential oil: a short review. Molecules 15:9252–9287

Min OJ, Sharma BR, Park CM, Rhyu DY (2011) Effect of *Stigma myadis* water extract on adipogenesis and blood glucose in 3T3-L1 adipocytes and db/db mice. Korean J Pharm 42: 201–208

Nessa F, Ismail Z, Mohamed N (2012) Antimicrobial activities of extracts and flavonoid glycosides of corn silk (*Zea mays* L). Int J Biotechnol Wellness Ind 1:115–121

Nguemfo EL, Dimo T, Azebaze AGB, Asongalem EA, Alaoui K, Dongmo AB, Cherrah Y, Kamtchouing P (2007) Anti-inflammatory and anti-nociceptive activities of the stem bark extracts from *Allanblackia monticola* STANER L.C. (Guttiferae). J Ethnopharmacol 114: 417–424

Nguyen T, Sherratt PJ, Pickett CB (2003) Regulatory mechanisms controlling gene expression mediated by the anti-oxidant response element. Annu Rev Pharm Toxicol 43:233–260

Nurhanan AR, Rosli WIW (2014) Nutritional compositions and antioxidative capacity of the silk obtained from immature and mature corn. J King Saud Univ Sci 26:119–127

Pan YX, Wang C, Chen ZQ, Li WW, Yuan GQ, Chen HX (2019) Physicochemical properties and anti-diabetic effects of a polysaccharide from corn silk in high-fat diet and streptozotocin-induced diabetic mice. Carbohydr Polym 164:370–378

Pietta PG (2000) Flavonoids as anti-oxidants. J Nat Prod 63:1035–1042

Pinheiro ACS, Pais AA, Tardivo ACB, Alves MJQF (2011) Effect of aqueous extract of corn silks (*Zea mays* L.) on the renal excretion of water and electrolytes and arterial pressure in anesthetized Wistar rats. Rev Brasil Plant Med 13:375–381

Preedy VR, Watson RR (2010) Glossary. In: Handbook of disease burdens and quality of life measures. Springer, New York, NY, p 4230

Ramasamy R, Vannucci SJ, Yan SSD, Herold K, Yan SF, Schmidt AM (2005) Advanced glycation end products and RAGE: a common thread in ageing, diabetes, neurodegeneration, and inflammation. Glycobiology 15:16R–28R

Ren SC, Liu ZL, Ding XL (2009) Isolation and identification of two novel flavone glycosides from corn silk (*Stigma maydis*). J Med Plants Res 32:1009–1015

Ren SH, Qiao QQ, Ding XL (2013) Anti-oxidant activity of five flavones glycosides from corn silk (*Stigma maydis*). Czech J Food Sci 2:148–155

Saheed S, Oladipipo AE, Abdulazeez AA, Olarewaju SA, Ismaila NO, Emmanuel IA, Fatimah QD, Aisha AY (2012) Toxicological evaluations of *Stigma maydis* (corn silk) aqueous extract on hematological and lipid parameters in Wistar rats. Toxicol Rep 2:638–644

Schreiner GE (1965) Toxic nephropathy: adverse renal effects caused by drugs and chemicals. JAMA 191:849–850

Senol FS, Orhan I, Yilmaz G, Cicek M, Sener B (2009) Acetylcholinesterase, butyrylcholinesterase, and tyrosinase inhibition studies and anti-oxidant activities of 33 Scutellaria L. taxa from Turkey. Food Chem Toxicol 48:781–788

Sepehri G, Derakhshanfar A, Zade FY (2011) Protective effects of corn silk extract administration on gentamicin induced nephrotoxicity in rat. Comp Clin Pathol 20:89–94

Snook ME, Widstrom NW, Gueldner RC (1989) Reversed-phase high-performance liquid chromatographic procedure for the determination of maysin in corn silks. J Chromatogr A 477:439–447

Steenkamp V (2003) Phytomedicines for the prostate. Fitoterapia 74:545–552

Stoecklin G, Lu M, Rattenbacher B, Moroni C (2003) A constitutive decay element promotes tumor necrosis factor alpha mRNA degradation via an AU-rich element-independent pathway. Mol Cell Biol 23:3506–3515

Surjee SMA, Zwain LAH (2015) Antibacterial activity of *Zea mays* L and *Urtica dioica* L extract on the isolated bacteria from children UTI in Erbil city. Int J Res Sci Technol 5:2249–2454

Umar MS (2016) Anti-diabetic potential of methanol extract of cooked corn silk (*Stigma maydis*) on alloxan-induced diabetes in albino mice. Pharm Chem J 3(4):68–72

Velazquez DVO, Xavier HS, Batista JEM, Castro-Chaves CD (2005) Zea mays L. extracts modify glomerular function and potassium urinary excretion in conscious rats. Phytomedicine 12:363–369

Vijitha TP, Saranya D (2017) Corn silk- a medicinal boon. Int J Chem Tech Res 10(10):129–137

Vladimir-Knežević S, Blažeković B, Štefan MB, Alegro A, Kőszegi T, Petrik J (2011) Antioxidant activities and polyphenolic contents of three selected micromeria species from Croatia. Molecules 16:1454–1470

Wan Rosli WI, Nurhanan AR, Mohsin SSJ, Farid CG (2008) Aqueous, alcoholic treated and proximate analysis of maydis stigma (*Zea mays* hairs). Ann Microsc 8:66–72

Wan Rosli WI, Nurhana AR, Farid CG, Mohsin SSJ (2010) Effect of sodium hydroxide (NaOH) and sodium hypochlorite (NaHClO) on morphology and mineral concentrations of *Zea mays* hairs (cornsilk). Ann Microsc 10:4–10

Wang C, Zhang T, Liu S, Zhang C, Wang E, Wang Z, Zhang Y, Liu J (2011a) Subchronic toxicity study of corn silk with rats. J Ethnopharmacol 137(1):36–43

Wang GQ, Xu T, Bu XM, Liu BY (2011c) Anti-inflammation effects of corn silk in a rat model of carrageenin-induced pleurisy. Inflammation 35:822–827

Wang GQ, Xu T, Bu XM, Liu BY (2012) Anti-inflammation effects of corn silk in a rat model of carrageenin-induced pleurisy. Inflammation 35:822–827

Wanrosli WI, Nurhanan AR, Solihah MA, Mohsin SSJ (2011) Corn silk improves nutrient content and physical characteristics of beef patties. Sains Malaysiana 40(2):155–161

Wu J, Ye M, Wang Z (2017) Extraction, purification and anti-hyperlipidemic activities of total flavonoids from corn silk. Pakistan Journal of Zoology, 49(6), 2173–2179

Xing F, Lei W, Tao MEI, Qing Z (2012) Studies on antimicrobial activity of aqueous extract of maize silk, vol 140. Trans Tech Publications, Switzerland, pp 426–430

Yadav R, Khan SH, Mada SB, Meena S, Kapila R, Kapila S (2019) Consumption of probiotic *Lactobacillus fermentum* MTCC: 5898-fermented milk attenuates dyslipidemia, oxidative stress and inflammation in male rats fed on cholesterol-enriched diet. Probiotics Antimicrob Proteins 11(2):509–518

Yan Z, Da-yun S, Jing-shu Z, Hong-li Z (2011) Microwave assisted extraction and anti-hyperlipidemic effect of total flavonoids from corn silk. Afr J Biotechnol 10(65):14583–14586

Yang J, Li X, Xue Y, Wang N, Liu W (2014) Anti-hepatoma activity and mechanism of corn silk polysaccharides in H22 tumor-bearing mice. Int J Biol Macromol 64:276–280

Yulina ES, Joseph IS, Ferdiana LA (2013) Study of kidney repair mechanism of corn silk (*Zea mays* L. hair)- Binahong (*Anredera cordifolia* (ten.) steenis) leaves in rat model of kidney failure. Int J Pharmacol 9(1):12–23

Zhang Y, Wu L, Ma Z, Cheng J, Liu J (2015) Anti-diabetic, anti-oxidant and anti-hyperlipidemic activities of flavonoids from corn silk on STZ-induced diabetic mice. Molecules 21(7):1–11

Zhao W, Yin Y, Yu Z, Liu J, Chen F (2012) Comparison of anti-diabetic effects of polysaccharides from corn silk on normal and hyperglycemia rats. Int J Biol Macromol 50(4):1133–1137

Zhu KX, Lian CX, Guo XN, Peng W, Zhou HM (2011) Antioxidant activities and total phenolic contents of various extracts from defatted wheat germ. Food Chem 126:1122–1126

Zilic S, Jankovic M, Basic Z, Vancetovic J, Maksimovic V (2016) Anti-oxidant activity, phenolic profile, chlorophyll and mineral matter content of corn silk (*Zea mays* L): comparison with medicinal herbs. J Cereal Sci 69:363–363

Chapter 8
Application of Metabolomics for the Discovery of Potent Antimicrobials from Plants

Pramod Barathe, Sagar Reddy, Kawaljeet Kaur, Varsha Shriram, and Vinay Kumar

Abstract One of the major goals of studying metabolome and metabolism has long been to find biomarkers for disease diagnosis and prognosis. The significance of metabolomics has been transformed from a straightforward biomarker identification tool to a technology for the detection of active biological process drivers, nevertheless. It is now understood that the metabolome modifies other "omics" levels, including as the genome, epigenome, transcriptome, and proteome, in order to influence cellular function. In this chapter, we highlight the strategies to use metabolomics to uncover the active function of metabolites in physiology and disease by understanding how the metabolome is useful in screening and to identify the active molecules from natural sources such as plants and their mode of action. The idea of using activity screens to find biologically active compounds using metabolomics, or what we call activity metabolomics, is already having a significant impact on biology.

Keywords Medicinal plants · Metabolomics · Antimicrobials · Antimicrobial resistance

8.1 Introduction

Plants always are a helping hand for humans as a natural remedy to treat bacteria and associated diseases or infections. It is worthy to note the impact of plant secondary metabolites for human health and well-being. Uncovered potential of antimicrobials from plants makes them a rich candidate as antimicrobials. It is reported that many

P. Barathe · S. Reddy · K. Kaur · V. Kumar (✉)
Department of Biotechnology, Modern College of Arts, Science and Commerce, Savitribai Phule Pune University, Pune, Maharashtra, India
e-mail: vinay.kumar@moderncollegegk.org

V. Shriram
Department of Botany, Prof. Ramkrishna More College, Savitribai Phule Pune University, Pune, Maharashtra, India

V. Kumar et al. (eds.), *Medicinal Plants and Antimicrobial Therapies*,
https://doi.org/10.1007/978-981-99-7261-6_8

antimicrobials from plants have antimicrobial activity like cinnamaldehyde from cinnamon, thymol and carvacrol from thyme, betulinic acid from ber tree (*Ziziphus jujuba*), allicin from garlic, curcumin from curcuma, and piperine from piper (Gorlenko et al. 2020). The diversity of plant constituents and their antimicrobial potential is nowadays being successfully studied with the application of metabolomics. Metabolomics can be defined as the large-scale study of small molecules, commonly known as metabolites, within cells, biofluids, tissues, or organisms (https://www.ebi.ac.uk). Collectively, these small molecules and their interactions within a biological system are known as the metabolome. Metabolites represent the activity of the genes at a specific time and in a specific environment since they are closely tied to biochemical, physiological, and pathological processes. The other global omics platforms, such as genomics, transcriptomics, and proteomics, are therefore complimentary to metabolomics (Mirsaeidi et al. 2015). The metabolomics analysis records the metabolome at a single time point, whereas its subtopic, fluxomics, keeps track of metabolites with stable isotope-labelled components and fully illustrates the metabolic changes. The understanding of changes occurring in the live organism under certain conditions is provided by metabolomics, which is assisted by chemometrics and bioinformatics. Nuclear magnetic resonance (NMR) and mass spectrometry (MS) coupled with liquid chromatography (LC-MS) and gas chromatography (GC-MS) are the two analytical methods utilized extensively in metabolomics. The metabolomic workflow, regardless of the method, necessitates spectrum processing with features detection, standardization, and deconvolution, followed by multivariate data analysis (Alonso et al. 2015). Statistical analysis is a component of bioinformatics calculations, but they also contain pathway and network analysis, which shows complex correlations within metabolite sets. All bioinformatics tools, software, and databases for metabolic data have been discussed in this chapter.

Metabolomics covers a number of topics in basic and applied research. It is used in antimicrobial research for the evaluation of plant phytochemicals, as well as bacterial responses to various environmental conditions and stress. It is suitable to explore various plant phytomolecules, essential oils, and antimicrobials from plants for their bioactivity-guided isolation, which guides fast-tracking and identification of molecules with desired biological potential (Rinschen et al. 2019). However, the ability of plant secondary metabolites to prevent the growth of microorganisms is an evolutionary strategy. The plant metabolome acts as an abundant source of phytochemicals offering numerous benefits for the humans (Baldwin 2010). Various compounds that plants use to interact with microbes might encourage the growth of synergistic bacteria while discouraging the opportunistic ones (Berg et al. 2017). Plants become a source of antimicrobial substances as a result of this natural process, which alters the metabolic processes of bacteria in various ways. Bacteria are sensitive to outside stimuli and need to adapt to a variety of environmental conditions in order to survive and grow. They can keep the cell system's homeostasis by detecting these changes and starting metabolic regulation (Chauhan et al. 2016). The transcriptional level of metabolic regulation often involves interactions between transcriptional factors and their target genes (Drapal et al. 2014). The end-products representing the metabolic status of the cell are adjusted production as a result of the subsequent upregulation or downregulation of enzyme activity (Mack et al. 2018). Metabolomics makes it possible to investigate the metabolic changes caused by plant secondary metabolites in bacterial cells. It exposes the modifications to metabolic pathways that result from bacteria's biochemical networks being reprogrammed (Zampieri et al. 2017). According to this perspective, metabolomics is useful in figuring out how cells behave differently when bacteria are subjected to metabolic stress brought on by secondary metabolites that inhibit their growth. The application of metabolomics in revealing bacterial response to plant-derived natural products still requires more attention and in-depth study. To date, metabolomics studies have extensively explored the bacterial response to changing environmental conditions (carbon source, pH, and starvation) (Drapal et al. 2014), the mechanisms of action of known antibiotics (Hoerr et al. 2016; Koen et al. 2018), or bacterial resistance mechanisms (Mack et al. 2018). Therefore, in this chapter, we attempt to focus on the metabolomics way to screen and select antimicrobial molecules from plants and determination of mode of actions.

8.2 Metabolomics: Principle and Techniques

Metabolomics has been derived from the word "metabolome" meaning "the complete set of metabolites in an organism," leading to the definition of metabolomics as "identification and quantification of all metabolites in a biological system" (Schripsema 2010). Metabolomics covers all the metabolome of all the metabolites present inside the cells and tissues and hence has wider applications in personalized medicines, identification of potent secondary metabolites and antimicrobials from plants, pharmacology, toxicology, etc. (Clish 2015). It allows thorough analysis of various low-molecular-weight molecules that operate as beginning, intermediate, and end products of metabolic processes in living organisms and complementing other global omics platforms such as genomes, transcriptomics, and proteomics (Sieniawska and Georgiev 2022). To identify these low-molecular-weight metabolites, metabolomics is divided into two primary analysis platforms: untargeted metabolomics and targeted metabolomics. Untargeted metabolomics, also known as hypothesis generating, emphasizes on gathering data from as many species as feasible, then annotating metabolites, and assessing both known and undiscovered metabolic alterations where data are further utilized for comparative quantification and providing hypothesis to further investigate using targeted approach (Schrimpe-Rutledge et al. 2016). Targeted metabolomics approaches by measuring specific chemically characterized and biologically annotated metabolites by analyzing it quantitatively or semi-quantitatively using standards, their kinetics, known as biochemical routes, and other pathways leading to identification and characterization of

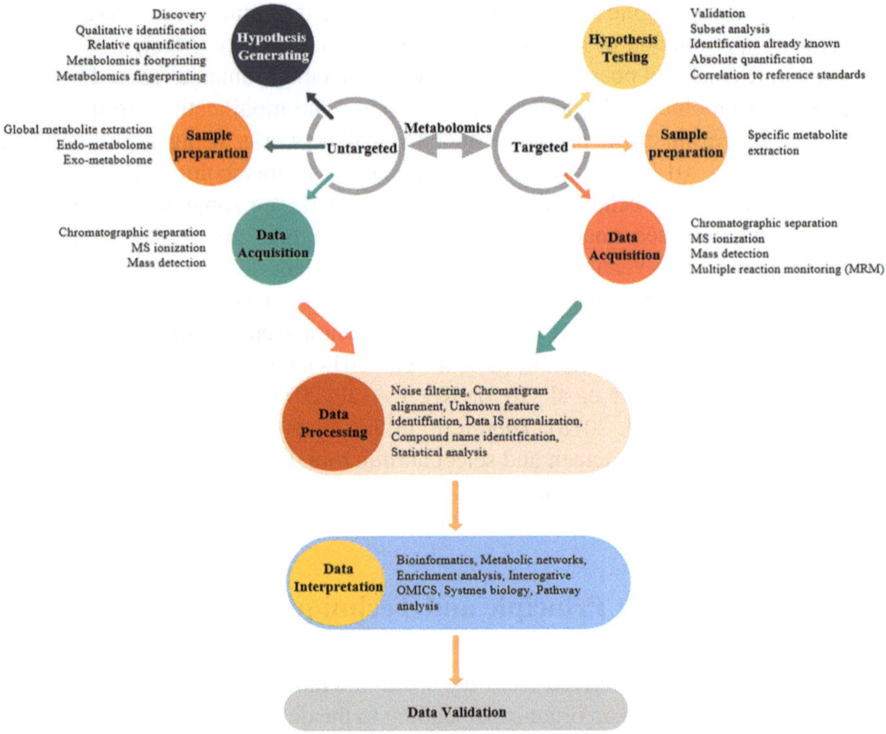

Fig. 8.1 Metabolomics analysis platforms: targeted and untargeted metabolomics

novel associations of metabolites (Costanzo et al. 2022; Roberts et al. 2012). Each type of platforms consists of their own strategies and techniques for sample preparation, data acquisition, data processing, data interpretation, and data validation of the metabolite as illustrated in Fig. 8.1 and Table 8.1.

8.3 Metabolism and Antimicrobial Resistance

Metabolomics deals with the study of metabolome in a biological sample and gives a glance of the utilized biochemical processes. Metabolomics includes the study of relationship between antimicrobial resistance (AMR) mechanisms and microbial metabolism. Key changes associated with AMR happen within the bacterial cell by metabolites that plays important role including cell-to-cell interactions in biofilms, cell envelope modifications, and cell energy modifications illustrated in Table 8.2.

Table 8.1 Techniques used in metabolomics

Sr. No.	Metabolomics technique	Metabolomics analysis platform	Sensitivity	Throughput	Comprehensiveness	Strengths	Challenges	References
1.	Gas chromatography–mass spectrometry (GC-MS)	Targeted and untargeted	High	High	High	• Identify and quantify small molecular metabolites • Identify and semi-quantify over 200 compounds • Detection of more than 300 additional unidentified signals	• Does not yield accurate mass data for compound identification • Has much slower scan rate than time-of-flight mass spectrometers • Severe matrix effects of biogenic amines	Fiehn (2016), Putri et al. (2022)
2.	Liquid chromatography–mass spectrometry (LC-MS)	Targeted and untargeted	Medium	High	High	• Multiple experimental, analytical, and computational steps • Identification and quantitation of metabolites indicative of biological/environmental perturbations	• Overwhelming number of unidentified peaks • Contextualize metabolites in pathways • Partial, incomplete metabolomes	Cui et al. (2018), Zhou et al. (2012)
3.	Nuclear magnetic resonance (NMR)	Targeted and untargeted	Low	Low–high	Low–high	• Reproducibility • Metabolite deconvolution • Dynamic nuclear polarization	• Lower signal-to-noise • Low natural abundant nuclei	Moco (2022), Nagana Gowda and Raftery (2021)

(continued)

Table 8.1 (continued)

Sr. No.	Metabolomics technique	Metabolomics analysis platform	Sensitivity	Throughput	Comprehensiveness	Strengths	Challenges	References
4.	Capillary electrophoresis–mass spectrometry (CE-MS)	Targeted and untargeted	High	Medium	High	• Analysis of (highly) polar and charged metabolites in a wide range of biological samples • High resolution	• Lack of standard operating procedures • Reproducibility studies	Ramautar et al. (2009, 2019)
5.	Direct-infusion mass spectrometry (DIMS)	Untargeted	High	High	High	• Characterizing molecular responses of organisms to disease, drugs, and the environment • Reproducibility • Evaluate novel data processing algorithms	• Coelution of metabolites into the mass spectrometer in the absence of chromatography	Kirwan et al. (2014)

Table 8.2 Studies indicating the role of metabolism in antimicrobial resistance (AMR) using metabolomics techniques

AMR mechanism	Main finding	Metabolic pathway	Analytical approach	Antibiotics	Species	Reference
Metabolic adaptation in energy production						
Nutrient supplementation	Antibiotic-resistant cells reduce activity in central carbon metabolism, which can be activated by nutrient supplementation	Glycolysis, TCA cycle	Untargeted GC–MS	Kanamycin	*Edwardsiella tarda*	Peng et al. (2015), Su et al. (2015)
Energy metabolism influences antibiotic efficacy.	Energy production is a better predictor of antibiotic efficacy than is growth rate	Nucleotides	Colorimetric	Ampicillin, carbenicillin, gentamicin, kanamycin, streptomycin, ciprofloxacin, levofloxacin, norfloxacin, cefsulodin	*Escherichia coli*	Lopatkin et al. (2019)
Metabolic adaptation to the microenvironment						
Biofilm formation	The population of the biofilm has a heterogeneous distribution of quorum-sensing molecules	Quorum sensing	Targeted MALDI-SIMS	NA	*P. aeruginosa, Staphylococcus aureus*	Vuong et al. (2000)
Metabolic adaptation in antibiotic-resistant cells	Adaptative evolution in the cystic fibrosis lung of antibiotic-resistant strains results in metabolic change	Nutrient consumption	Targeted LC-UV and LC-MS	NA	*Pseudomonas aeruginosa*	Yang et al. (2017)
Cell wall modifications						
Cell wall disruption and synthesis	By driving the energy production flow to glucose utilization and shuttled TCA cycle, colistin therapy drives metabolic flux towards cell wall repair	Glycolysis, lipids, TCA cycle	Untargeted GC–MS	Colistin	*Acinetobacter baumannii*	Pamp et al. (2008)
	The loss of envelope and membrane biogenesis processes result in complete lipid reconstruction,	Amino acids, lipids, pentose	Untargeted LC-MS	Colistin	*Klebsiella pneumoniae*	Li et al. (2020)

(continued)

Table 8.2 (continued)

AMR mechanism	Main finding	Metabolic pathway	Analytical approach	Antibiotics	Species	Reference
	including changes in lipid A moiety, resulting in the energy metabolic switch to glycolysis	phosphate, TCA cycle				
Multidrug resistance (MDR) overexpression	Increased MDR efflux pump activity triggers metabolic rewiring toward anaerobic respiration	Nitrates and oxygen	Colorimetric oximeter	NA	*P. aeruginosa*	Olivares Pacheco et al. (2017)

CE-MS capillary electrophoresis–mass spectrometry, *GC–MS* gas chromatography with mass spectrometry, *LC* liquid chromatography, *LC-UV* liquid chromatography with ultraviolet detection, *MALDI-SIMS* matrix-assisted laser desorption ionization—secondary ion mass spectrometry, *NMR* nuclear magnetic resonance, *TCA* tricarboxylic acid

8.3.1 Cell Energy Modifications

Activation of cellular functional responses or dormancy to avoid eradication of microbes by antibiotics results from the activity of energy-producing metabolic pathways. Metabolic pathway that is most convenient to microbe is glycolysis and tricarboxylic acid cycle (TCA) for the production of energy through the electron transport chain (ETC). Bacteria switches their metabolism from aerobic to less efficient anaerobic fermentative energy production pathway (Passalacqua et al. 2016). This is to evade the host defense system and play a role in aminoglycoside tolerance. Targeting such pathway is novel act to resensitize some of the antibiotics like aminoglycosides. This can be done by the use of variety of carbon sources to induce aerobic conditions in microbes. Metabolomics study has demonstrated that these stimulations change the activity of TCA and ETC. Increasing the passive entry of charged molecules can increase the efficacy of aminoglycosides. Increased electric transmembrane potential induced by ETC stimulation facilitates positively charged aminoglycoside influx mediated by proton motive force (PMF) (Lobritz et al. 2015; Meylan et al. 2017; Peng et al. 2015; Su et al. 2015) This would result in decreasing the survival of antibiotic-resistant strains. Other antibiotic classes such as beta lactams and fluoroquinolones induce redox disbalance as secondary antibiotic effect that depends on cellular respiration. Metabolic specialization of targeted bacterial cell decides the antibiotic uptake and its secondary effect that is influenced by nutrient supplement.

8.3.2 Modification of the Cell Envelope

Polymyxin antibiotics are the last resort of antibiotics that targets lipid A in lipopolysaccharides (LPS) of Gram-negative bacteria. Modifications to this LPS can lead to resistance to polymixins. This modification may be intrinsic, acquired, chromosomally encoded, or plasmid-mediated adaptations (Baron et al. 2016; Haeili et al. 2017; Li et al. 2020). This supports fatty acid synthesis, which demands high energy. Metabolomics study of polymyxin-resistant strain demonstrated that the modifications in the lipid modifications, in another pathway, while TCA cycle was shunted upper carbon flux in the glycolysis pathway, was elevated (Koen et al. 2018). This suggested that there is a switch to glucose fermenting metabolism for energy production in polymyxin-resistant cells. This forced us to focus on the biosynthetic routes for lipopolysaccharides (LPS).

8.3.3 Cell–Cell Interactions in Biofilm

Biofilm decreases antibiotic activity by the production of extracellular polymeric substances, which costs efficient cellular communication and metabolic adaptations.

Quorum sensing mechanisms used by bacteria for the cell communication produces small molecules, which are essential for the formation of biofilm. Thus, these processes are essential to target for the treatment of biofilm infections. This requires identification and characterization of quorum sensing (QS) molecules. Chronic bacterial infections are frequently accompanied with mature, well-developed biofilms and have lower antibiotic potency. In particular, mature, well-developed biofilms are linked to significant nourishing gradients that are brought on by the structure of the biofilm. The ability of pathogens to switch metabolically to other food sources is frequently a determinant of the biofilm maturation process. Analyses of the metabolites produced during central carbon metabolism in real time showed that the biofilm's depth and time allowed for metabolic adjustments to anaerobic fermentation pathways. Increased aminoglycoside eradication was achieved by redirecting metabolism in *Pseudomonas aeruginosa* biofilms by the addition of TCA cycle carbon sources, demonstrating the possibility of food supplementation to reduce metabolically induced tolerance in biofilms (Koeva et al. 2017; Meylan et al. 2017).

8.4 Screening and Selection of Antimicrobial Molecules from Plants and Determination of Mode of Action

The rapid emergence and reemergence of bacterial resistance to frequently prescribed and even to the last-resort antibiotics call for the development of novel, highly potent antibacterial medications. Standard taxonomy and bioactivity-based methods are useful for the identification of antibacterial compounds, but they are insufficient and frequently lead to the rediscovery of already well-known compounds. Therefore, it is advised to use a holistic approach for the preliminary screening of numerous novel compounds acting synergistically or as a single active ingredient. Metabolomics has the ability to identify bioactive metabolites in the first fractionation stage, eliminating the need for subsequent bioassays and increasing the likelihood of isolating targeted natural products at higher yields. Various methods or techniques involved in metabolomics include nuclear magnetic resonance (NMR), high-performance liquid chromatography (HPLC), liquid chromatography coupled to mass spectroscopy (LC-MS), and gas chromatography coupled to mass spectroscopy (GC-MS). All of these techniques are selective and sensitive but generate large amounts of complex data. Chemometric is the application of mathematical and statistical techniques, which is used to retrieve valuable information from a complex dataset. It contributes to correlating the metabolic data with the biological activity of several samples simultaneously. There are many chemometric analyses, but the most commonly used are principal component analysis (PCA), partial least square (PLS), and orthogonal projections to latent structures discriminant analysis (OLPS-DA), a modification of the nonlinear iterative partial least squares (NIPALS) algorithm. The combination of metabolomics and chemometric analysis as illustrated in Fig. 8.2a

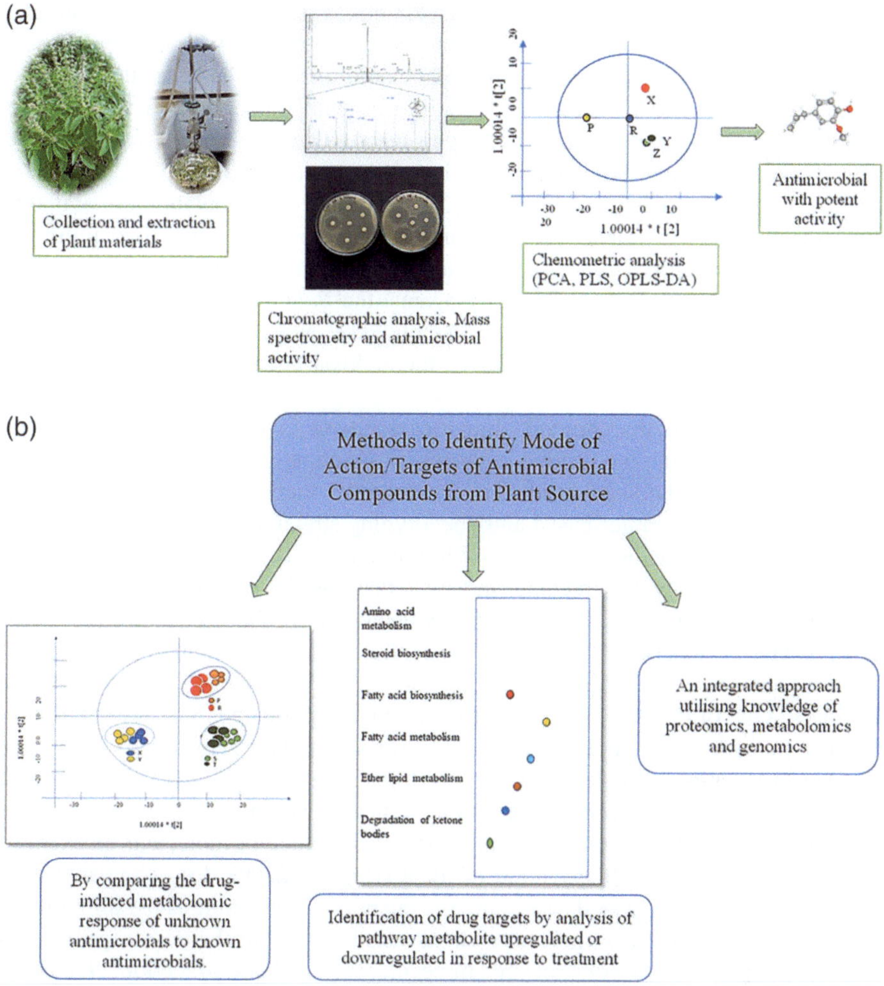

Fig. 8.2 (**a**) Identification of antimicrobial compounds from plants using metabolomics approaches. (**b**) Metabolomics approach to identify action/target of antimicrobial compounds from plant sources

offers several advantages over conventional approaches and facilitates the screening of a large number of potent compounds, such as essential oils, plant extracts, and other bioactive substances. Maree et al. (2014) successfully used chemometric analysis in conjunction with GC-MS to assess the antimicrobial activity and chemistry of essential oils, and they identified eugenol as a biomarker of efficacy responsible for the strong antimicrobial activity of samples of essential oils against all of the tested bacterial isolates. On the other hand, limonene, α-pinene, sabinene, α-phellandrene, and limonene were identified as compounds responsible for the samples' poorer antibacterial activity. Similarly, by OLPS-DA analysis, dos Santos et al. (2018) discovered palmitic acid, 7,8-epoxy-1-octene cis-α-bergamotene, methyl linolelaidate, alloaromadendrene, and veridiflorol as key biomarkers responsible for the good antibacterial activity of volatile oils. On the other hand, bornyl acetate was shown to be related to volatile oils with moderate activity, whereas 4-terpineol was found to be associated with inactive volatile oils, solving the cost-related issue regarding the isolation and identification of individual components of volatile or essential oils. Carvalho et al. (2021) identified five metabolites, similar to the essential oils, namely cis-3-hexenyl β-primeveroside, macahydantoin A, lepidiline A, uvaricin, and 15-oxo-18Z-tetracosenoic acid from *Lepidium meyenii* extract that shows antibacterial activity against the *Bacillus cereus* using a combination of metabolomic approach and chemometric analysis such as partial least squares-discriminant analysis (PLS-DA), hierarchical cluster analysis (HCA), and variable analysis.

Another problem that hinders drug development is the lack of an effective approach to determine the in vivo mechanism of action and effectiveness of novel therapeutic leads (Halouska et al. 2012). Currently, there are two main drug lead discovery strategies, target-based and phenotyping strategies. Target-based strategy begins with the predefined target whose inhibition results in the desired phenotypic effects, while phenotypic screening starts with cell-based assay that monitors growth inhibition (da Cunha et al. 2021). Both the methods have their own advantages and disadvantages. The ability to use both molecular and chemical knowledge to investigate specific molecular hypotheses is one of the benefits of the target-based approach. One of the disadvantages of target-based approach is that the solution to the specific molecular hypothesis may not be relevant to pathophysiology or provide an adequate therapeutic index. One such example includes inability of chemical lead to penetrate the bacterium and establish the efficacy. The major strength of phenotypic screening is that it does not require prior knowledge of molecular mechanism of action and has more likely hood of identifying candidate drug compared with target-based approach. The disadvantage of phenotypic screening is the difficulty of optimizing candidate drug molecular characteristics without the design parameters offered by prior knowledge of the molecular mechanism of action. To better understand the efficacy of innovative drug candidates, thorough knowledge of the mechanism of action is required. One approach is based on the idea that drugs with similar mechanisms of action or therapeutic goals will affect the metabolome similarly. As a result, the method of action of a novel chemical lead can be inferred from drugs with established biological targets. Vincent et al. (2016) identified mechanism of action of eight compounds using untargeted liquid chromatography–mass spectrometry approach. Thymidylate kinase, isoprenoid biosynthesis, acyltransferase, and deoxyribonucleic acid (DNA) metabolism were identified as the potent targets for the novel antimicrobials 1-[3-fluoro-4-(5-methyl-2,4-dioxo-pyrimidin-1-yl)phenyl]-3-[2- 26 (trifluoromethyl)phenyl]urea (AZ1), fosmidomycin, inhibitor of LpxC in lipid A biosynthesis (CHIR-090), and 2-(cyclobutylmethoxy)-5′-deoxyadenosine. Chen et al. (2020) identified metabolic pathway targeted by the essential oil of *Cinnamomum camphora* against methicillin-resistant *Staphylococcus*, similar to the antibiotics, using GC-MS coupled with chemometric analysis. The disruption of the TCA cycle as the principal mechanism of action of essential oils was indicated by a considerable drop in two upstream (citrate and malate) and downstream metabolites of the TCA cycle. This highlights the potent use of metabolomics approach for the rapid and efficient identification of mechanism of action of novel antimicrobial from natural sources as illustrated in Fig. 8.2b.

8.5 Bioinformatic Tools for Metabolomics

The main goals of bioinformatics analyses utilized in metabolic investigations include accurate identification of metabolic products, quantitative evaluation of metabolites, contribution to metabolic reconstruction, disclosure of metabolic features, and identification of metabolic effects of environmental variables. A wide variety of MS and NMR spectra are used to represent data from metabolic analyses. For routine investigations to perform data processing, peak annotation, metabolite identification, and statistical and pathway analysis, metabolomics hardware typically comes with the software already installed. However, these qualities might not be sufficient for more sophisticated metabolomics applications. In addition, because the software and data formats are proprietary, there is a problem with interoperability between various processes and software versions. Thus, there is an increasing number of free-to-use open-source software tools that can perform one or more of these tasks. A survey was conducted by Weber and others for the prevalence of utilization of open-source and free-to-use software tools (Weber et al. 2017). Commercial software is bundled with instruments by the metabolomics community. Among the open sources, 70% of them used XCMS (Tautenhahn et al. 2012) online processes and provides complete metabolomics work flow including detection, retention time correction, alignment, statistical analysis, and annotation. Agilent ChemStation is one of the most popular commercial programs used, while AMDIS (http://amdis.net) was the most popular open-source software.

Skyline is the vendor-independent open-source software tool for targeted MS (Adams et al. 2020). The authors claimed that Skyline has been developed into a flexible and potent open-source software solution for quantitative MS data workflows, capable of supporting both proteomics and metabolomics data processing in a vendor-neutral environment, enhancing transparency, and facilitating the exchange of data and methods. 70% of the annotations used CAMERA (Kuhl et al. 2012), which is a free program for Windows, Mac OS, and Linux operating systems and is available from the Bioconductor repository. A wider range of software, including Bioconductor's (https://bioconductor.org/) Metlin (https://metlin.scripps.edu https://massconsortium.com), MetFrag (https://msbi.ipb-halle.de/MetFrag), XCMS, and RMassBank (https://bioconductor.org/packages/release/bioc/html/RMassBank.html), was used to annotate the data generated by Multistage Mass Spectrometry (MS^n). The NIST Mass Spectrometry Data Centre and AMDIS were the two most popular options for annotating GC-MS data (Weber et al. 2017). Chenomx's NMR suite and AMIX from Bruker (https://www.bruker.com/en/

products-and-solutions/mr/nmr-software/amix.html), which each have 39% of users, are the two commercial software programs most commonly used for annotating NMR data (Weber et al. 2017). The Birmingham Metabolite Library (Ludwig et al. 2012) and further data-mining tools like BATMAN (Hao et al. 2012) and rNMR, which have respective user bases of 22%, 17%, and 9%, are examples of open-source software programs for this purpose. Incompatibility between software from raw data to reporting formats, it needs closer integration of metabolomics with other omics data to generate a better understanding of biological processes. Omics Discovery Index (Omics DI) integrates data from genomics, proteomics, transcriptomics, and metabolomics (https://www.omicsdi.org).

8.6 Metabolomics in Preclinical Studies

Antimicrobials once chosen for the further research must undergo preclinical testing using animal models and other in vitro systems to ensure their safety and efficacy before being evaluated in clinical trials (Lu and Chen 2017). However, poor pharmacokinetic profile and unanticipated toxicity are some of the significant reasons behind chemical failures that lead to inadequate effectiveness of the antimicrobial in preclinical studies. To overcome these drawbacks, metabolomics is used as effective tool for attaining these aims (Kumar et al. 2014). Apart from preclinical and clinical research, metabolomics has become a vital tool in drug discovery and development. Antimicrobial identification and their characterization have been a major part of any drug development. Novel antimicrobial metabolites (O-glycoside-kaempferol, luteolin glucoside, caffeoylquinic acid, and kaempferol) were identified from untargeted metabolomic profiling of *Sphagnum fallax* using mass spectrometry (MS) systems and 1H nuclear magnetic resonance (1H NMR) spectroscopy (Fudyma et al. 2019). Further, preclinical studies on the kaempferol polyphenol in human promyelocytic leukemia (HL-60) cells revealed its antitumor properties by affecting proliferation and apoptosis of cancer cells (Kluska et al. 2021). GC-MS analysis of nonpolar extract of the halotolerant marine fungus *Penicillium chrysogenum* MZ945518 isolated from the Mediterranean Sea in Egypt detected 20 metabolites, mainly (Z)-18-octadec-9-enolide (36.28%) and 1,2-benzenedicarboxylic acid (26.73%) with antibacterial activity against *Proteus mirabilis* ATCC 29906 and *Micrococcus luteus* ATCC 9341, which further molecular-docked with target proteins such as DNA gyrase, glutathione S-transferase, and acetylcholinesterase and confirmed their antimicrobial activity and effectivity for further investigations (El-Sayed et al. 2023). Further, GC-MS combined with chemometrics for the identification of antimicrobial compounds from selected commercial essential oils reveals eugenol compound with the highest antimicrobial activity against *Candida albicans* and other Gram-positive bacteria (Maree et al. 2014). Apart from drug and antimicrobial identification in drug discovery, metabolomics is also used in the discovery of roles of various enzymes, drug metabolism, and determining underlying resistant mechanisms of pathogenic bacteria using experimental animal models such as knockout mice (Lu and Chen 2017). Recent studies on metabolomics analysis in tumor metabolism reprogramming have helped scientists develop amino acid metabolism reprogramming in tumor growth, lipid metabolic reprogramming in cancer cells, glucose metabolism reprogramming in cancer progression, and metabolomic markers in cancer progression using renal clear cell carcinoma cells, kidney cancer cells, Michigan Cancer Foundation-7 (MCF-7) breast cancer cells, and pancreatic ductal adenocarcinoma cells, respectively (Han et al. 2021). Recently, drug repurposing and reprogramming has emerged in medicine field where metabolomics has established itself as an excellent tool for large data screening leading to the development of metabolomics databases and mapping of large datasets in the field (Alarcon-Barrera et al. 2022). Studies on biochemical alterations in the treatment of drugs to differentiated human-induced pluripotent stem cell (hiPSC)-derived models for studying human central nervous system (CNS) disorders has helped "humanize" the process of CNS drug discovery (Silva and Haggarty 2020). The approach of integrating metabolomics in various drug repurposing, drug reprogramming, and drug metabolism studies has effectively investigated many drug-induced alterations and hence is growing as a vital tool in drug discovery.

8.7 Metabolomics in Clinical Trials

Clinical trials are prospective assessments of the effects of human treatments in which the investigator, typically in accordance with a protocol, assigns the intervention (treatment) (Harlan 2016). They are used to determine the beneficial and harmful effects of new biomedical or behavioral interventions and inform society, policymakers, and health-care providers about the treatment decision (Wright 2017). Metabolomics has become vital in clinical trials in recent years, offering precise, reliable, qualitative, and quantitative data (Wishart 2008). Metabolomics can have a direct use in assessing the toxicity and identification of metabolic markers of efficacy (Lu and Chen 2017), for instance, a system-wide assessment of metabolic responses to the administration of combined metabolic activator (CMA) with N-acetyl-L-cysteine (NAC) or cysteine as a metabolic activator. Evidence suggests that plasma nicotinamide adenine dinucleotide (NAD+) and glutathione (GSH) deficiency play an important role in the development of metabolic diseases. The GSH is produced from the primary amino acids, glutamate, glycine, and cysteine and is essential for maintaining intracellular redox balance. The GSH deficiency causes pathogenesis with an oxidative burden. Yang et al. (2023) investigated the effect of CMA, which includes NAD+ and glutathione (GSH) precursor, on boosting deficient metabolic pathways, and efficacy in treating the disease. It was discovered in this study that supplementing with CMA considerably improved the metabolic pathways associated with disease avoidance (especially the serine metabolism pathway). Furthermore, it has been demonstrated that supplementing CMA with NAC or cysteine is safe for human consumption.

In addition to establishing the efficacy and safety of medications, metabolomics analysis can detect patient nonadherence to therapy, for example, the analysis of prevalence of nonadherence to antihypertensive therapies using direct detection of blood pressure (BP)-lowering drug in urine samples analyzed by HPLC–tandem mass spectrometry (MS/MS) analytical tools (Tomaszewski et al. 2014). When compared to the existing approaches, this urine-based screening tool for nonadherence to antihypertensive therapy has significant advantages. First and foremost, this procedure is simple, precise, and fully noninvasive, preventing unnecessary additional harm and treatments. It is also effective for improved patient stratification prior to expensive, irreversible operations and therapy escalation (Tomaszewski et al. 2014). Metabolomics also plays a role in the study of patients' dietary compliance and in the identification of metabolic markers that indicate a diet rich in fruits, fish, cocoa-containing items, and vegetables (Andersen et al. 2014). This can be used in clinical trials to assess the effect of eating pattern on treatment efficacy. Drug misuse and relapse are two other topics that metabolomics might shed light on. A quick, noninvasive LC-MS/MS approach was used by Koster et al. (2014) for detecting drug abuse in human hair. With the help of this technique, 17 different substances or their metabolites, including methadone, cocaine, morphine, codeine, heroin, nicotine, and delta-9-tetrahydrocannabinol, can be detected. This may provide physicians or other medical professionals crucial data that can be used to decide on patient-specific treatments or patient stratification prior to the start of clinical trials (Koster et al. 2014). Similarly, Vincent et al. (2016) untargeted liquid chromatography–mass spectrometry approach and identified thymidylate kinase, isoprenoid biosynthesis, acyl-transferase, and DNA metabolism as the potent targets of the novel antimicrobials AZ1, fosmidomycin, CHIR-090, and 2-(cyclobutylmethoxy)-5′-deoxyadenosine. These approaches, however, were found to be less effective in the case of drugs that do not work through metabolic pathways, such as proton carrier carbonyl cyanide m-chlorophenyl hydrazone (CCCP), and have no discernable impact on metabolomes. As a result, an integrated approach based on proteomic, transcriptomic, metabolomic, and genomic knowledge is recommended, for instance, the study by Gao et al. (2022) in which they first identified the mechanism of action of epigallocatechin gallate by in vitro assays such as biofilm formation, hemolytic activity, and growth inhibition. Further, reported downregulation of proteins involved in cell wall synthesis, DNA replication, and virulence. In addition, identified ATP-binding cassette (ABC) transporters, glycolysis/gluconeogenesis, and aminoacyl-transfer ribonucleic acid (tRNA) perturbed pathways in response to epigallocatechin gallate treatment, effectively combining proteomics, metabolomics, and traditional activity-based approach for determining the mechanism of action of antimicrobials.

8.8 Conclusion

Metabolomics finds new application in searching of new antimicrobial agents from the plant sources and to understand their mode on action. This chapter gives a glance on the screening and mode of action of plant antimicrobials against bacteria using metabolomics study. Metabolomics can enhance our understanding of the mechanisms underlying bacterial resistance and may help us to build more potent future therapies, such as a chance of regaining bacterial susceptibility to antibiotics. Synergistic approach of antibiotics and natural compounds from plants may reveal some potential new targets in bacteria. Metabolomics alone does not explore bacterial response to stress conditions. This can be achieved by the combination of gene transcription data (transcriptomic and translation quantification data) and proteomics with the accumulated metabolites. This advocates the need to focus on approaches combining transcriptomics, genomics, and proteomics. Bioinformatics is a developing field that provides tools to monitor metabolites in bacteria. Combination study of omics data further gives possible enzymatic targets in bacteria cells with a light shed on bacterial response to plant antimicrobials. As a result, an integrative, all-encompassing strategy to studying antimicrobial development and their effect on human health should be part of the expected future growth of metabolomics.

Acknowledgments The authors acknowledge the funding support under DBT-BUILDER (BT/INF/22/SP45363/2022) from the Department of Biotechnology (DBT), Government of India, and the DST-FIST (SR/FST/COLLEGE-/19/568) from the Department of Science and Technology (DST), Government of India, implemented at Modern College, Ganeshkhind, Pune, India.

References

Adams KJ, Pratt B, Bose N, Dubois LG, St. John-Williams L, Perrott KM, Ky K, Kapahi P, Sharma V, MacCoss MJ, Moseley MA, Colton CA, MacLean BX, Schilling B, Thompson JW (2020) Skyline for small molecules: a unifying software package for quantitative metabolomics. J Proteome Res 19(4):1447–1458. https://doi.org/10.1021/acs.jproteome.9b00640

Alarcon-Barrera JC, Kostidis S, Ondo-Mendez A, Giera M (2022) Recent advances in metabolomics analysis for early drug development. Drug Discov Today 27:1763–1773. https://doi.org/10.1016/j.drudis.2022.02.018

Alonso A, Marsal S, Julià A (2015) Analytical methods in untargeted metabolomics: state of the art in 2015. Front Bioeng Biotechnol 3:1–20. https://doi.org/10.3389/fbioe.2015.00023

Andersen MS, Rinnan Å, Manach C, Poulsen SK, Pujos-guillot E, Larsen TM, Astrup A, Dragsted LO (2014) Compliance to a dietary pattern. J Proteome Res 13:1405–1418

Baldwin IT (2010) Plant volatiles. Curr Biol 20(9):R392–R397. https://doi.org/10.1016/j.cub.2010.02.052

Baron S, Hadjadj L, Rolain J-M, Olaitan AO (2016) Molecular mechanisms of polymyxin resistance: knowns and unknowns. Int J Antimicrob Agents 48(6):583–591. https://doi.org/10.1016/j.ijantimicag.2016.06.023

Berg G, Köberl M, Rybakova D, Müller H, Grosch R, Smalla K (2017) Plant microbial diversity is suggested as the key to future biocontrol and health trends. FEMS Microbiol Ecol 93(5). https://doi.org/10.1093/femsec/fix050

Carvalho FV, Fonseca Santana L, Diogenes A, da Silva V, Costa SL, Zambotti-Villelae L, Colepicolo P, Ferraz CG, Ribeiro PR (2021) Combination of a multiplatform metabolite profiling approach and chemometrics as a powerful strategy to identify bioactive metabolites in Lepidium meyenii (Peruvian maca). Food Chem 364:130453. https://doi.org/10.1016/j.foodchem.2021.130453

Chauhan R, Ravi J, Datta P, Chen T, Schnappinger D, Bassler KE, Balázsi G, Gennaro ML (2016) Reconstruction and topological characterization of the sigma factor regulatory network of mycobacterium tuberculosis. Nat Commun 7(1):11062. https://doi.org/10.1038/ncomms11062

Chen J, Tang C, Zhang R, Ye S, Zhao Z, Huang Y, Xu X, Lan W, Yang D (2020) Metabolomics analysis to evaluate the antibacterial activity of the essential oil from the leaves of Cinnamomum camphora (Linn.) Presl. J Ethnopharm 253:112652. https://doi.org/10.1016/j.jep.2020.112652

Clish CB (2015) Metabolomics: an emerging but powerful tool for precision medicine. Mol Case Studies 1(1):a000588. https://doi.org/10.1101/mcs.a000588

Costanzo M, Caterino M, Ruoppolo M (2022) Targeted metabolomics. In: Metabolomics perspectives. Elsevier, pp 219–236. https://doi.org/10.1016/B978-0-323-85062-9.00006-4

Cui L, Lu H, Lee YH (2018) Challenges and emergent solutions for LC-MS/MS based untargeted metabolomics in diseases. Mass Spectrom Rev 37(6):772–792. https://doi.org/10.1002/mas.21562

da Cunha BR, Zoio P, Fonseca LP, Calado CRC (2021) Technologies for high-throughput identification of antibiotic mechanism of action. Antibiotics 10(5):1–20. https://doi.org/10.3390/antibiotics10050565

dos Santos FA, Sousa IP, Furtado NAJC, Da Costa FB (2018) Combined OPLS-DA and decision tree as a strategy to identify antimicrobial biomarkers of volatile oils analyzed by gas chromatography–mass spectrometry. Rev Bras 28(6):647–653. https://doi.org/10.1016/j.bjp.2018.08.006

Drapal M, Perez-Fons L, Wheeler PR, Fraser PD (2014) The application of metabolite profiling to Mycobacterium spp.: determination of metabolite changes associated with growth. J Microbiol Methods 106:23–32. https://doi.org/10.1016/j.mimet.2014.07.037

El-Sayed H, Hamada MA, Elhenawy AA et al (2023) Acetylcholine esterase inhibitory effect, antimicrobial, antioxidant, metabolomic profiling, and an in silico study of non-polar extract of the halotolerant marine fungus Penicillium chrysogenum MZ945518. Microorganisms 11:769. https://doi.org/10.3390/microorganisms11030769

Fiehn O (2016) Metabolomics by gas chromatography–mass spectrometry: combined targeted and untargeted profiling. Curr Protoc Mol Biol 114(1):30.4.1. https://doi.org/10.1002/0471142727.mb3004s114

Fudyma JD, Lyon J, Amini Tabrizi R et al (2019) Untargeted metabolomic profiling of Sphagnum fallax reveals novel antimicrobial metabolites. Plant Direct 3:e00179. https://doi.org/10.1002/pld3.179

Gao T, Ye F, Tan Y, Peng M, Yuan F, Liu Z, Zhou D, Yang K, Liu W, Guo R, Zhang T, Zheng L, Zhou R, Tian Y (2022) Metabolomics and proteomics analyses revealed mechanistic insights on the antimicrobial activity of epigallocatechin gallate against Streptococcus suis. Front Cell Infect Microbiol 12:1–14. https://doi.org/10.3389/fcimb.2022.973282

Gorlenko CL, Kiselev HY, Budanova EV, Zamyatnin AA, Ikryannikova LN (2020) Plant secondary metabolites in the battle of drugs and drug-resistant bacteria: new heroes or worse clones of antibiotics? Antibiotics 9(4):170. https://doi.org/10.3390/antibiotics9040170

Haeili M, Javani A, Moradi J, Jafari Z, Feizabadi MM, Babaei E (2017) MgrB alterations mediate Colistin resistance in Klebsiella pneumoniae isolates from Iran. Front Microbiol 8:2470. https://doi.org/10.3389/fmicb.2017.02470

Halouska S, Fenton RJ, Barletta RG, Powers R (2012) Predicting the in vivo mechanism of action for drug leads using NMR metabolomics. ACS Chem Biol 7(1):166–171. https://doi.org/10.1021/cb200348m

Han J, Li Q, Chen Y, Yang Y (2021) Recent metabolomics analysis in tumor metabolism reprogramming. Front Mol Biosci 8:763902. https://doi.org/10.3389/fmolb.2021.763902

Hao J, Astle W, De Iorio M, Ebbels TMD (2012) BATMAN—an R package for the automated quantification of metabolites from nuclear magnetic resonance spectra using a Bayesian model. Bioinformatics 28(15):2088–2090. https://doi.org/10.1093/bioinformatics/bts308

Harlan WR (2016) Clinical trials. In: International encyclopedia of public health, vol 1, 2nd edn. Elsevier. https://doi.org/10.1016/B978-0-12-803678-5.00078-3

Hoerr V, Duggan GE, Zbytnuik L, Poon KKH, Große C, Neugebauer U, Methling K, Löffler B, Vogel HJ (2016) Characterization and prediction of the mechanism of action of antibiotics through NMR metabolomics. BMC Microbiol 16(1):82. https://doi.org/10.1186/s12866-016-0696-5

Kirwan JA, Weber RJM, Broadhurst DI, Viant MR (2014) Direct infusion mass spectrometry metabolomics dataset: a benchmark for data processing and quality control. Sci Data 1(1): 140012. https://doi.org/10.1038/sdata.2014.12

Kluska M, Juszczak M, Żuchowski J et al (2021) Kaempferol and its glycoside derivatives as modulators of etoposide activity in HL-60 cells. Int J Mol Sci 22:3520. https://doi.org/10.3390/ijms22073520

Koen N, van Breda SV, Loots DT (2018) Elucidating the antimicrobial mechanisms of colistin sulfate on mycobacterium tuberculosis using metabolomics. Tuberculosis 111:14–19. https://doi.org/10.1016/j.tube.2018.05.001

Koeva M, Gutu AD, Hebert W, Wager JD, Yonker LM, O'Toole GA, Ausubel FM, Moskowitz SM, Joseph-McCarthy D (2017) An antipersister strategy for treatment of chronic Pseudomonas aeruginosa infections. Antimicrob Agents Chemother 61(12):e00987. https://doi.org/10.1128/AAC.00987-17

Koster RA, Alffenaar JC, Greijdanus B, Vandernagel JEL, Uges DRA (2014) Fast and highly selective LC-MS/MS screening for THC and 16 other abused drugs and metabolites in human hair to monitor patients for drug abuse. Ther Drug Monit 36(2):234–243

Kuhl C, Tautenhahn R, Böttcher C, Larson TR, Neumann S (2012) CAMERA: an integrated strategy for compound spectra extraction and annotation of liquid chromatography/mass spectrometry data sets. Anal Chem 84(1):283–289. https://doi.org/10.1021/ac202450g

Kumar B, Prakash A, Ruhela RK, Medhi B (2014) Potential of metabolomics in preclinical and clinical drug development. Pharm Rep 66:956–963. https://doi.org/10.1016/j.pharep.2014.06.010

Li B, Yin F, Zhao X, Guo Y, Wang W, Wang P, Zhu H, Yin Y, Wang X (2020) Colistin resistance gene mcr-1 mediates cell permeability and resistance to hydrophobic antibiotics. Front Microbiol 10:3015. https://doi.org/10.3389/fmicb.2019.03015

Lobritz MA, Belenky P, Porter CBM, Gutierrez A, Yang JH, Schwarz EG, Dwyer DJ, Khalil AS, Collins JJ (2015) Antibiotic efficacy is linked to bacterial cellular respiration. Proc Natl Acad Sci 112(27):8173–8180. https://doi.org/10.1073/pnas.1509743112

Lopatkin AJ, Stokes JM, Zheng EJ, Yang JH, Takahashi MK, You L, Collins JJ (2019) Bacterial metabolic state more accurately predicts antibiotic lethality than growth rate. Nat Microbiol 4(12):2109–2117. https://doi.org/10.1038/s41564-019-0536-0

Lu Y, Chen C (2017) Metabolomics: bridging chemistry and biology in drug discovery and development. Curr Pharm Rep 3:16–25. https://doi.org/10.1007/s40495-017-0083-4

Ludwig C, Easton JM, Lodi A, Tiziani S, Manzoor SE, Southam AD, Byrne JJ, Bishop LM, He S, Arvanitis TN, Günther UL, Viant MR (2012) Birmingham metabolite library: a publicly accessible database of 1-D 1H and 2-D 1H J-resolved NMR spectra of authentic metabolite standards (BML-NMR). Metabolomics 8(1):8–18. https://doi.org/10.1007/s11306-011-0347-7

Mack SG, Turner RL, Dwyer DJ (2018) Achieving a predictive understanding of antimicrobial stress physiology through systems biology. Trends Microbiol 26(4):296–312. https://doi.org/10.1016/j.tim.2018.02.004

Maree J, Kamatou G, Gibbons S et al (2014) The application of GC–MS combined with chemometrics for the identification of antimicrobial compounds from selected commercial essential oils. Chemom Intell Lab Syst 130:172–181. https://doi.org/10.1016/j.chemolab.2013.11.004

Meylan S, Porter CBM, Yang JH, Belenky P, Gutierrez A, Lobritz MA, Park J, Kim SH, Moskowitz SM, Collins JJ (2017) Carbon sources tune antibiotic susceptibility in Pseudomonas aeruginosa via tricarboxylic acid cycle control. Cell Chem Biol 24(2):195–206. https://doi.org/10.1016/j.chembiol.2016.12.015

Mirsaeidi M, Banoei MM, Winston BW, Schraufnagel DE (2015) Metabolomics: applications and promise in mycobacterial disease. Ann Am Thorac Soc 12(9):1278–1287. https://doi.org/10.1513/AnnalsATS.201505-279PS

Moco S (2022) Studying metabolism by NMR-based metabolomics. Front Mol Biosci 9:882487. https://doi.org/10.3389/fmolb.2022.882487

Nagana Gowda GA, Raftery D (2021) NMR-based metabolomics. Adv Exp Med Biol 1280:19–37. https://doi.org/10.1007/978-3-030-51652-9_2

Olivares Pacheco J, Alvarez-Ortega C, Alcalde Rico M, Martínez JL (2017) Metabolic compensation of fitness costs is a general outcome for antibiotic-resistant Pseudomonas aeruginosa mutants overexpressing efflux pumps. MBio 8(4):e00500. https://doi.org/10.1128/mBio.00500-17

Pamp SJ, Gjermansen M, Johansen HK, Tolker-Nielsen T (2008) Tolerance to the antimicrobial peptide colistin in Pseudomonas aeruginosa biofilms is linked to metabolically active cells, and depends on the pmr and mexAB-oprM genes. Mol Microbiol 68(1):223–240. https://doi.org/10.1111/j.1365-2958.2008.06152.x

Passalacqua KD, Charbonneau M-E, O'Riordan MXD (2016) Bacterial metabolism shapes the host–pathogen interface. Microbiol Spectr 4(3). https://doi.org/10.1128/microbiolspec.VMBF-0027-2015

Peng B, Su Y, Li H, Han Y, Guo C, Tian Y, Peng X (2015) Exogenous alanine and/or glucose plus Kanamycin kills antibiotic-resistant bacteria. Cell Metab 21(2):249–262. https://doi.org/10.1016/j.cmet.2015.01.008

Putri SP, Ikram MMM, Sato A, Dahlan HA, Rahmawati D, Ohto Y, Fukusaki E (2022) Application of gas chromatography-mass spectrometry-based metabolomics in food science and technology. J Biosci Bioeng 133(5):425–435. https://doi.org/10.1016/j.jbiosc.2022.01.011

Ramautar R, Somsen GW, de Jong GJ (2009) CE-MS in metabolomics. Electrophoresis 30(1):276–291. https://doi.org/10.1002/elps.200800512

Ramautar R, Somsen GW, de Jong GJ (2019) CE-MS for metabolomics: developments and applications in the period 2016-2018. Electrophoresis 40(1):165–179. https://doi.org/10.1002/elps.201800323

Rinschen MM, Ivanisevic J, Giera M, Siuzdak G (2019) Identification of bioactive metabolites using activity metabolomics. Nat Rev Mol Cell Biol 20(6):353–367. https://doi.org/10.1038/s41580-019-0108-4

Roberts LD, Souza AL, Gerszten RE, Clish CB (2012) Targeted metabolomics. Curr Protoc Mol Biol 98(1). https://doi.org/10.1002/0471142727.mb3002s98

Schrimpe-Rutledge AC, Codreanu SG, Sherrod SD, McLean JA (2016) Untargeted metabolomics strategies—challenges and emerging directions. J Am Soc Mass Spectrom 27(12):1897–1905. https://doi.org/10.1007/s13361-016-1469-y

Schripsema J (2010) Application of NMR in plant metabolomics: techniques, problems and prospects. Phytochem Anal 21(1):14–21. https://doi.org/10.1002/pca.1185

Sieniawska E, Georgiev MI (2022) Metabolomics: towards acceleration of antibacterial plant-based leads discovery. Phytochem Rev 21(3):765–781. https://doi.org/10.1007/s11101-021-09762-4

Silva MC, Haggarty SJ (2020) Human pluripotent stem cell–derived models and drug screening in CNS precision medicine. Ann N Y Acad Sci 1471:18–56. https://doi.org/10.1111/nyas.14012

Su Y, Peng B, Han Y, Li H, Peng X (2015) Fructose restores susceptibility of multidrug-resistant Edwardsiella tarda to Kanamycin. J Proteome Res 14(3):1612–1620. https://doi.org/10.1021/pr501285f

Tautenhahn R, Patti GJ, Rinehart D, Siuzdak G (2012) XCMS online: a web-based platform to process untargeted metabolomic data. Anal Chem 84(11):5035–5039. https://doi.org/10.1021/ac300698c

Tomaszewski M, White C, Patel P, Masca N, Damani R, Hepworth J, Samani NJ, Gupta P, Madira W, Stanley A, Williams B (2014) High rates of non-adherence to antihypertensive treatment revealed by high-performance liquid chromatography-tandem mass spectrometry (HP LC-MS/MS) urine analysis. Heart 100(11):855–861. https://doi.org/10.1136/heartjnl-2013-305063

Vincent IM, Ehmann DE, Mills SD, Perros M, Barrett P (2016) Untargeted metabolomics to ascertain antibiotic modes of action. Antimicrob Agents Chemother 60:2281. https://doi.org/10.1128/AAC.02109-15

Vuong C, Saenz HL, Götz F, Otto M (2000) Impact of the agr quorum-sensing system on adherence to polystyrene in Staphylococcus aureus. J Infect Dis 182(6):1688–1693. https://doi.org/10.1086/317606

Weber RJM, Lawson TN, Salek RM, Ebbels TMD, Glen RC, Goodacre R, Griffin JL, Haug K, Koulman A, Moreno P, Ralser M, Steinbeck C, Dunn WB, Viant MR (2017) Computational tools and workflows in metabolomics: an international survey highlights the opportunity for harmonisation through galaxy. Metabolomics 13(2):12. https://doi.org/10.1007/s11306-016-1147-x

Wishart DS (2008) Applications of metabolomics in drug discovery and development. Drugs R D 9(5):307–322. https://doi.org/10.2165/00126839-200809050-00002

Wright B (2017) Introduction to clinical trials. In: A comprehensive and practical guide to clinical trials. Elsevier Inc., London. https://doi.org/10.1016/B978-0-12-804729-3.00001-8

Yang JH, Bhargava P, McCloskey D, Mao N, Palsson BO, Collins JJ (2017) Antibiotic-induced changes to the host metabolic environment inhibit drug efficacy and alter immune function. Cell Host Microbe 22(6):757–765.e3. https://doi.org/10.1016/j.chom.2017.10.020

Yang H, Li X, Jin H, Turkez H, Ozturk G, Doganay HL, Zhang C, Nielsen J, Uhlén M, Borén J, Mardinoglu A (2023) Longitudinal metabolomics analysis reveals the acute effect of cysteine and NAC included in the combined metabolic activators. Free Radic Biol Med 204:347–358. https://doi.org/10.1016/j.freeradbiomed.2023.05.013

Zampieri M, Zimmermann M, Claassen M, Sauer U (2017) Nontargeted metabolomics reveals the multilevel response to antibiotic perturbations. Cell Rep 19(6):1214–1228. https://doi.org/10.1016/j.celrep.2017.04.002

Zhou B, Xiao JF, Tuli L, Ressom HW (2012) LC-MS-based metabolomics. Mol BioSyst 8(2):470–481. https://doi.org/10.1039/C1MB05350G

Chapter 9
Phytonanotechnologies for Addressing Antimicrobial Resistance

Rupali Srivastava, Ananya Padmakumar, Paloma Patra, Sushma V. Mudigunda, and Aravind Kumar Rengan

Abstract Antibiotic resistance has become a major issue, resulting in hundreds of thousands of deaths each year. This necessitates the use of more toxic, expensive, and inadequate medications. Antibiotic resistance is caused by a number of factors, including antibiotic overuse in people and cattle. In this circumstance, experts are looking for innovative ways to combat this alarming condition. Several nanoparticle-based antibacterial techniques are now being developed, offering essential information about the structural features that assume critical roles in the development of antimicrobial nanotherapeutic medicines. Plant-based nanoparticles will be less harmful than their synthetic counterparts in this setting, making them intriguing options for avoiding extensive microbiome damage associated with the present techniques. The goal of this chapter is to summarize what is currently known about phytonanotherapeutic solutions for tackling antibiotic resistance and their potential to substitute antibiotics, as well as their adjuvant property in the treatment of multidrug-resistant bacterial illnesses.

Keywords Antimicrobials · Drug resistance · Biogenic nanoparticles · Plant nanotechnology · Nanotherapeutics

Abbreviations

A. baumannii	*Acinetobacter baumannii*
Ag NPs	Silver nanoparticles
Ag	Silver
Amp	Ampicillin
Au NPs	Gold nanoparticles
Au	Gold
B. subtilis	*Bacillus subtilis*

R. Srivastava · A. Padmakumar · P. Patra · S. V. Mudigunda · A. K. Rengan (✉)
Department of Biomedical Engineering, Indian Institute of Technology, Hyderabad, India
e-mail: aravind@bme.iith.ac.in

V. Kumar et al. (eds.), *Medicinal Plants and Antimicrobial Therapies*,
https://doi.org/10.1007/978-981-99-7261-6_9

Ce	Cerium
Co	Cobalt
Cu	Copper
CuO	Copper oxide
E. coli	*Escherichia coli*
E. faecalis	*Enterococcus faecalis*
Fe	Iron
Fe_2O_3	Iron oxide
K. aerogenes	*Klebsiella aerogenes*
MDROs	Multidrug-resistant organisms
MNPs	Metallic nanoparticles
Ni	Nickel
NPs	Nanoparticles
P. aeruginosa	*Pseudomonas aeruginosa*
Pd	Palladium
Pt	Platinum
ROS	Reactive oxygen species
S. aureus	*Staphylococcus aureus*
S. typhimurium	*Salmonella typhimurium*
SCC	Silver–carbene complexes
Se	Selenium
Te	Tellurium
Zn	Zinc
ZnO	Zinc oxide

9.1 Introduction to Phytonanotechnology

The core concept of nanotechnology is the control of individual atoms and molecules at the nanoscale. Nanotechnology is an intriguing field of research in the modern sciences, and it provides a wide range of products, including nanorods, nanoclusters, and nanotubes of varying sizes. These nanoscale objects provide certain functions like shifting between crystalline, amorphous, and solid phases as well as altering their morphological and chemical properties. As a result of their configuration on the nanoscale, nanoparticles (NPs) possess distinct characteristics that differentiate them from their bulk counterparts. The production of nanoparticles was achieved by the use of a variety of proficient methods. Chemical and physical processes are two examples of older technologies that waste a lot of energy and require costly and sometimes dangerous materials. However, there is a pressing need for safer, more environmentally friendly ways of synthesis and manufacture of nanomaterials due to the safety issues associated with these procedures. Utilizing plant sources in nanoparticle synthesis develops links between plant sciences and nanotechnology that are economically advantageous. This intersection of plant biology and nanotechnology is sometimes referred to as "green nanotechnology." Fungi, bacteria, viruses, and plants are all utilized as reducing agents in biological synthesis. The green synthesis approach has garnered the most attention due to its relative simplicity. Increased infections and the development of antibiotic-resistant bacteria pose a major threat to public health. It is critical to develop cutting-edge antibiotic detection methods and new antimicrobial medicines in order to counteract the evolutionary resistance mechanisms by which pathogenic bacteria have evolved. Nanomaterials have been shown to be very efficient at combating bacterial resistance. Plants provide a better alternative due to their ecological friendliness and the range of compounds they contain (Anand et al. 2022; Aromal and Philip 2012; Iravani 2011; Kuppusamy et al. 2016; Mittal et al. 2013). Apart from the secondary metabolites in plant extracts, the proteins isolated from plants and less well-known legumes like peanuts and chickpeas are used for the synthesis of biogenic nanoparticles. Because of the great stability and rapid rate of biosynthesis of plant-based NPs, the biosynthesis of metal NPs sparked interest in the understanding and characterization of the processes of metal ion acquisition and bioreduction by plants.

9.2 Plant-Driven Biosynthesis of Nanoparticles

Plant extract refers to the natural product derived from distinct plant parts such as leaves, stems, flowers, roots, and peel. Recently, plant-mediated nanomaterials (PMNs) of varying sizes and shapes derived from plant tissues and organs of plants have gained greater attention due to their enhanced physicochemical properties, efficiency, and low toxicity (Peralta-Videa et al. 2016; Nguyen et al. 2022; Hano and Abbasi 2022). Owing to the occurrence of several secondary metabolites and coenzymes, plant extracts act as reducing and stabilizing agents (Muddapur et al. 2022) in the bioreduction reaction involved in the synthesis of metallic nanoparticles (MNPs), thus exhibiting the potential to serve as precursors for the green synthesis of nanomaterials (Kuppusamy et al. 2016; Hano and Abbasi 2022). Accordingly, plant-mediated synthesis approaches have given rise to multiple bio-based nanoparticles such as gold (Au), silver (Ag), platinum (Pt), nickel (Ni), cobalt (Co), selenium (Se), palladium (Pd), copper (Cu), iron (Fe), tellurium (Te), and cerium (Ce) (Jain et al. 2021; Rahmanpour et al. 2022; Chen et al. 2022). It has also been reported that biosynthesis can be performed by live plants in addition to plant extracts. Some plants, namely *Brassica juncea*, *Medicago sativa*, and *Helianthus annuus*, have shown promise to accumulate Ag in substantial quantities when it occurs in the substrate. Fe, Co, Cu, and Ni nanoparticles have also been synthesized in living plants (Gardea-Torresdey et al. 2003; Prathna et al. 2011). Fascinatingly, it was pointed out that plant-driven biosynthesis could produce nanoparticles of a wide variety of sizes and shapes compared with other routes of green synthesis (Iravani 2011; Makarov et al. 2014) as mentioned in Fig. 9.1.

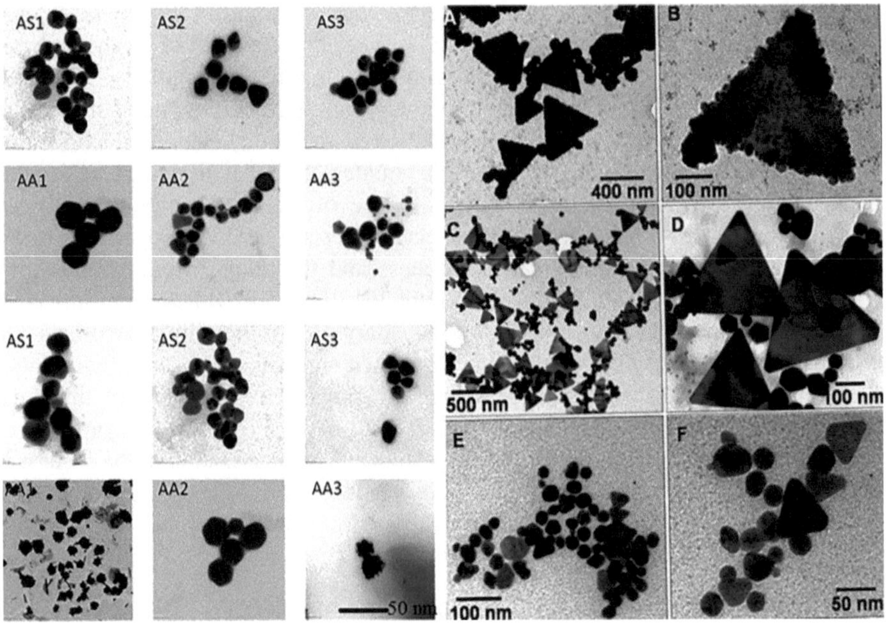

Fig. 9.1 Distinct shapes and sizes of plant-synthesized biogenic nanoparticles. (Reproduced with permission from (Kuppusamy et al. 2016))

9.2.1 Mechanism of Plant-Mediated Preparation of Nanoparticles

The notion of "green chemistry" has recently been expansively explored. Due to rising awareness regarding the toxicity caused by chemicals, the fabrication of green nanoparticles requires (1) an eco-friendly solvent (e.g., water and ethanol), (2) a suitable nontoxic reductant, and (3) a safe stabilization agent. Undeniably, extensive synthetic pathways are fabricated for nanoparticle development, with the most prevalent physical, chemical, and biosynthetic routes. The chemical approach entails the utilization of hazardous and highly toxic chemicals, which further causes a variety of environmental menaces. Contrastingly, the green synthesis is a benign, biocompatible, and environmentally sound alternative to producing nanoparticles. It is mediated by living organisms such as fungi, algae, bacteria, and plants. Plant extract-mediated synthesis confers certain benefits because various phytochemicals in them serve as biocapping and bioreducing agents in the fabrication, thus instigating increased particle stability (Fig. 9.2). Rafique et al. (2019) exhibited that nanoparticle synthesis through a plant-mediated route exhibited enhanced efficiency, ease of handling, and safer and more rapid process compared with biological methods. In addition, Singh et al. (2019b) confirmed that the antioxidant activity of the phytochemicals present in the plant extract drives the development of copper oxide nanoparticles (CuO NPs). Similarly, in another study, it was demonstrated that

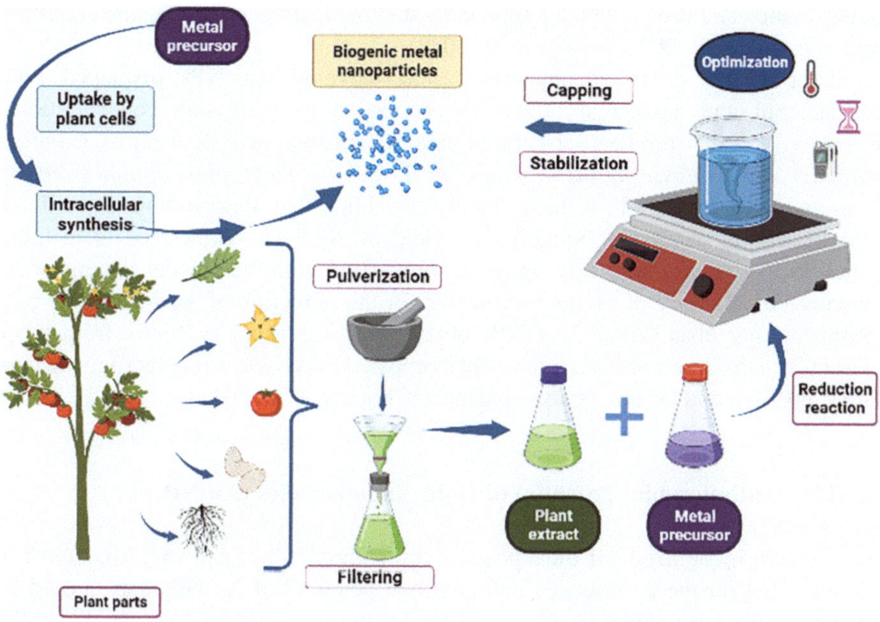

Fig. 9.2 Schematic representation of the mechanism involved in the plant-driven green synthesis of biogenic nanoparticles

amino acids, alkaloids, polysaccharides, and reducing sugar compounds participated in reducing Ag^+ to Ag^0 (Kesharwani et al. 2009). To conclude, these phytochemicals are proficient in altering the metallic precursors into corresponding nanoparticles.

9.2.2 *Different Types of Biogenic Nanoparticles and Their Antimicrobial Activity*

9.2.2.1 Silver Nanoparticles (Ag NPs) and Their Antimicrobial Potential

Since ancient times, silver has been extensively utilized as a potential antibacterial agent, especially in treating wounds (Rafique et al. 2017; Oves et al. 2022; Deivanathan and Prakash 2022). Prolonged exposure and undesirable use of antibiotics have resulted in antibiotic resistance among various bacteria. This is facilitated by the bacterial defense strategies due to mutations in the bacterial genome, which eventually aid in inactivating antibiotics (Murray et al. 2022). Hence, researchers have strived to address the concern of antibiotic resistance in bacteria. Leveraging the antimicrobial potential of biogenic nanoparticles is one among them. Of the several kinds, Ag NPs have demonstrated promise in combating antimicrobial resistance in bacteria (Nakazato et al. 2019; Singh et al. 2019a; Deivanathan and Prakash 2022). Furthermore, Ag NPs are used for the nanocarrier-mediated delivery of antibiotics and drugs, thereby enhancing the therapeutic activity against resistant microbes (Ni et al. 2022; Rabiee et al. 2022).

Zhang et al. (2016) demonstrated that biogenic Ag NPs displayed both antibacterial and antifungal effects in proportion to increasing concentrations. Besides, Ag NPs produced by fungi had heightened activity against *Candida albicans* and *Malassezia furfur* (Roy et al. 2019), while those synthesized by actinomycetes appeared to have methylene blue dye degradation capabilities (Mechouche et al. 2022). Notably, the yield of Ag NPs synthesized from tuber extract of *Curcuma longa* was higher in comparison with the powder, owing to the enhanced availability of agents responsible for the reduction of Ag ions in Ag NPs (Sathishkumar et al. 2010). Ag NPs obtained by the rapid synthesis from tuber extracts of *Dioscorea bulbifera* have been reposted to possess strong activity against Gram-positive and Gram-negative bacteria (Chopade et al. 2012).

9.2.2.2 Antimicrobial Potential of Gold Nanoparticles (Au NPs)

Au NPs are recognized for their potency, biocompatibility, and low toxicity. It is reported that for the pronounced antimicrobial potential of Au NPs they should be coupled with other biomolecules. Gelatin, chitosan, collagen, etc. facilitate the binding of easy tagging of Au NPs with the biomolecules (Rajendran et al. 2018). Significantly, Au NPs show promise as nanocarriers for the delivery of antimicrobial therapeutics, thereby providing a synergistic effect. For instance, as reported by Shittu et al. (2017), Au NPs synthesized from the leaf extract of *Piper guineense* aided the better delivery of lincomycin. Further, vancomycin-coupled Au NPs displayed 50 times improved antibacterial properties against vancomycin-resistant *Escherichia* and *Enterococci* (Gu et al. 2003). Photosensitizer and antibody-conjugated Au NPs revealed strong antibacterial properties via photothermal and photodynamic therapy (Savas et al. 2018; ElZorkany et al. 2019).

Au NPs obtained from the leaf extract of *Cymbopogon citratus* boost the predation efficiency of copepod *Mesocyclops aspericornis* against malaria and dengue mosquitoes at lower doses (Murugan et al. 2015). Recently, Aljabali et al. (2018), Vijayan et al. (2018), and Singh et al. (2018) verified the use of *Indigofera suffruticosa*, *Ziziphus jujuba*, and *Euphrasia officinalis* leaf extracts for the development of potent Au NPs where the phytoconstituents acted as capping and stabilizing agents.

9.2.2.3 Antimicrobial Property of Zinc Oxide Nanoparticles (ZnO NPs)

Owing to its robustness, biocompatibility, and photo-oxidizing and photocatalytic effects, ZnO NPs are now preferred as antimicrobial agents (Sur and Mukhopadhyay 2019). Besides, the high conductivity of zinc oxide (ZnO) allows ultraviolet (UV) light absorption, resulting in a greater interaction of ZnO NPs with bacteria. Many reactive oxygen species (ROS) are desorbed from the exterior due to interaction with UV light. Through the phototoxic effects of ZnO NPs, substantial ROS can be evidenced when treating bacterial broth with them and illuminating it with UV light (Wang 2004). These species facilitate better entry and killing of bacterial cells by ZnO NPs. This is also reinforced by unique physicochemical properties and high surface area-to-volume ratios offered by ZnO NPs. Further, the biocompatibility of ZnO NPs in human cells was established and is still being meticulously investigated. ZnO NPs could be efficient against both Gram-negative and Gram-positive bacteria (Raghupathi et al. 2011). In comparison with other biogenic nanoparticles of magnesium oxide, titanium oxide (TiO_2), cupric oxide, etc., ZnO NPs present enhanced bactericidal activity on *Staphylococcus aureus* (Zhang et al. 2008; Raghupathi et al. 2011). The green synthesis of zinc oxide nanoparticles by the biological method using aqueous stem extract of *Ruta graveolens* appeared to be productive against *Klebsiella aerogenes*, *Pseudomonas aeruginosa*, *Escherichia coli*, and Gram-positive strain including *S. aureus* (Lingaraju et al. 2016).

9.2.2.4 Bactericidal Properties of Iron Oxide Nanoparticles (Fe_2O_3 NPs)

Similar to Ag NPs and Au NPs, nanosized biogenic iron oxide nanoparticles have demonstrated great promise as antimicrobial agents. Biogenic Fe_2O_3 NPs of sizes 21 nm and 32 nm prepared from the leaf extract of *Gardenia jasminoides* and *Lawsonia inermis*, respectively, had a repressive effect against *S. aureus*, *Salmonella enterica*, *Proteus mirabilis*, *and E. coli* (Naseem and Farrukh 2015). *P. aeruginosa* was known to be inhibited by 100-nm-sized semi-crystalline biogenic Fe_2O_3 NPs produced from *Tridax procumbens* extract (Senthil and Ramesh 2012). Alam et al. (2019) reported biogenic Fe_2O_3 NPs of sizes ranging from 56 nm to 350 nm to have disrupted the cell wall of a plant pathogen *Ralstonia solanacearum*. These Fe_2O_3 NPs were synthesized from the leaf extract of *Skimmia laureola*. Recently, in another study by Aisida et al. (2020), Fe_2O_3 NPs synthesized from *Moringa oleifera* displayed strong antibacterial potency against *P. aeruginosa*, *E. coli*, *S. aureus*, *Salmonella typhimurium*, and *Pasteurella multocida*.

9.2.2.5 Other Metal Nanoparticles and Their Antimicrobial Properties

Antimicrobial effect has also been evidenced for biogenic Pd, Se, Ce, and Te nanoparticles, respectively. A methanolic extract of *Moringa oleifera* peel was used to generate Pd nanoparticles with strong activity against *E. coli* and *Staphylococcus aureus* (Surendra et al. 2016). In 2016, Surendra and Roopan (2016) obtained cerium dioxide (CeO_2) nanoparticles (CNPs) with antibacterial properties from the peel extract of *M. oleifera*. Similarly, CNPs synthesized by the bioreduction from *Gloriosa superba* L. leaf showed improved activity toward various Gram-negative and Gram-positive bacteria including *Pseudomonas*, *Klebsiella*, and *Streptococcus* (Arumugam et al. 2015). Recently, palladium nanoparticles (Pd NPs) were obtained harnessing the aqueous fruit extract of *Couroupita guianensis Aubl.* The as-synthesized Pd NPs appeared to have remarkable bactericidal capacity against *Salmonella typhimurium, Vibrio cholerae, E. coli, and Klebsiella pneumonia* (Gnanasekar et al. 2018). TiO_2 nanoparticles from the root, stem, leaves, seeds, and flower extracts of plants including *Myristica fragrans*, *Acorus calamus*, and *Ledebouria revoluta* possess phytocompounds that act as both capping and reducing agents (Aswini et al. 2021; Sagadevan et al. 2021; Ansari et al. 2022; Sunny et al. 2022). These nanoparticles showed great promise when evaluated for antibacterial activity against various Gram-positive and Gram-negative bacteria. Similarly, there are other biogenic nanoparticles synthesized from different parts of the plant having antimicrobial property against different microbes, mentioned in Table 9.1.

9.3 Antimicrobial Mechanism of Action of Nanoparticles

Drug resistance is a type of evolutionary mechanism by which bacteria evade drug killing and continue to multiply within their hosts. This happens when bacteria mutate to take certain drugs. This is because only microorganisms become resistant, not humans or animals. The misuse of antibiotics may increase selection pressure and genetic relatedness, hence developing antibiotic resistance and pathogenicity in organisms by means of numerous adaptations.

Antimicrobial resistance is a global concern and causes hundreds of thousands of deaths each year. Expression of enzymes can modify or degrade antibacterial agents such as aminoglycosides and beta-lactamases, cell wall modifications, ribosomal subunit mutations, 16S ribosomal ribonucleic acid (rRNA) methylation, changes in RNA polymerase, and carbapenamase subunits that confer resistance to various antibiotics, and finally, overexpression of drug-specific efflux pumps is only part of the mechanism that microbes use to suppress the effects of antibiotics. Regardless of the fact that antimicrobial therapy of bacteria can result in the creation of biofilms, other more complicated traits such as quorum sensing and biofilms are not correlated to antimicrobial agents. Antimicrobial resistance can be acquired naturally, by random mutation (de novo) or through horizontal transmission from donor bacteria, viral vectors, or plasmid deoxyribonucleic acid (DNA) to recipient microbes. Nanoparticles have been extensively explored as a possible therapy due to the absence of new antibacterial agents and the inadequacy of regularly used antibiotics to treat persistent microbial infections. It is widely recognized that antimicrobial nanomaterials have different ways of combating antibiotic resistance. Nanoparticles can work in different ways to combat drug resistance. For microbes to be able to avoid nanoparticle processes, each cell must go through many genetic changes at the same time. However, it is rare for more than one biological gene mutation to happen at the same time in the same cell (Baptista et al. 2018; Sultan et al. 2018; Sharma et al. 2016; Pelgrift and Friedman 2013; Smerkova et al. 2020).

Furthermore, antimicrobial nanomaterials have been employed to fight methicillin-resistant *Staphylococcus aureus* (MRSA) and other multidrug-resistant

Table 9.1 Various plant-synthesized metallic nanoparticles and their antimicrobial activities

Sl. No.	Plants	Plant part used	Biogenic nanoparticle	Size (nm) and shapes	Metabolite facilitating bioreduction	Microbe affected	References
1	*Phyllanthus reticulatus*	Leaf	Cupric oxide	4–14, spherical	Saponins, tannin	*Escherichia coli*, *Staphylococcus aureus*, and *Klebsiella pneumoniae*	Potbhare et al. (2019)
2	*Hyptis suaveolens*	Leaf	Cu	5–8.7, spherical	Terpenoids	*Bacillus cereus* and *Pseudomonas fluorescens*	Shubhashree et al. (2022)
3	*Digera muricata*	Leaf	Au	40 ± 5, spherical	–	*Vibrio cholera*, *Staphylococcus pyrogen*, *Klebsiella*, *Citrobacter*, and *Enterobacter*	Shah et al. (2020)
4	*Crescentia cujete* L.	Leaf	Au	32.89, cubic crystalline	Flavonoids	*E. coli*, *P. aeruginosa V. cholerae*, *S. typhimurium*, *Shigella flexneri* f., and *Bacillus subtilis*	Seetharaman et al. (2017)
5	*Coleus forskohlii*	Root	Au	25–40, triangular	Forskolin	*E. coli*, *P. aeruginosa*, and *S. aureus*	Naraginti and Sivakumar (2014)
6	*Galaxaura elongata*	Whole algae	Au	3–77, rod, triangular, hexagonal	Acetogenins, phlorotannins	*S. aureus*, methicillin-resistant *S. aureus* (MRSA), *E. coli*, and *P. aeruginosa*	Abdel-Raouf et al. (2017)
7	*Pistacia integerrima*	Gall	Au	20–200, granular	Terpenoids	*K. pneumonia*, *Bacillus subtilis*, *S. aureus*, *Alternaria solani*, *Aspergillus niger*, and *Aspergillus flavus*	Islam et al. (2019)
8	*Fagonia indica*	Callus	Ag	40 nm	Phenolic compounds	*E. coli*, *Citrobacter amalonaticus*, *Shigella sonnei*, *S. typhimurium*	Adil et al. (2019)
9	*Brillantaisia owariensis*, *Crossopteryx febrifuga*, *Senna siamea*	Leaf	Ag	45 to 110 nm, spherical	Alkaloids	*S. aureus*, *E. coli*, and *P. aeruginosa*	Kambale et al. (2020)
10	*Ocimum tenuiflorum*	Leaf	Ni	12 and 36 nm, spherical	Terpenoids	*E. coli*, *K. pneumoniae*, *S. typhimurium*, *B. subtilis*, *Staphylococcus epidermidis*, *C. albicans*, *Candida tropicalis*,	Jeyaraj Pandian et al. (2016)

(continued)

Table 9.1 (continued)

Sl. No.	Plants	Plant part used	Biogenic nanoparticle	Size (nm) and shapes	Metabolite facilitating bioreduction	Microbe affected	References
						Aspergillus fumigatus, *Aspergillus clavatus*, and *A. niger*	
11	*Ruta graveolens*	Stem	Zinc oxide	28 nm, spherical	–	*K. aerogenes*, *P. aeruginosa*, *E. coli*, and *Staphylococcus aureus*	Lingaraju et al. (2016)
12	*Fagonia indica*	Callus	Ag	40 nm	Phenolic compounds	*E. coli*, *Citrobacter amalonaticus*, *S. sonnei*, *S. typhimurium*	Adil et al. (2019)
13	*Brillantaisia owariensis*, *Crossopteryx febrifuga*, *Senna siamea*	Leaf	Ag	45 to 110 nm, spherical	Alkaloids	*S. aureus*, *E. coli*, and *P. aeruginosa*	Kambale et al. (2020)
14	*Fagonia indica*	Callus	Ag	40 nm	Phenolic compounds	*E. coli*, *Citrobacter amalonaticus*, *S. sonnei*, *S. typhimurium*	Adil et al. (2019)
15	*Ocimum tenuiflorum*	Leaf	Ni	12 and 36 nm, spherical	Terpenoids	*E. coli*, *K. pneumoniae*, *S. typhimurium*, *B. subtilis*, *S. epidermidis*, *C. albicans*, *C. tropicalis*, *Aspergillus fumigatus*, *A. clavatus*, and *A. niger*	Jeyaraj Pandian et al. (2016)
16	*Ruta graveolens*	Stem	Zinc oxide	28 nm, spherical	–	*K. aerogenes*, *P. aeruginosa*, *E. coli*, *Staphylococcus aureus*	Lingaraju et al. (2016)

bacteria. We are hopeful about the prophylactic application of nanoelements given their efficacy as disinfectants. It has been demonstrated that their activity inhibits the growth of indoor mold, diseases, and fungus. Nanostructured materials have the potential to carry antimicrobial agents, assist in the delivery of novel medications, and even possess long-lasting antimicrobial activity. In addition, due to their antibacterial capabilities, nanoparticles (such as metallic, organic, and carbon nanotubes) can avoid bacterial drug resistance mechanisms and hinder biofilm formation and other critical processes. In vitro antibacterial studies show that metal nanoparticles are effective in blocking multiple microbial species. Through various processes, metallic nanoparticles have the ability to prevent or overcome the development of biofilms and multidrug resistance. The material used to manufacture the nanoparticles and their particle size are two important factors that determine how effective metal nanoparticles are against microorganisms. To permanently colonize surfaces and colonize with peptidoglycan coats, bacteria release substances such as proteins, leading to biofilm formation. Biofilm formation renders bacteria resistant to and unresponsive to antibiotics (Zhao et al. 2017; Riga et al. 2017; Kaur et al. 2021; Singh et al. 2014, 2018, 2019; Ogar et al. 2015; Baptista et al. 2018; Rex et al. 2019; Watnick and Kolter 2000; Høiby et al. 2010).

The nanoparticles are capable of acting as antibacterial agents, including direct interaction with the cell walls of bacteria, the prevention of the formation of biofilm, the induction of innate and adaptive immune responses from the host, the generation of reactive oxygen species (ROS), and the effects that occur inside the cells (Baptista et al. 2018). For example, one of the mechanisms of action for nanoparticles is that they cause the production of reactive oxygen species (ROS). These radicals are responsible for oxidative stress and have an effect on membrane lipids. They do this by modifying the structure of cellular proteins and mitochondrial DNA, which may sometimes result in lipid mutagenesis or oxidation. This may disturb the inflammatory process, leading to an increase in apoptosis and disease. Although the nanoparticles have some properties that are toxic to human cells, like cell membrane damage and increased oxidative stress, they do not lead to the development of resistance mechanisms as they do not have the same mechanism of action as conventional antibiotics. Therefore, nanoparticles appear to be effective against pathogenic bacteria and may be very effective against multidrug-resistant organism (MDRO) (Bahadar et al. 2016; Beyth et al. 2015; Trifan et al. 2020; Baptista et al. 2018; Kaur et al. 2021).

The physicochemical characteristics of nanoparticles have a role in determining their antibacterial activity. However, it is essential to keep in mind that other variables, such as environmental conditions and bacterial strains, play key roles in determining the degree to which microorganisms interact with nanoparticles. Previous research has shown that the temperature and pH of the test medium have an effect on the in vitro dissolution of nanoparticles as well as their antibacterial activity. By lowering the pH, the nanoparticles become more soluble, which in turn boosts the effectiveness of their antibacterial activity. As a result, the solubility of nanoparticles is enhanced when they are present in an acidic environment. During the testing of silver nanoparticles (Ag NPs) and gold nanoparticles (Au NPs), poisonous and highly diffusible ions were released into the medium, which led researchers to conclude that a diffusion technique would be the most appropriate method to use with Ag NPs. Instead, there is no release of gold ions into the media, and the antibacterial activity of the Au NPs is brought about by their physical interactions with bacteria. Nanoparticles and other nanoscale compounds have the ability to interact with bacterial cells, alter the permeability of cell membranes, and disrupt molecular signaling pathways. They may also pass through the membranes of bacteria (Lee et al. 2019; Westmeier et al. 2018; Smerkova et al. 2020; Peretyazhko et al. 2014; Saliani et al. 2015). NPs, especially Ag NPs, can generally easily enter nontransformed cells either by endocytosis or by macropinocytosis. These hypotheses indicate that nanoparticles have a dual effect on bacterial membranes. First, nanoparticles disrupt membrane potential and integrity and then trigger the formation of ROS. In addition, currently accepted methods include the generation of intracellular effects that restrict RNA and protein synthesis, biofilm development, and the induction of host immune responses. Tellurium nanoparticles (Te NPs) have a stronger ability to reduce *E. coli* plaques than *Staphylococcus aureus* and reduce biomass by more than 90%. Alternative approaches are also being considered, such as the combination of nanoparticles and herbal antimicrobials to address toxicity issues (Anand et al. 2022; Gómez-Gómez et al. 2020; Chandra et al. 2017; Ruddaraju et al. 2020). Different modes of action of biogenic nanoparticles against microbes are represented in Fig. 9.3.

Gram-positive and Gram-negative bacteria have distinct membrane proteins and architectures, as well as diverse adsorption methods for nanoparticles. There may be a weak correlation between changes in bacterial outer membrane permeability and the overexpression of efflux transporters in bacteria that are resistant to nanoparticles. The disruptions caused by nanoparticles result in aberrant cellular respiration, improper transport across membranes, and an energy deficit that may potentially cause these cells to lyse. It is believed that smaller nanoparticles are more effective owing to the increased surface area of the nanoparticles themselves. However, the precise mechanism of action is not yet understood (Lesniak et al. 2013; Finley et al. 2015; Zhao and Jiang 2013; Anand et al. 2022; Panáček et al. 2006). The two negative effects of increasing surface area that often inhibit antibacterial action are the development of aggregates more quickly and the production of smaller molecules. Before making the move to using nanoparticles, it is important to do in-depth research on a number of different concentrations to determine which ones have biological effects that are only favorable to certain microbes, fungi, or viruses. In addition, the problem of bacterial resilience to the nanoparticles that are being employed has to be looked into. In addition, there are assertions that nanoparticles have effects on systems other than the respiratory chain alone. Several studies have pointed out that silver has a strong affinity for sulfur and nitrogen, which may influence the structural makeup of proteins. In addition, since silver is a photocatalyst, it may cause cytotoxicity by inducing reactive oxygen species (ROS), inducing apoptosis, and damaging cellular components. These effects can be traced back to the photocatalytic capabilities of silver (Bartłomiejczyk

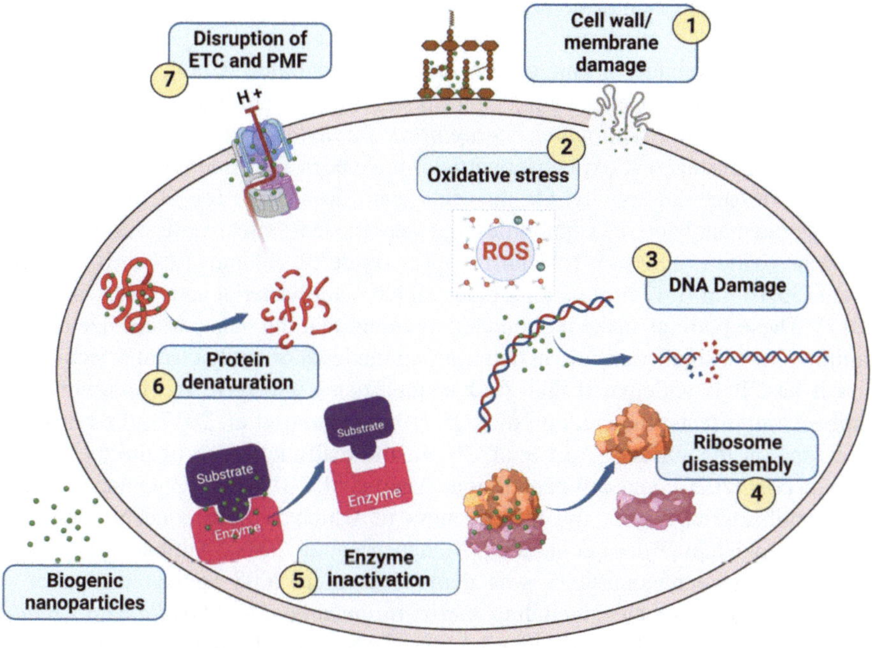

Fig. 9.3 Various modes of action of biogenic nanoparticles to combat microbes. *ETC* electron transport chain, *PMF* proton motive force

et al. 2013; Drake and Hazelwood 2005; Tolaymat et al. 2010; Choi et al. 2008; Carlson et al. 2008; Ninganagouda et al. 2014; Piao et al. 2011; Greulich et al. 2011; Luther et al. 2011). Silver is the most significant inorganic nanoparticle since it has potent antibacterial, antifungal, antiviral, as well as anti-inflammatory activity. A review of the literature revealed the following approaches to characterization of the antibacterial potency of silver nanoparticles. The bacterial outer membrane is denatured, the bacterial cell membrane is fragmented, and metabolic activity is disrupted by the interaction of Ag NPs with disulfide or sulfhydryl groups of enzymes, resulting in cell death (Zinjarde 2012; Lok et al. 2006; Iavicoli et al. 2013; Yun et al. 2013; Egger et al. 2009). The truncated triangular nanoparticles are inherently more reactive due to the enhanced antimicrobial activity on their surface (Tak et al. 2015). Owing to their nontoxicity, special functionalization ability, multivalent effects, and photothermal activity, the fabrication of Au nanoparticles is very useful for the development of effective antibacterial agents (Lima et al. 2013; Tiwari et al. 2011; Zhou et al. 2012). However, the antibacterial properties of gold nanoparticles are independent of the activity, leading to the formation of reactive oxygen species. Researchers attempted to study the antibacterial ability of Au nanoparticles by attaching the nanoparticles to bacterial membranes and tuning the membrane potential to reduce adenosine triphosphate (ATP) levels. Furthermore, this compound prevented transfer ribonucleic acid (tRNA) from attaching to ribosomes (Cui et al. 2012).

Nanoparticles made of zinc oxide (ZnO), copper oxide (CuO), and iron oxide (Fe_2O_3) are effective in eliminating Gram-positive and Gram-negative bacteria, such as *Escherichia coli*, *Pseudomonas aeruginosa*, *Staphylococcus aureus*, and *Bacillus subtilis*. As a result, zinc oxide nanoparticles have been shown to possess the highest level of antibacterial activity. On the other hand, iron oxide (Fe_2O_3) nanoparticles demonstrated antibacterial capabilities that were the least effective. It was discovered that zinc oxide (19.89 nm–1.43 nm), copper oxide (29.11 nm–1.61 nm), and iron oxide (35.16 nm–1.47 nm) nanoparticles all have antibacterial activity (Azam et al. 2012). These findings make it abundantly evident that the nanoparticle size of each sample plays a significant role in determining the level of antibacterial effectiveness that it has. It is anticipated that ZnO nanoparticles will have the following four antibacterial effects: the creation of ROS (Abdel-Raouf et al. 2017); the release of zinc ions on the surface (Adil et al. 2019); the malfunctioning of the membrane; (Aisida et al. 2020); and cell penetration (Alam et al. 2019). The ZnO nanoparticles have antibacterial action, the effectiveness of which is both concentration- and surface-dependent (Buzea et al. 2007; Mahapatra et al. 2008). Antibacterial activity of copper oxide nanoparticles was demonstrated in 2008 against a number of different bacterial species, including *Klebsiella pneumoniae*, *Shigella*, and *Pseudomonas aeruginosa*. They discovered that copper oxide nanoparticles had an antibacterial activity that corresponded to the microorganisms that they tested them against. Copper has the ability to interact with amine and carboxyl groups that are found in microbes like *Bacillus subtilis*. Furthermore, large quantities of Cu^{2+} ions have the potential to create ROS (Singh et al. 2018). It was previously believed that nanoparticles might penetrate the membranes of bacterial cells, causing damage to essential enzymes in the organism and ultimately leading to the death of the cells.

For instance, compared with chemically manufactured nanoparticles and commercial nanoparticles, the antimicrobial activity of nanoparticles made using environmentally friendly methods is much higher. This is because of the medicinal qualities of the plants that are utilized in the creation of nanoparticles, such as *Ocimum sanctum* (also known as tulsi) and *Azadirachta indica* (also known as neem) (Ramteke et al. 2013; Verma and Mehata 2016). For instance, as compared to traditional silver nanoparticles, silver nanoparticles created using environmentally friendly methods demonstrated significant and efficient clearance zones against a wide variety of bacterial strains (Velmurugan et al. 2017). However, bimetallic Au–Pt NPs show antibacterial actions against multidrug-resistant *Escherichia coli* by decreasing the potential of the bacterial membrane and raising the quantity of adenosine triphosphate (ATP) (Baptista et al. 2018). In addition to this, research on the antibacterial capabilities of reduced iron and iron oxide nanoparticles has also been carried out. These nanoparticles cause harm to bacterial cells by rupturing the membrane of the bacterial cell and producing oxidative stress inside the cell. Ag, Au, and ZnO NPs, together with antibiotics, have been shown to be effective against *S. aureus*, *Enterococcus faecalis*, *Escherichia coli*, *Acinetobacter baumannii*, and

Pseudomonas aeruginosa. These microorganisms do this by piercing the bacterial cell membrane and interrupting essential molecular processes. There is now reason to suppose that bacteria are less prone than Ag NPs to evolving resistance to antibiotics. This is supported by the fact that there is evidence to support this hypothesis (Hemeg 2017; Baranwal et al. 2018; Chernousova and Epple 2013; Leid et al. 2012).

When nanoparticles interact with bacteria and enter bacteria, the efficiency of the several modes of action identified for nanoparticles relies on a wide variety of factors, mainly particle size, shape, and surface coating. In this respect, a number of instances that were previously reported in the scientific literature have been utilized to highlight the many antimicrobial modes of action that may be demonstrated by nanoparticles. When creating nanoparticles, monodispersity is an additional significant characteristic to take into consideration. Altering the circumstances under which the reaction occurs also allows for the control of this parameter. For instance, chemicals like stabilizers and reducing agents can be used as part of the various synthetic processes to increase or decrease their reactivity. Using potato tuber extracts, scientists have been able to produce reduced silver nanoparticles in the shapes of spherical, triangular, and polygonal (Ghosh et al. 2012). It is suggested that effective control of nanoparticles' form, size, and monodispersity be examined in order to evaluate the effects of the reaction conditions now undergoing synthesis optimization. The use of organisms with a high production capacity and regulated reaction circumstances has the potential to speed up and improve the reliability of the manufacturing of well-characterized nanoparticles in comparison with chemical and physical approaches. Many industries, such as those involved in food, health, cosmetics, and pharmaceuticals, might benefit from this environmentally beneficial technique (Iravani et al. 2014). While the sensitivity of *B. subtilis* to antimicrobial nanoparticles is highly size-dependent, the sensitivity of *E. coli* to antimicrobial nanoparticles was rather constant over a variety of particle sizes, suggesting that the efficiency of Au NPs was comparable across this size spectrum. These results further demonstrate that NP size is crucial (Hayden et al. 2012). The form of the nanoparticles has also been exemplified by the finding that spherical Ag NPs are more effective at improving *Klebsiella pneumoniae's* sensitivity to antibiotics than rod-like Ag NPs (Acharya et al. 2018). When it comes to the coating, the compounds secreted by the bacteria serve as organic shields and stabilizers for the Ag NPs. These molecules stop aggregation and ensure the Ag NPs' long-term stability while also contributing to their anticancer and antioxidant qualities. Biofunctionalization, which can also refer to these ultra-thin coatings, is a powerful way to boost NPs' efficacy. Coating biofunctionalized ZrO_2 NPs with glutamate, as suggested by Khan et al. 2020, can control the NPs' size, dispersion, and yield and may also have antibacterial properties (Singh et al. 2015). Highly efficient Ag NP-coated gloves were coated ex situ with *Corymbia citriodora* (*Eucalyptus citriodora*), a member of the *Myrtaceae* family, to further guard against the spread of nosocomial disease (Paosen et al. 2021). It is possible that his work on synthesizing Ag NPs from ampicillin (Amp-Ag NPs) demonstrates how biofunctionalization may be used to create molecules with combined antibacterial and silver properties. Interventions that spark the creation of new chemicals (Khatoon et al. 2019) results from the study by Khatoon et al. (2019) corroborate the lack of resistance in the tested bacterial strains after 15 cycles of exposure to the Amp-Ag NPs. Evidence suggests that this material is much more efficient than chemical NPs and antibiotics in killing off *Escherichia coli* and *Staphylococcus aureus*. Numerous studies have shown clear results, lending confidence to the vast potential of NP applications. Nano-Ag and ZnO and their effects on mesophilic and halophilic bacteria in cell cultures were emphasized, as were the various instances of antibacterial processes happening in nanoparticles (Sinha et al. 2011). This study implies that the effects may differ depending on the kind of NP and the bacterium, since nanotoxicity was seen to be more prominent against Gram-negative bacteria. It was found that Ag NPs are harmful to *E. coli* by studying proton leak events across cell membranes (Sondi and Salopek-Sondi 2004; Sharma et al. 2009). Consistent with what Sinha et al. (2011) found, the charge carried by the cell membrane is crucial to the efficacy of metal nanoparticles. Morones et al. (2005) and Feng et al. (2013a, b) found that charged Ag and ZnO NPs interact electrostatically with negatively charged cell surfaces, altering membrane permeability and therefore affecting cytotoxicity.

In contrast to Gram-positive cells, which have a thicker peptidoglycan layer that is more immune to NP penetration, Gram-negative cells have a cell membrane that is rich in lipopolysaccharide. Gram-negative cells thus possess a stronger negative charge than Gram-positive cells. Bacterial cell membranes are the source of nanotoxicity. This interaction is significantly influenced by the structure (Sanpui et al. 2008). It has been hypothesized that the alteration in the composition of the cell wall has a substantial role in the resistance of Gram-positive bacteria to Ag NPs. Instead of peptidoglycan, Gram-negative halophilic bacteria have a thinner exoskeleton made of charged cardiolipin. As a result, they are less resistant to metallic NPs than Gram-positive bacteria, which have better resistance. Environments with seawater include halophilic microorganisms (Ventosa et al. 1998).

In their paper, Sinha et al. (2011) presented the findings of an in-depth research investigation that focused on the electrostatically mediated sensitivity of nanoparticles. We were able to demonstrate that zinc oxide (ZnO) was toxic to both the Gram-positive EMB4 strain and the Gram-negative *Marinobacter* species. This had a detrimental impact on the growth limitation that was produced by the buildup of nanoparticles in the cytoplasm. Ag NPs, on the other hand, did not have any effect on EMB4 events, which suggests that they may not penetrate bacteria. In point of fact, the electrostatic confinement of NPs may also be mediated by interaction with proteins, as was shown in the bacterium *Staphylococcus aureus* and the fungus *Candida albicans*. Platinum nanoparticles (Pt NPs) have an effect on cells by causing disruptions in the cytoplasmic and plasma membranes, but the behavior and accumulation of Au NP molecules are distinct from those of Ag NPs. According to Chwalibog et al. (2010), even though they interacted with the bacteria without actually coming into contact with them, they were still able to destroy the fungal cells. In addition to metallic nanoparticles, nonactive organic carbon (NOC)-based nanoparticles synthesized by commonly used combustion methods are also useful for antimicrobial activity against resistant bacterial strains. Organic carbon-based nanoparticles induce lipid peroxidation and increase membrane permeability through the formation of altered fullerenes. Against this background, four silver–carbene complexes (SCCs) at low doses (0.5 mg/L–90 mg/L) are potent against MDRO-resistant strains such as *Acinetobacter baumannii*, *Yersinia pestis*, and *Bacillus anthracis*. Application of 10 mg/L of Ag NPs triggered cell death through the formation of reactive oxygen species (ROS), in addition to the broader antibacterial activities of NPs discussed above. These findings agree with those of research conducted on *E. coli*, which was published by Zhou et al. (2012). Therefore, studies on the useful features of nanoparticles are always developing, even if more information is required on the mechanisms underlying nanoparticles' antibacterial activity (Rusciano et al. 2009; Zhang et al. 2013; Leid et al. 2012; Aruguete et al. 2013).

Apart from metallic and organic nanoparticles, chitosan-containing nanoparticles are synthesized using *Pelargonium graveolens* leaf extract (Naggae et al. 2022). Due to its positive charge, chitosan binds with DNA in bacterial and fungal cells, thereby inhibiting transcription of mRNA, resulting in protein translation, and it also decreases the activities of metalloproteins (Singh et al. 2018).

9.4 Various Green Synthetic Methods for the Preparation of Antimicrobial Phytonanoparticles

Infectious diseases caused by pathogenic microorganisms are a serious worldwide health concern, particularly in the underdeveloped nations due to a lack of understanding regarding the usage of antibiotic and preventive vaccination (Sirelkhatim et al. 2015). This has led to the emergence of multidrug-resistant (MR) strains. To treat infectious diseases, there seems to be an urgent need to develop drugs that can combat multiple harmful bacteria, especially MR variants. In this regard, the researchers have been designing nanoparticles (NPs) that are hazardous against pathogenic bacterial strains (Wigginton et al. 2007; Sharma et al. 2020). NPs are the most basic, minuscule, and distinguishable particles wherein atoms are linked together to create three-dimensional structures in the range varying from 1 to 100 nm (Daraee et al. 2016; Moreno-Vega et al. 2012). A variety of materials have been employed for the fabrication of nanoparticles, where the morphology, size range, and chemical properties can influence their effects (Ali et al. 2016). Physical, chemical, and biological approaches are utilized to manufacture NPs. Two primary procedures that comprise physical methods are evaporation–condensation and laser ablation. In chemical procedures, nanoparticles are generated primarily through the chemical reduction of metals using inorganic and organic reducing agents. Biological approaches employ living entities ranging from bacteria, fungal cells, plants, algae, and even viral particles to synthesize NPs (Mittal et al. 2013). It is argued that chemical procedures are extremely costly and environmentally unfriendly.

Therefore, it becomes necessary to design eco-friendly and economically reliable strategies (Kumar and Yadav 2009). Green synthesis, wherein biocompatible living organisms are employed to synthesize nanoparticles, represents one of the approaches (Roy et al. 2013). This unique bio-based method doesn't require elevated temperature, pressure, or energy conditions.

The techniques utilized for the green synthesis of NPs rely mostly on hydrophobic organic solvents and capping reagents. The essential requirements for the green synthesis of NPs are selection of an eco-friendly reducing agent, a solvent system, and nonhazardous stabilizing agents (Raveendran et al. 2003; Dakal et al. 2016). Some examples of plant-mediated green synthesized NPs include gold, magnesium, silver, titanium, zinc, copper, and alginate. Among these, silver NPs and nanocomposites have established themselves as one of the most efficient antibacterial agents against a variety of microorganisms (Logeswari et al. 2015). It has recently been extensively documented that several conventional drugs are ineffective against multidrug-resistant strains of bacterial pathogens comprising the genera *Pseudomonas*, *Staphylococcus*, *Klebsiella*, *Enterococcus*, and *Enterobacter* among others, contributing to rising cases of mortality (Ismail et al. 2018; Kuo et al. 2018; Ting et al. 2018; Tsao et al. 2018). Several of these multidrug-resistant pathogens have been demonstrated to be adversely affected by NPs containing appropriate antimicrobial agents (Natan and Banin 2017; Chen et al. 2014). These NPs also interfere with the biofilm forming capability of bacteria besides having a bactericidal effect (Almaaytah et al. 2017). Antibiofilm function and antimicrobial activity of these NPs clearly propose a novel application in the treatment of multidrug-resistant pathogenic diseases. Due to their significance, we shall focus primarily on plant-based green synthetic approaches of NP synthesis in the next sections.

9.4.1 Sonochemical/Ultrasonication Method

Ultrasonication is an effective, an easy, a convenient, and a completely safe method for producing NPs. By triggering new chemical reactions, this technique yields unanticipated chemical species (Cravotto and Cintas 2006; Ziarati et al. 2013). Initially, nickel molybdate NPs were synthesized using this method. The advantage of high-intensity ultrasonic energy is that it generates a substantial output of innovative materials at a reduced temperature and pressure, as well as at a shorter reaction time, compared with traditional procedures (Kianpour et al. 2013; Xu et al. 2013; Xu and Sun 2013). Figure 9.4 depicts the high-intensity ultrasonic horn typically employed in laboratories (Sharma et al. 2021). This technique was also employed to produce silver nanoparticles (Ag NPs) and calcium carbonate ($CaCO_3$) NPs, where it was discovered that the application of ultrasonic irradiation in batch-carbonation reactions induced swift nucleation, a substantial improvement in calcium ion supersaturation, and an improvement in solute transport (Darroudi et al. 2012; He et al. 2005). In a study reported by Faried et al. (2016), preparation of silver

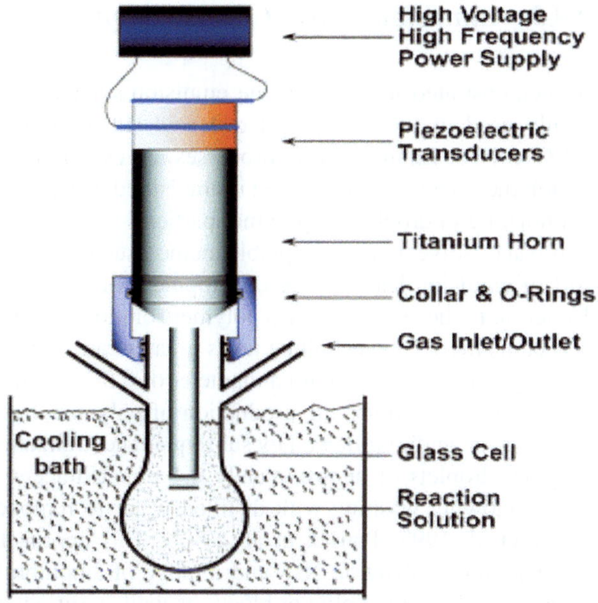

Fig. 9.4 A typical laboratory ultrasonic horn. (Reprinted with permission from Sharma et al. (2021))

nanoparticles (Ag NPs) was carried out under the influence of ultrasound as the reducing agent and *Kappaphycus alvarezii* (*K. alvarezii*) as the natural bio-media. The *K. alvarezii*/Ag NPs exhibited antimicrobial effects against *S. aureus*, *E. coli*, *Salmonella choleraesuis*, and *Bacillus subtilis* (Faried et al. 2016). Other documented studies of Ag NPs synthesized via this technique demonstrated bactericidal effects against a variety of microbial species (Jansirani et al. 2016; Bagherzade et al. 2017). The nanoparticles of silver and iron oxide were also generated from fenugreek seed extract, where the latter served as a stabilizer, a capping agent, and a reductant. NPs prepared by ultrasonic technique possessed superior antioxidant and antibacterial activities and high stability over those produced by magnetic stirring (Deshmukh et al. 2019). In another study carried out by Nouri et al. (2020), Ag NPs were fabricated using Mentha aquatica leaf extract under ultrasonic stimulation. Monodisperse and smaller sized NPs were obtained with ultrasonic assisted synthesis when compared to the hydrothermal method. The smaller size also enabled the Ag NPs to show enhanced antimicrobial efficacy (Nouri et al. 2020). Another very recent study carried out by Khatamifar and Fatemi (2021) utilized the aqueous galls extract of *Quercus infectoria* as a reductant and a capping agent coupled with ultrasonic irradiation to prepare pure copper oxide NPs. These NPs exhibited antimicrobial effects against both Gram-positive and Gram-negative bacterial species (Khatamifar and Fatemi 2021). Thus, the green synthesis method assisted by ultrasound proves to be an effective combination against pathogenic microbes.

9.4.2 Emulsion-Solvent Evaporation Process

As demonstrated in Fig. 9.5, the emulsion-solvent evaporation technique is commonly used to produce poly-lactic acid (PLA) and poly(lactic-co-glycolic acid) (PLGA) NPs. Emulsification processes are usually of three types: First, a solution of polymer in an organic solvent is emulsified using an oil-in-water (o/w) emulsifier, culminating in droplets of polymer particles scattered in the aqueous phase. For the removal of solvent, any acceptable method such as stripping may be employed. This sort of emulsion leads to the development of droplets with a diameter of at least 1 micron. In the second form, a polymer solution is emulsified with water without an o/w emulsifier. This technique takes greater care and control than the first type and yields polymer droplets with a diameter of 0.8–1.0 μm or greater. The third form of emulsification involves the production of polymers having functional groups (usually basic or acidic) followed by subsequent neutralization with water. This produces polymer droplets of approximately 0.1 μm. Ultracentrifugation is done to collect NPs, following washing with distilled water to remove any free additive residues (Song et al. 1997; Patrick et al. 1997). This approach is widely utilized for the fabrication of polymeric NPs and a range of amphiphilic block copolymers incorporating hydrophilic polymer blocks as well as surfactant micelles with a core-shell structure (Yabu et al. 2005). In a work carried out by El-Sherbiny et al. (2016), poly (ε-caprolactone)/curcumin/grape leaf extract–silver hybrid NPs were synthesized using emulsion-solvent evaporation technique. The synthesis could be performed under mild aqueous conditions, and the resultant NPs showed good antimicrobial activity. Likewise, antimicrobial activities of hesperidin-loaded PLGA NPs (El-Sherbiny et al. 2016) and chitosan-coated PLGA NPs loaded with *Peganum*

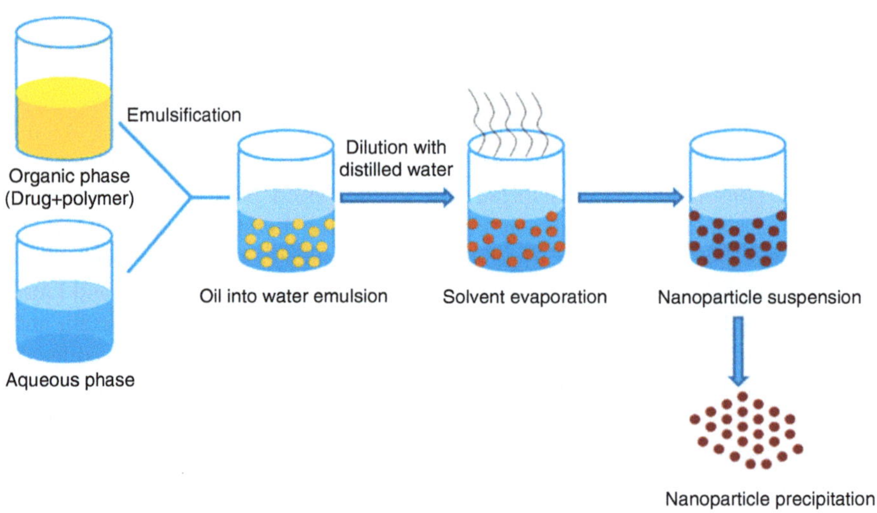

Fig. 9.5 Schematic depicting the steps involved in emulsion-solvent evaporation method. (Reprinted with permission from Sharma et al. (2021))

harmala alkaloids (Balakrishnan et al. 2021) were also explored after their synthesis via emulsion-solvent evaporation technique.

9.4.3 Hydrothermal Method

Hydrothermal approach is adding reactants to water or an organic solvent inside an autoclave such that a reaction can occur under high pressure and temperature. This approach is termed hydrothermal when water is employed as a reaction media and solvothermal when nonaqueous solvents are involved (Sharma et al. 2021). Figure 9.6 depicts a schematic of the hydrothermal process for the synthesis of nanoparticles (Rao et al. 2001). This method can be combined with ultrasonication, mechanochemical, microwave-assisted, and electrochemical procedures to improve reaction kinetics and increase the ability to develop novel materials (Tatarchuk et al. 2018). Previous reports have documented using aloe vera leaf extract toward the hydrothermal production of Ag NPs. These green-synthesized NPs exhibited bactericidal activity against the pathogens *Staphylococcus epidermidis* and *Pseudomonas aeruginosa* (Cho et al. 1995; Tippayawat et al. 2016). Using the hydrothermal technique, Maheswari et al. (2018) produced titanium dioxide NPs from ginger, turmeric, and garlic extracts. These green-synthesized NPs displayed antibacterial activity against *E. coli*, *Streptococcus mutans*, *S. aureus*, *P. aeruginosa*, and *K. pneumoniae* (Maheswari et al. 2018).

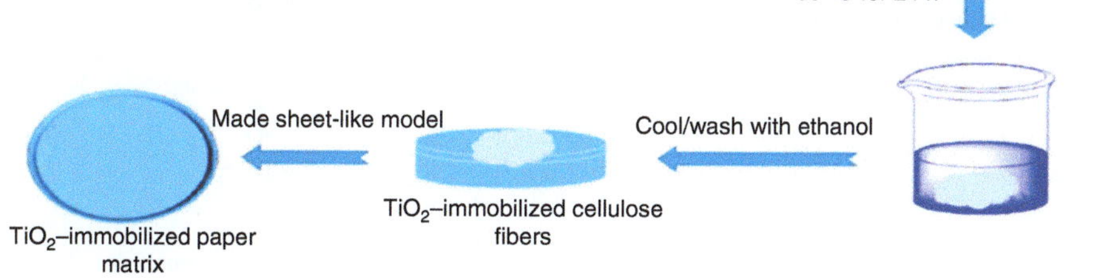

Fig. 9.6 The hydrothermal method of nanoparticle synthesis. (Reprinted with permission from Sharma et al. (2021))

9.4.4 Microwave-Assisted Synthetic Method

The microwave-assisted approach is useful for decreasing particle size and enhancing solubility. Due to their direct bonding with molecules and accelerated thermal conductivity, microwaves lead to extensive production of organic nanoparticles. Microwaves typically interact with polar compounds, resulting in their vibrational and rotational motion followed by the production of heat. This method binds charged molecules in a reaction mixture to electromagnetic radiation at a specified frequency. This technique's key benefit is homogenous heating, which improves polymer–drug binding and affects structural variances. This approach is helpful for binding polymers and drugs and modifying pharmaceuticals. This ecological and successful technology is utilized to make tablets, gel beads, microspheres, nanosystems, solid dispersion components, and film coatings (Gupta 2017; Kappe and Dallinger 2006; Jaiswal et al. 2018). Figure 9.7 depicts the synthesis of NPs using microwave-assisted technique. Microwave-assisted synthesis of bimetallic as well as metallic silver and copper NPs and their alloy NPs from the biopolymer starch have been reported. These NPs mediated by starch exhibited antibacterial activity against both *S. aureus* and *E. coli* (Valodkar et al. 2011). Using a microwave-assisted technique, extracts of tea and coffee were utilized to manufacture copper oxide NPs, which exhibited good antibacterial activity against various human pathogenic organisms (*Streptococcus pneumoniae*, *E. coli*, *S. aureus*, and *Vibrio cholerae*) (Sutradhar et al. 2014). Using chitosan, ascorbic acid, and microwave irradiation, Cu and Ag NPs were produced. These demonstrated antibacterial activities toward *E. coli* and *B. subtilis*. Synthesized Ag NPs from a variety of plant extracts such as *Phyllanthus niruri L.*, Juglans regia leaves, and mulberry leaves showed antibacterial efficacy against *Salmonella typhimurium*, *E. coli*, *P. aeruginosa*, *S. aureus*, *K. pneumoniae*, *C. albicans*, *B. subtilis*, and *Aspergillus niger* (Haris et al. 2017; Eshghi et al. 2018;

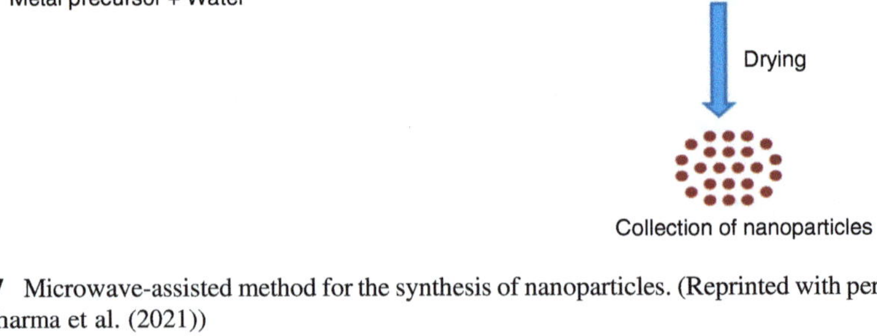

Fig. 9.7 Microwave-assisted method for the synthesis of nanoparticles. (Reprinted with permission from Sharma et al. (2021))

Liem and Nguyen 2019). Thus, green synthesis-based production of nanomedicine has proven to be an effective strategy in tackling several microbial infections including the very notorious multidrug-resistant (MDR) strains as well.

9.5 Conclusion

The use of plant-based derived nanoparticles has strong rationale advantages and purposes in medicine and human health care. In particular, research suggests that nanoparticles can elicit antibacterial effects independently or in conjunction with antibiotics, thereby significantly lowering the core problem of acquired resistance triggered by antibiotic misuse or overuse. Despite this potential use, future work should focus on two aspects: evaluating the possible risks of leveraging plant-driven synthesis of nanoparticles and minimizing the impact of their fabrication process. To render plant-based nanoparticles a viable alternative, more elaborate synthesis approaches are essential, as is the investigation into the modes of action that influence the antibacterial effect of nanoparticles.

Developing biogenic metallic nanomaterials with antimicrobial properties is a simple as well as cost-effective process, and these nanoparticles inhibit multiple pathogenic microorganisms excellently. Certain such green-synthesized nanomaterials inhibited different bacterial species that had evolved high resistance to existing therapeutic drugs. Consequently, some nanomaterials facilitate the delivery of antimicrobial agents systemically. It is expected that these biogenic nanoparticles will efficaciously substitute the currently offered drugs against which the bacteria have gained resistance. Aside from biomedical applications, these environmentally friendly nanoparticles have the potential to alleviate the issue of microbial contamination of potable water as well.

Acknowledgments The authors would like to thank Department of Biotechnology (DBT) under grants DBT/CE/FO64/2020-21/G307, MHRD IMPRINT (4291), ICMR (No.35/1/2020-GIA/Nano/BMS), DST-INSPIRE (DST/INSPIRE/04/2015/000377), DST-AMT (DST/TDT/AMT/2017/227), SERB-CRG (CRG/2020/005069), ICMR-CoE, and IITH/BME/SOCH3. RS would like to acknowledge DBT-AMRflows under the grant DBT/CE/FO64/2020-21/G307. AP appreciates and thanks Ministry of Education (MoE) for their financial support of her fellowship. PP would like to thank MoE-PMRF (ID 2001697) for funding her fellowship. SM would like to thank MoE for her fellowship.

References

Abdel-Raouf N, Al-Enazi NM, Ibraheem IBM (2017) Green biosynthesis of gold nanoparticles using Galaxaura elongata and characterization of their antibacterial activity. Arab J Chem 10: S3029–S3039. https://doi.org/10.1016/j.arabjc.2013.11.044

Acharya D, Singha KM, Pandey P, Mohanta B, Rajkumari J, Singha LP (2018) Shape dependent physical mutilation and lethal effects of silver nanoparticles on bacteria. Sci Rep 8:201. https://doi.org/10.1038/s41598-017-18590-6

Adil M, Khan T, Aasim M, Khan AA, Ashraf M (2019) Evaluation of the antibacterial potential of silver nanoparticles synthesized through the interaction of antibiotic and Aqueous callus extract of Fagonia indica. AMB Express 9:75. https://doi.org/10.1186/s13568-019-0797-2

Aisida SO, Madubuonu N, Alnasir MH, Ahmad I, Botha S, Maaza M, Ezema FI (2020) Biogenic synthesis of iron oxide nanorods using Moringa oleifera leaf extract for antibacterial applications. Appl Nanosci (Switzerland) 10:305–315. https://doi.org/10.1007/s13204-019-01099-x

Alam T, Khan RAA, Ali A, Sher H, Ullah Z, Ali M (2019) Biogenic synthesis of iron oxide nanoparticles via Skimmia laureola and their antibacterial efficacy against bacterial wilt pathogen Ralstonia solanacearum. Mater Sci Eng C 98:101–108. https://doi.org/10.1016/j.msec.2018.12.117

Ali A, Zafar H, Zia M, ul Haq I, Phull AR, Ali JS, Hussain A (2016) Synthesis, characterization, applications, and challenges of iron oxide nanoparticles. Nanotechnol Sci Appl 9:49. https://doi.org/10.2147/NSA.S99986

Aljabali AAA, Akkam Y, Al Zoubi MS, Al-Batayneh KM, Al-Trad B, Alrob OA, Alkilany AM, Benamara M, Evans DJ (2018) Synthesis of gold nanoparticles using leaf extract of Ziziphus zizyphus and their antimicrobial activity. Nano 8:174. https://doi.org/10.3390/nano8030174

Almaaytah A, Mohammed GK, Abualhaijaa A, Al-Balas Q (2017) Development of novel ultrashort antimicrobial peptide nanoparticles with potent antimicrobial and antibiofilm activities against multidrug-resistant bacteria. Drug Des Dev Ther 11:3159. https://doi.org/10.2147/DDDT.S147450

Anand U, Carpena M, Kowalska-Góralska M, Garcia-Perez P, Sunita K, Bontempi E, Simal-Gandara J (2022) Safer plant-based nanoparticles for combating antibiotic resistance in bacteria: a comprehensive review on its potential applications, recent advances, and future perspective. Sci Total Environ 821:153472

Ansari A, Siddiqui VU, Rehman WU, Akram MK, Siddiqi WA, Alosaimi AM, Hussein MA, Rafatullah M (2022) Green synthesis of TiO2 nanoparticles using Acorus calamus leaf extract and evaluating its photocatalytic and in vitro antimicrobial activity. Catalysts 12:181. https://doi.org/10.3390/catal12020181

Aromal SA, Philip D (2012) Green synthesis of gold nanoparticles using Trigonella foenumgraecum and its size-dependent catalytic activity. Spectrochim Acta A Mol Biomol Spectrosc 97:1–5. https://doi.org/10.1016/j.saa.2012.05.083

Aruguete DM, Kim B, Hochella MF, Ma Y, Cheng Y, Hoegh A, Liu J, Pruden A (2013) Antimicrobial nanotechnology: its potential for the effective management of microbial drug resistance and implications for research needs in microbial nanotoxicology. Environ Sci Process Impacts 15:93–102. https://doi.org/10.1039/c2em30692a

Arumugam A, Karthikeyan C, Haja Hameed AS, Gopinath K, Gowri S, Karthika V (2015) Synthesis of cerium oxide nanoparticles using Gloriosa superba L. leaf extract and their structural, optical and antibacterial properties. Mater Sci Eng C 49:408–415. https://doi.org/10.1016/j.msec.2015.01.042

Aswini R, Murugesan S, Kannan K (2021) Bio-engineered TiO2 nanoparticles using Ledebouria revoluta extract: larvicidal, histopathological, antibacterial and anticancer activity. Int J Environ Anal Chem 101:2926–2936. https://doi.org/10.1080/03067319.2020.1718668

Azam A, Ahmed AS, Oves M et al (2012) Antimicrobial activity of metal oxide nanoparticles against gram-positive and gram-negative bacteria: a comparative study. Int J Nanomedicine 7: 6003–6009. https://doi.org/10.2147/IJN.S35347

Bagherzade G, Tavakoli MM, Namaei MH (2017) Green synthesis of silver nanoparticles using aqueous extract of saffron (Crocus sativus L.) wastages and its antibacterial activity against six bacteria, Asian Pacific. J Trop Biomed 7:227–233. https://doi.org/10.1016/j.apjtb.2016.12.014

Bahadar H, Maqbool F, Niaz K, Abdollahi M (2016) Toxicity of nanoparticles and an overview of current experimental models. Iran Biomed J 20(1):1–11. https://doi.org/10.7508/ibj.2016.01.001. Epub 2015 Aug 19. PMID: 26286636; PMCID: PMC4689276

Balakrishnan K, Casimeer SC, Ghidan AY, Al Antary TM, Singaravelu A (2021) Exploration of antioxidant, antibacterial activities of green synthesized hesperidin loaded PLGA nanoparticles. Biointerface Res Appl Chem 11(6):14520–14528

Baptista PV, McCusker MP, Carvalho A, Ferreira DA, Mohan NM, Martins M, Fernandes AR (2018) Nano-strategies to fight multidrug resistant bacteria—"a Battle of the titans". Front Microbiol 9:1441

Baranwal A, Srivastava A, Kumar P, Bajpai VK, Maurya PK, Chandra P (2018) Prospects of nanostructure materials and their composites as antimicrobial agents. Front Microbiol 9:422. https://doi.org/10.3389/fmicb.2018.00422

Bartłomiejczyk T, Lankoff A, Kruszewski M, Szumiel I (2013) Silver nanoparticles—allies or adversaries? Ann Agric Environ Med 20:48–54

Beyth N, Houri-Haddad Y, Domb A, Khan W, Hazan R (2015) Alternative antimicrobial approach: nano-antimicrobial materials. Evid Based Complement Alternat Med. 2015:246012. https://doi.org/10.1155/2015/246012. Epub 2015 Mar 16. PMID: 25861355; PMCID: PMC4378595

Buzea C, Pacheco II, Robbie K (2007) Nanomaterials and nanoparticles: sources and toxicity. Biointerphases 2:MR17–MR71

Carlson C, Hussein SM, Schrand AM, Braydich-Stolle LK, Hess KL, Jones RL, Schlager JJ (2008) Unique cellular interaction of silver nanoparticles: size-dependent generation of reactive oxygen species. J Phys Chem B 112:13608–13619. https://doi.org/10.1021/jp712087m

Chandra H, Bishnoi P, Yadav A, Patni B, Mishra AP, Nautiyal AR (2017) Antimicrobial resistance and the alternative resources with special emphasis on plant-based antimicrobials—a review. Plan Theory 6:16. https://doi.org/10.3390/plants6020016

Chen CW, Hsu CY, Lai SM, Syu WJ, Wang TY, Lai PS (2014) Metal nanobullets for multidrug resistant bacteria and biofilms. Adv Drug Deliv Rev 78:88–104. https://doi.org/10.1016/j.addr.2014.08.004

Chen P, Yang Z, Mai Z, Huang Z, Bie Y, Wu S, Dong X, Fu X, Ko F, Zhang S, Zheng W, Zhang S, Zhou W (2022) Electrospun nanofibrous membrane with antibacterial and antiviral properties decorated with Myoporum bontioides extract and silver-doped carbon nitride nanoparticles for medical masks application. Sep Purif Technol 298:121565. https://doi.org/10.1016/j.seppur.2022.121565

Chernousova S, Epple M (2013) Silver as antibacterial agent: ion, nanoparticle, and metal. Angew Chem Int Ed 52:1636–1653. https://doi.org/10.1002/anie.201205923

Cho WS, Yashima M, Kakihana M, Kudo A, Sakata T, Yoshimura M (1995) Room temperature preparation of the highly crystallized luminescent CaWO4 film by an electrochemical method. Appl Phys Lett 66:1027–1029. https://doi.org/10.1063/1.113563

Choi O, Deng KK, Kim NJ, Ross L, Surampalli RY, Hu Z (2008) The inhibitory effects of silver nanoparticles, silver ions, and silver chloride colloids on microbial growth. Water Res 42:3066–3074. https://doi.org/10.1016/j.watres.2008.02.021

Chopade B, Ghosh P, Ahire K, Jabgunde K, Pardesi C, Bellare D (2012) Synthesis of silver nanoparticles using Dioscorea bulbifera tuber extract and evaluation of its synergistic potential in combination with antimicrobial agents. Int J Nanomedicine 1:483. https://doi.org/10.2147/ijn.s24793

Chwalibog A, Sawosz E, Hotowy A, Szeliga J, Mitura S, Mitura K, Grodzik M, Orlowski P, Sokolowska A (2010) Visualization of interaction between inorganic nanoparticles and bacteria or fungi. Int J Nanomedicine 5:1085–1094. https://doi.org/10.2147/IJN.S13532

Cravotto G, Cintas P (2006) Power ultrasound in organic synthesis: moving cavitational chemistry from academia to innovative and large-scale applications. Chem Soc Rev 35:180–196. https://doi.org/10.1039/B503848K

Cui Y, Zhao Y, Tian Y et al (2012) The molecular mechanism of action of bactericidal gold nanoparticles on Escherichia coli. Biomaterials 33:2327–2333. https://doi.org/10.1016/j.biomaterials.2011.11.057

Dakal TC, Kumar A, Majumdar RS, Yadav V (2016) Mechanistic basis of antimicrobial actions of silver nanoparticles. Front Microbiol 7:1831. https://doi.org/10.3389/fmicb.2016.01831

Daraee H, Eatemadi A, Abbasi E, Fekri Aval S, Kouhi M, Akbarzadeh A (2016) Application of gold nanoparticles in biomedical and drug delivery. Artif Cells Nanomed Biotechnol 44:410–422. https://doi.org/10.3109/21691401.2014.955107

Darroudi M, Zak AK, Muhamad MR, Huang NM, Hakimi M (2012) Green synthesis of colloidal silver nanoparticles by sonochemical method. Mater Lett 66:117–120. https://doi.org/10.1016/j.matlet.2011.08.016

Deivanathan SK, Prakash JTJ (2022) Green synthesis of silver nanoparticles using aqueous leaf extract of Guettarda speciosa and its antimicrobial and anti-oxidative properties. Chem Data Collect 38:100831. https://doi.org/10.1016/j.cdc.2022.100831

Deshmukh AR, Gupta A, Kim BS (2019) Ultrasound assisted green synthesis of silver and iron oxide nanoparticles using fenugreek seed extract and their enhanced antibacterial and antioxidant activities. Biomed Res Int 2019:1714358. https://doi.org/10.1155/2019/1714358

Drake PL, Hazelwood KJ (2005) Exposure-related health effects of silver and silver compounds: a review. Ann Occup Hyg 49:575–585. https://doi.org/10.1093/annhyg/mei019

Egger S, Lehmann RP, Height MJ et al (2009) Antimicrobial properties of a novel silver-silica nanocomposite material. Appl Environ Microbiol 75:2973–2976. https://doi.org/10.1128/AEM.01658-08

El-Naggar NEA, Saber WIA, Zweil AM et al (2022) An innovative green synthesis approach of chitosan nanoparticles and their inhibitory activity against phytopathogenic Botrytis cinerea on strawberry leaves. Sci Rep 12:3515. https://doi.org/10.1038/s41598-022-07073-y

El-Sherbiny IM, El-Shibiny A, Salih E (2016) Photo-induced green synthesis and antimicrobial efficacy of poly (ε-caprolactone)/curcumin/grape leaf extract-silver hybrid nanoparticles. J Photochem Photobiol B Biol 160:355–363. https://doi.org/10.1016/J.JPHOTOBIOL.2016.04.029

ElZorkany HES, Youssef T, Mohamed MB, Amin RM (2019) Photothermal versus photodynamic treatment for the inactivation of the bacteria Escherichia coli and Bacillus cereus: an in vitro study. Photodiagn Photodyn Ther 27:317–326. https://doi.org/10.1016/j.pdpdt.2019.06.020

Eshghi M, Vaghari H, Najian Y, Najian MJ, Jafarizadeh-Malmiri H, Berenjian A (2018) Microwave-assisted green synthesis of silver nanoparticles using Juglans regia leaf extract and evaluation of their physico-chemical and antibacterial properties. Antibiotics 7:68. https://doi.org/10.3390/antibiotics7030068

Faried M, Shameli K, Miyake M, Hajalilou A, Kalantari K, Zakaria Z, Hara H, Khairudin NBA (2016) Synthesis of silver nanoparticles via green method using ultrasound irradiation in seaweed Kappaphycus alvarezii media. Res Chem Intermed 42:7991–8004. https://doi.org/10.1007/s11164-016-2574-z

Feng Y, Cui X, He S, Dong G, Chen M, Wang J, Lin X (2013a) Erratum: addition to the role of metal nanoparticles in influencing arbuscular mycorrhizal fungi effects on plant growth. Environ Sci Technol 47(16):9496–9504. https://doi.org/10.1021/es402109n

Feng Y, Cui X, He S, Dong G, Chen M, Wang J, Lin X (2013b) The role of metal nanoparticles in influencing Arbuscular mycorrhizal fungi effects on plant growth. Environ Sci Technol 47:9496–9504. https://doi.org/10.1021/es403789p

Finley PJ, Norton R, Austin C, Mitchell A, Zank S, Durham P (2015) Unprecedented silver resistance in clinically isolated Enterobacteriaceae: major implications for burn and wound management. Antimicrob Agents Chemother 59:4734–4741. https://doi.org/10.1128/AAC.00026-15

Gardea-Torresdey JL, Gomez E, Peralta-Videa JR, Parsons JG, Troiani H, Jose-Yacaman M (2003) Alfalfa sprouts: a natural source for the synthesis of silver nanoparticles. Langmuir 19:1357–1361. https://doi.org/10.1021/la020835i

Ghosh S, Patil S, Ahire M, Kitture R, Kale S, Pardesi K, Cameotra SS, Bellare J, Dhavale DD, Jabgunde A, Chopade BA (2012) Synthesis of silver nanoparticles using Dioscorea bulbifera tuber extract and evaluation of its synergistic potential in combination with antimicrobial agents. Int J Nanomedicine 7:483–496. https://doi.org/10.2147/ijn.s24793

Gnanasekar S, Murugaraj J, Dhivyabharathi B, Krishnamoorthy V, Jha PK, Seetharaman P, Vilwanathan R, Sivaperumal S (2018) Antibacterial and cytotoxicity effects of biogenic palladium nanoparticles synthesized using fruit extract of Couroupita guianensis Aubl. J Appl Biomed 16:59–65. https://doi.org/10.1016/j.jab.2017.10.001

Gómez-Gómez B, Sanz-Landaluce J, Pérez-Corona MT, Madrid Y (2020) Fate and effect of in-house synthesized tellurium based nanoparticles on bacterial biofilm biomass and architecture. Challenges for nanoparticles characterization in living systems. Sci Total Environ 719: 137501. https://doi.org/10.1016/j.scitotenv.2020.137501

Greulich C, Diendorf J, Geßmann J, Simon T, Habijan T, Eggeler G, Schildhauer TA, Epple M, Köller M (2011) Cell type-specific responses of peripheral blood mononuclear cells to silver nanoparticles. Acta Biomater 7:3505–3514. https://doi.org/10.1016/j.actbio.2011.05.030

Gu H, Ho PL, Tong E, Wang L, Xu B (2003) Presenting vancomycin on nanoparticles to enhance antimicrobial activities. Nano Lett 3:1261–1263. https://doi.org/10.1021/nl034396z

Gupta N (2017) Microwave-assisted synthesis: review of recent developments. Int J Adv Res Sci Engin 6(3)

Hano C, Abbasi BH (2022) Plant-based green synthesis of nanoparticles: production, characterization and applications. Biomol Ther 12:31

Haris M, Kumar A, Ahmad A, Abuzinadah MF, Basheikh M, Khan SA, Mujeeb M (2017) Microwave-assisted green synthesis and antimicrobial activity of silver nanoparticles derived from a supercritical carbon dioxide extract of the fresh aerial parts of Phyllanthus niruri L. Trop J Pharm Res 16:e2967. https://doi.org/10.4314/tjpr.v16i12.22

Hayden SC, Zhao G, Saha K, Phillips RL, Li X, Miranda OR, Rotello VM, El-Sayed MA, Schmidt-Krey I, Bunz UHF (2012) Aggregation and interaction of cationic nanoparticles on bacterial surfaces. J Am Chem Soc 134:6920–6923. https://doi.org/10.1021/ja301167y

He M, Forssberg E, Wang Y, Han Y (2005) Ultrasonication-assisted synthesis of calcium carbonate nanoparticles. Chem Eng Commun 192:1468–1481. https://doi.org/10.1080/009864490896025

Hemeg HA (2017) Nanomaterials for alternative antibacterial therapy. Int J Nanomedicine 12: 8211–8225. https://doi.org/10.2147/IJN.S132163

Høiby N, Bjarnsholt T, Givskov M, Molin S, Ciofu O (2010) Antibiotic resistance of bacterial biofilms. Int J Antimicrob Agents 35:322–332. https://doi.org/10.1016/j.ijantimicag.2009.12.011

Iavicoli I, Fontana L, Leso V, Bergamaschi A (2013) The effects of nanomaterials as endocrine disruptors. Int J Mol Sci 14:16732–16801. https://doi.org/10.3390/ijms140816732

Iravani S (2011) Green synthesis of metal nanoparticles using plants. Green Chem 13:2638–2650. https://doi.org/10.1039/c1gc15386b

Iravani S, Korbekandi H, Mirmohammadi SV, Zolfaghari B (2014) Synthesis of silver nanoparticles: chemical, physical and biological methods. Res Pharm Sci 9:385–406

Islam NU, Jalil K, Shahid M, Muhammad N, Rauf A (2019) Pistacia integerrima gall extract mediated green synthesis of gold nanoparticles and their biological activities. Arab J Chem 12: 2310–2319. https://doi.org/10.1016/j.arabjc.2015.02.014

Ismail B, Shafei MN, Harun A, Ali S, Omar M, Deris ZZ (2018) Predictors of polymyxin B treatment failure in gram-negative healthcare-associated infections among critically ill patients. J Microbiol Immunol Infect 51:763–769. https://doi.org/10.1016/j.jmii.2017.03.007

Jain N, Jain P, Rajput D, Patil UK (2021) Green synthesized plant-based silver nanoparticles: therapeutic prospective for anticancer and antiviral activity. Micro Nano Syst Lett 9:5

Jaiswal KK, Manikandan D, Murugan R, Ramaswamy AP (2018) Microwave-assisted rapid synthesis of Fe3O4/poly (styrene-divinylbenzene-acrylic acid) polymeric magnetic composites and investigation of their structural and magnetic properties. Eur Polym J 98:177–190. https://doi.org/10.1016/j.eurpolymj.2017.11.005

Jansirani D, Karthick Raja N, Hariprasanth RJ, Sweetin Preethi S, Sorna Kumar RSA (2016) Synthesis of colloidal starched silver nanoparticles by sonochemical method and evaluation of its antibacterial activity. J Chem Pharm Sci 9:177–179. https://www.researchgate.net/publication/301655641

Jeyaraj Pandian C, Palanivel R, Dhanasekaran S (2016) Screening antimicrobial activity of nickel nanoparticles synthesized using Ocimum sanctum leaf extract. J Nanoparticles 2016:1–13. https://doi.org/10.1155/2016/4694367

Kambale EK, Nkanga CI, Mutonkole BPI, Bapolisi AM, Tassa DO, Liesse JMI, Krause RWM, Memvanga PB (2020) Green synthesis of antimicrobial silver nanoparticles using aqueous leaf extracts from three Congolese plant species (Brillantaisia patula, Crossopteryx febrifuga and Senna siamea). Heliyon 6:e04493. https://doi.org/10.1016/j.heliyon.2020.e04493

Kappe CO, Dallinger D (2006) The impact of microwave synthesis on drug discovery. Nat Rev Drug Discov 5:e51–e63. https://doi.org/10.1038/nrd1926

Kaur K, Reddy S, Barathe P, Shriram V, Anand U, Proćków J, Kumar V (2021) Combating drug-resistant bacteria using photothermally active nanomaterials: a perspective review. Front Microbiol 12:747019. https://doi.org/10.3389/fmicb.2021.747019

Kesharwani J, Yoon KY, Hwang J, Rai M (2009) Phytofabrication of silver nanoparticles by leaf extract of Datura metel: hypothetical mechanism involved in synthesis. J Bionanoscience 3:39–44. https://doi.org/10.1166/jbns.2009.1008

Khan M, Shaik MR, Khan ST, Adil SF, Kuniyil M, Khan M, Al-Warthan AA, Siddiqui MRH, Nawaz Tahir M (2020) Enhanced antimicrobial activity of biofunctionalized zirconia nanoparticles. ACS Omega 5:1987–1996. https://doi.org/10.1021/acsomega.9b03840

Khatamifar M, Fatemi SJ (2021) Green synthesis of pure copper oxide nanoparticles using Quercus infectoria galls extract, thermal behavior and their antimicrobial effects. Part Sci Technol 40:18. https://doi.org/10.1080/02726351.2021.1901810

Khatoon N, Alam H, Khan A, Raza K, Sardar M (2019) Ampicillin silver nanoformulations against multidrug resistant bacteria. Sci Rep 9:6848. https://doi.org/10.1038/s41598-019-43309-0

Kianpour G, Salavati- Niasari M, Emadi H (2013) Sonochemical synthesis and characterization of NiMoO4 nanorods. Ultrason Sonochem 20:418–424. https://doi.org/10.1016/j.ultsonch.2012.08.012

Kumar V, Yadav SK (2009) Plant-mediated synthesis of silver and gold nanoparticles and their applications. J Chem Technol Biotechnol 84:151–157. https://doi.org/10.1002/jctb.2023

Kuo AJ, Shu JC, Liu TP, Lu JJ, Lee MH, Wu TS, Su LH, Wu TL (2018) Vancomycin resistant Enterococcus faecium at a university hospital in Taiwan, 2002-2015: fluctuation of genetic populations and emergence of a new structure type of the Tn1546-like element. J Microbiol Immunol Infect 51:821–828. https://doi.org/10.1016/j.jmii.2018.08.008

Kuppusamy P, Yusoff MM, Maniam GP, Govindan N (2016) Biosynthesis of metallic nanoparticles using plant derivatives and their new avenues in pharmacological applications—an updated report. Saudi Pharm J 24:473–484. https://doi.org/10.1016/j.jsps.2014.11.013

Lee J, Adegoke O, Park EY (2019) High-performance biosensing systems based on various nanomaterials as signal transducers. Biotechnol J 14:1800249. https://doi.org/10.1002/biot.201800249

Leid JG, Ditto AJ, Knapp A, Shah PN, Wright BD, Blust R, Christensen L, Clemons CB, Wilber JP, Young GW, Kang AG, Panzner MJ, Cannon CL, Yun YH, Youngs WJ, Seckinger NM, Cope EK (2012) In vitro antimicrobial studies of silver carbene complexes: activity of free and nanoparticle carbene formulations against clinical isolates of pathogenic bacteria. J Antimicrob Chemother 67:138–148. https://doi.org/10.1093/jac/dkr408

Lesniak A, Alvati S, Santos Mortinez MJ, Radomski MW, Dawson KA, Åberg C (2013) Nanoparticle adhesion to the cell membrane and its effect on nanoparticle up-take efficiency. J Am Chem Soc 135:1438–1444. https://doi.org/10.1021/ja309812z

Liem LN, Nguyen D (2019) Microwave assisted green synthesis of silver nanoparticles using mulberry leaves extract and silver nitrate solution. Technologies 7:7. https://doi.org/10.3390/technologies7010007

Lima E, Guerra R, Lara V, Guzmán A (2013) Gold nanoparticles as efficient antimicrobial agents for Escherichia coli and Salmonella typhi. Chem Cent J 7:11. https://doi.org/10.1186/1752-153x-7-11

Lingaraju K, Raja Naika H, Manjunath K, Basavaraj RB, Nagabhushana H, Nagaraju G, Suresh D (2016) Biogenic synthesis of zinc oxide nanoparticles using Ruta graveolens (L.) and their antibacterial and antioxidant activities. Appl Nanosci (Switzerland) 6:703–710. https://doi.org/10.1007/s13204-015-0487-6

Logeswari P, Silambarasan S, Abraham J (2015) Synthesis of silver nanoparticles using plants extract and analysis of their antimicrobial property. J Saudi Chem Soc 19:311–317. https://doi.org/10.1016/j.jscs.2012.04.007

Lok C, Ho C, Chen R et al (2006) Proteomic analysis of the mode of antibacterial action of silver nanoparticles. J Proteome Res 5:916–924. https://doi.org/10.1021/pr0504079

Luther EM, Koehler Y, Diendorf J, Epple M, Dringen R (2011) Accumulation of silver nanoparticles by cultured primary brain astrocytes. Nanotechnology 22:375101. https://doi.org/10.1088/0957-4484/22/37/375101

Mahapatra O, Bhagat M, Gopalakrishnan C, Arunachalam KD (2008) Ultrafine dispersed CuO nanoparticles and their antibacterial activity. J Exp Nanosci 3:185–193. https://doi.org/10.1080/17458080802395460

Maheswari P, Ponnusamy S, Harish S, Ganesh MR, Hayakawa Y (2018) Hydrothermal synthesis of pure and bio modified TiO2: characterization, evaluation of antibacterial activity against gram positive and gram-negative bacteria and anticancer activity against KB oral cancer cell line. Arab J Chem 13:3484–3497. https://doi.org/10.1016/j.arabjc.2018.11.020

Makarov VV, Love AJ, Sinitsyna OV, Makarova SS, Yaminsky IV, Taliansky ME, Kalinina NO (2014) "Green" nanotechnologies: synthesis of metal nanoparticles using plants. Acta Naturae 6:35

Mechouche MS, Merouane F, Messaad CEH, Golzadeh N, Vasseghian Y, Berkani M (2022) Biosynthesis, characterization, and evaluation of antibacterial and photocatalytic methylene blue dye degradation activities of silver nanoparticles from Streptomyces tuirus strain. Environ Res 204:112360. https://doi.org/10.1016/j.envres.2021.112360

Mittal AK, Chisti Y, Banerjee UC (2013) Synthesis of metallic nanoparticles using plant extracts. Biotechnol Adv 31:346–356. https://doi.org/10.1016/j.biotechadv.2013.01.003

Moreno-Vega AI, Gomez-Quintero T, Nunez-Anita RE, Acosta-Torres LS, Castano V (2012) Polymeric and ceramic nanoparticles in biomedical applications. J Nanotechnol 2012:1. https://doi.org/10.1155/2012/936041

Morones JR, Elechiguerra JL, Camacho A, Holt K, Kouri JB, Ramírez JT, Yacaman MJ (2005) The bactericidal effect of silver nanoparticles. Nanotechnology 16:2346–2353. https://doi.org/10.1088/0957-4484/16/10/059

Muddapur UM, Alshehri S, Ghoneim MM, Mahnashi MH, Alshahrani MA, Khan AA, Shakeel Iqubal SM, Bahafi A, More SS, Shaikh IA, Mannasaheb BA, Othman N, Maqbul MS, Ahmad MZ (2022) Plant-based synthesis of gold nanoparticles and theranostic applications: a review. Molecules 27:1391

Murray CJ, Ikuta KS, Sharara F, Swetschinski L, Robles Aguilar G, Gray A, Han C, Bisignano C, Rao P, Wool E, Johnson SC, Browne AJ, Chipeta MG, Fell F, Hackett S, Haines-Woodhouse G, Kashef Hamadani BH, Kumaran EAP, McManigal B, Agarwal R, Akech S, Albertson S, Amuasi J, Andrews J, Aravkin A, Ashley E, Bailey F, Baker S, Basnyat B, Bekker A, Bender R, Bethou A, Bielicki J, Boonkasidecha S, Bukosia J, Carvalheiro C, Castañeda-Orjuela C, Chansamouth V, Chaurasia S, Chiurchiù S, Chowdhury F, Cook AJ, Cooper B, Cressey TR, Criollo-Mora E, Cunningham M, Darboe S, Day NPJ, de Luca M, Dokova K, Dramowski A, Dunachie SJ, Eckmanns T, Eibach D, Emami A, Feasey N, Fisher-Pearson N, Forrest K, Garrett D, Gastmeier P, Giref AZ, Greer RC, Gupta V, Haller S, Haselbeck A, Hay SI, Holm M, Hopkins S, Iregbu KC, Jacobs J, Jarovsky D, Javanmardi F, Khorana M, Kissoon N, Kobeissi E, Kostyanev T, Krapp F, Krumkamp R, Kumar A, Kyu HH, Lim C, Limmathurotsakul D, Loftus MJ, Lunn M, Ma J, Mturi N, Munera-Huertas T, Musicha P,

Mussi-Pinhata MM, Nakamura T, Nanavati R, Nangia S, Newton P, Ngoun C, Novotney A, Nwakanma D, Obiero CW, Olivas-Martinez A, Olliaro P, Ooko E, Ortiz-Brizuela E, Peleg AY, Perrone C, Plakkal N, Ponce-de-Leon A, Raad M, Ramdin T, Riddell A, Roberts T, Robotham JV, Roca A, Rudd KE, Russell N, Schnall J, Scott JAG, Shivamallappa M, Sifuentes-Osornio J, Steenkeste N, Stewardson AJ, Stoeva T, Tasak N, Thaiprakong A, Thwaites G, Turner C, Turner P, van Doorn HR, Velaphi S, Vongpradith A, Vu H, Walsh T, Waner S, Wangrangsimakul T, Wozniak T, Zheng P, Sartorius B, Lopez AD, Stergachis A, Moore C, Dolecek C, Naghavi M (2022) Global burden of bacterial antimicrobial resistance in 2019: a systematic analysis. Lancet 399:629–655. https://doi.org/10.1016/S0140-6736(21)02724-0

Murugan K, Benelli G, Panneerselvam C, Subramaniam J, Jeyalalitha T, Dinesh D, Nicoletti M, Hwang JS, Suresh U, Madhiyazhagan P (2015) Cymbopogon citratus-synthesized gold nanoparticles boost the predation efficiency of copepod Mesocyclops aspericornis against malaria and dengue mosquitoes. Exp Parasitol 153:129–138. https://doi.org/10.1016/j.exppara.2015.03.017

Nakazato G, Oves M, Prasad R, Qiu W, Li B, Mmi M, Mahidul Islam Masum M, Mahfuja Siddiqa M, Arif Ali K, Zhang Y, Abdallah Y, Ibrahim E, Yan C (2019) Biogenic synthesis of silver nanoparticles using Phyllanthus emblica fruit extract and its inhibitory action against the pathogen Acidovorax oryzae strain RS-2 of rice bacterial brown stripe. Front Microbiol 10:820. https://doi.org/10.3389/fmicb.2019.00820

Naraginti S, Sivakumar A (2014) Eco-friendly synthesis of silver and gold nanoparticles with enhanced bactericidal activity and study of silver catalyzed reduction of 4-nitrophenol. Spectrochim Acta A Mol Biomol Spectrosc 128:357–362. https://doi.org/10.1016/j.saa.2014.02.083

Naseem T, Farrukh MA (2015) Antibacterial activity of green synthesis of iron nanoparticles using Lawsonia inermis and Gardenia jasminoides leaves extract. J Chem 2015:1. https://doi.org/10.1155/2015/912342

Natan M, Banin E (2017) From nano to micro: using nanotechnology to combat microorganisms and their multidrug resistance. FEMS Microbiol Rev 41:302–322. https://doi.org/10.1093/femsre/fux003

Nguyen NTT, Nguyen LM, Nguyen TTT, Nguyen TT, Nguyen DTC, Van TT (2022) Formation, antimicrobial activity, and biomedical performance of plant-based nanoparticles: a review. Environ Chem Lett 20(4):2531

Ni C, Zhong Y, Wu W, Song Y, Makvandi P, Yu C, Song H (2022) Co-delivery of nano-silver and vancomycin via silica nanopollens for enhanced antibacterial functions. Antibiotics (Basel) 11(5):685. https://doi.org/10.3390/antibiotics11050685

Ninganagouda S, Rathod V, Singh D, Hiremath J, Singh AK, Mathew J, Ul-Haq M (2014) Growth kinetics and mechanistic action of reactive oxygen species released by silver nanoparticles from Aspergillus niger on Escherichia coli. Biomed Res Int 2014:53419. https://doi.org/10.1155/2014/753419

Nouri A, Tavakkoli Yaraki M, Lajevardi A, Rezaei Z, Ghorbanpour M, Tanzifi M (2020) Ultrasonic-assisted green synthesis of silver nanoparticles using Mentha aquatica leaf extract for enhanced antibacterial properties and catalytic activity, colloid interface. Sci Commun 35:100252. https://doi.org/10.1016/J.COLCOM.2020.100252

Ogar A, Tylko G, Turnau K (2015) Antifungal properties of silver nanoparticles against indoor mould growth. Sci Total Environ 521–522:305–314. https://doi.org/10.1016/j.scitotenv.2015.03.101

Oves M, Ahmar Rauf M, Aslam M, Qari HA, Sonbol H, Ahmad I, Sarwar Zaman G, Saeed M (2022) Green synthesis of silver nanoparticles by Conocarpus lancifolius plant extract and their antimicrobial and anticancer activities. Saudi J Biol Sci 29:460–471. https://doi.org/10.1016/j.sjbs.2021.09.007

Panáček A, Kvítek L, Prucek R, Kolář M, Večeřová R, Pizúrová N, Sharma VK, Nevěčná T, Zbořil R (2006) Silver colloid nanoparticles: synthesis, characterization, and their antibacterial activity. J Phys Chem B 110:16248–16253. https://doi.org/10.1021/jp063826h

Paosen S, Lethongkam S, Wunnoo S, Lehman N, Kalkornsurapranee E, Septama AW, Voravuthikunchai SP (2021) Prevention of nosocomial transmission and biofilm formation on novel biocompatible antimicrobial gloves impregnated with biosynthesized silver nanoparticles synthesized using Eucalyptus citriodora leaf extract. Biotechnol J 16:2100030. https://doi.org/10.1002/biot.202100030

Patrick DR, Findon G, Miller TE (1997) Residual moisture determines the level of touch-contact-associated bacterial transfer following hand washing. Epidemiol Infect 119:319–325. https://doi.org/10.1017/s0950268897008261

Pelgrift RY, Friedman AJ (2013) Nanotechnology as a therapeutic tool to combat microbial resistance. Adv Drug Deliv Rev 65:1803–1596

Peralta-Videa JR, Huang Y, Parsons JG, Zhao L, Lopez-Moreno L, Hernandez-Viezcas JA, Gardea-Torresdey JL (2016) Plant-based green synthesis of metallic nanoparticles: scientific curiosity or a realistic alternative to chemical synthesis? Nanotechnol Environ Engineer 1:1. https://doi.org/10.1007/s41204-016-0004-5

Peretyazhko TS, Zhang Q, Colvin VL (2014) Size-controlled dissolution of silver nanoparticles at neutral and acidic pH conditions: kinetics and size changes. Environ Sci Technol 48:11954–11961. https://doi.org/10.1021/es5023202

Piao MJ, Kang KA, Lee IK, Kim HS, Kim S, Choi JY, Choi J, Hyun JW (2011) Silver nanoparticles induce oxidative cell damage in human liver cells through inhibition of reduced glutathione and induction of mitochondria-involved apoptosis. Toxicol Lett 201:92–100. https://doi.org/10.1016/j.toxlet.2010.12.010

Potbhare AK, Chaudhary RG, Chouke PB, Yerpude S, Mondal A, Sonkusare VN, Rai AR, Juneja HD (2019) Phytosynthesis of nearly monodisperse CuO nanospheres using Phyllanthus reticulatus/Conyza bonariensis and its antioxidant/antibacterial assays. Mater Sci Eng C 99: 783–793. https://doi.org/10.1016/j.msec.2019.02.010

Prathna TC, Chandrasekaran N, Raichur AM, Mukherjee A (2011) Biomimetic synthesis of silver nanoparticles by Citrus limon (lemon) aqueous extract and theoretical prediction of particle size. Colloids Surf B: Biointerfaces 82:152–159. https://doi.org/10.1016/j.colsurfb.2010.08.036

Rabiee N, Ahmadi S, Akhavan O, Luque R (2022) Materials silver and gold nanoparticles for antimicrobial purposes against multi-drug resistance bacteria. Materials (Basel) 15(5):1799. https://doi.org/10.3390/ma15051799

Rafique M, Sadaf I, Rafique MS, Tahir MB (2017) A review on green synthesis of silver nanoparticles and their applications. Artif Cells Nanomed Biotechnol 45:1272–1291

Rafique M, Sadaf I, Tahir MB, Rafique MS, Nabi G, Iqbal T, Sughra K (2019) Novel and facile synthesis of silver nanoparticles using Albizia procera leaf extract for dye degradation and antibacterial applications. Mater Sci Eng C 99:1313–1324. https://doi.org/10.1016/j.msec.2019.02.059

Raghupathi KR, Koodali RT, Manna AC (2011) Size-dependent bacterial growth inhibition and mechanism of antibacterial activity of zinc oxide nanoparticles. Langmuir 27:4020–4028. https://doi.org/10.1021/la104825u

Rahmanpour A, Farahpour MR, Shapouri R, Jafarirad S, Rahimi P (2022) Synthesis and characterization of alumina-based nanocomposites of TiO_2/Al_2O_3/chitosan with antibacterial properties accelerate healing of infected excision wounds. Colloids Surf A Physicochem Eng Asp 644: 128839. https://doi.org/10.1016/j.colsurfa.2022.128839

Rajendran NK, Kumar SSD, Houreld NN, Abrahamse H (2018) A review on nanoparticle based treatment for wound healing. J Drug Deliv Sci Technol 44:421–430

Ramteke C, Chakrabarti T, Sarangi BK, Pandey R (2013) Synthesis of silver nanoparticles from the aqueous extract of leaves of Ocimum sanctum for enhanced antibacterial activity. Hindawi Publ Corp J Chem 2013:1–8. https://doi.org/10.1155/2013/278925

Rao CNR, Govindaraj A, Deepak FL, Gunari NA, Nath M (2001) Surfactant assisted synthesis of semiconductor nanotubes and nanowires. Appl Phys Lett 78:e1853–e1855. https://doi.org/10.1063/1.1359145

Raveendran P, Fu J, Wallen SL (2003) Completely "green" synthesis and stabilization of metal nanoparticles. J Am Chem Soc 125:13940–13941. https://doi.org/10.1021/ja029267j
Rex JH, Fernandez Lynch H, Cohen IG, Darrow JJ, Outterson K (2019) Designing development programs for non-traditional antibacterial agents. Nat Commun 10:3416. https://doi.org/10.1038/s41467-019-11303-9
Riga EK, Vöhringer M, Widyaya VT, Lienkamp K (2017) Polymer-based surfaces designed to reduce biofilm formation: from antimicrobial polymers to strategies for long-term applications. Macromol Rapid Commun 38:1700216. https://doi.org/10.1002/marc.201700216
Roy N, Gaur A, Jain A, Bhattacharya S, Rani V (2013) Green synthesis of silver nanoparticles: an approach to overcome toxicity. Environ Toxicol Pharmacol 36:807–812. https://doi.org/10.1016/j.etap.2013.07.005
Roy A, Bulut O, Some S, Mandal AK, Yilmaz MD (2019) Green synthesis of silver nanoparticles: biomolecule-nanoparticle organizations targeting antimicrobial activity. RSC Adv 9:2673–2702
Ruddaraju LK, Pammi SVN, Guntuku GS, Padavala VS, Kolapalli VRM (2020) A review on antibacterials to combat resistance: from ancient era of plants and metals to present and future perspectives of green nano technological combinations. Asian J Pharm Sci 15:42–59. https://doi.org/10.1016/j.ajps.2019.03.002
Rusciano G, De Luca AC, Pesce G, Sasso A (2009) On the interaction of nano-sized organic carbon particles with model lipid membranes. Carbon 47:2950–2957. https://doi.org/10.1016/j.carbon.2009.06.042
Sagadevan S, Anita Lett J, Vennila S, Varun Prasath P, Saravanan Kaliaraj G, Fatimah I, Léonard E, Mohammad F, Al-Lohedan HA, Alshahateet SF, Lee CT (2021) Photocatalytic activity and antibacterial efficacy of titanium dioxide nanoparticles mediated by Myristica fragrans seed extract. Chem Phys Lett 771:138527. https://doi.org/10.1016/j.cplett.2021.138527
Saliani M, Jalal R, Goharshadi EK (2015) Effects of pH and temperature on antibacterial activity of zinc oxide nanofluid against Escherichia coli O157: H7 and Staphylococcus aureus. Jundishapur J Microbiol 8:1–6. https://doi.org/10.5812/jjm.17115
Sanpui P, Murugadoss A, Prasad PVD, Ghosh SS, Chattopadhyay A (2008) The antibacterial properties of a novel chitosan-Ag-nanoparticle composite. Int J Food Microbiol 124:142–146. https://doi.org/10.1016/j.ijfoodmicro.2008.03.004
Sathishkumar M, Sneha K, Yun YS (2010) Immobilization of silver nanoparticles synthesized using Curcuma longa tuber powder and extract on cotton cloth for bactericidal activity. Bioresour Technol 101:7958–7965. https://doi.org/10.1016/j.biortech.2010.05.051
Savas S, Ersoy A, Gulmez Y, Kilic S, Levent B, Altintas Z (2018) Nanoparticle enhanced antibody and DNA biosensors for sensitive detection of Salmonella. Materials 11:1541. https://doi.org/10.3390/ma11091541
Seetharaman P, Chandrasekaran R, Gnanasekar S, Mani I, Sivaperumal S (2017) Biogenic gold nanoparticles synthesized using Crescentia cujete L. and evaluation of their different biological activities. Biocatal Agric Biotechnol 11:75–82. https://doi.org/10.1016/j.bcab.2017.06.004
Senthil M, Ramesh C (2012) Biogenic synthesis of Fe_3O_4 nanoparticles using tridax procumbens leaf extract and its antibacterial activity on Pseudomonas aeruginosa. Digest J Nanomater Biostruct 7:4
Shah R, Shah SA, Shah S, Faisal S, Ullah F (2020) Green synthesis and antibacterial activity of gold nanoparticles of Digera muricata. Indian J Pharm Sci 82(2):374
Sharma VK, Yngard RA, Lin Y (2009) Silver nanoparticles: green synthesis and their antimicrobial activities. Adv Colloid Interf Sci 145:83–96. https://doi.org/10.1016/j.cis.2008.09.002
Sharma VK, Johnson N, Cizmas L, McDonald TJ, Kim H (2016) A review of the influence of treatment strategies on antibiotic resistant bacteria and antibiotic resistance genes. Chemosphere 150:702–714. https://doi.org/10.1016/j.chemosphere.2015.12.084
Sharma M, Thakur P, Saini RV, Kumar R, Torino E (2020) Unveiling antimicrobial and anticancerous behavior of AuNPs and AgNPs moderated by rhizome extracts of Curcuma longa from diverse altitudes of Himalaya. Sci Rep 10:1–11. https://doi.org/10.1038/s41598-020-67673-4

Sharma D, Shandilya P, Saini NK, Singh P, Thakur VK, Saini RV, Mittal D, Chandan G, Saini V, Saini AK (2021) Insights into the synthesis and mechanism of green synthesized antimicrobial nanoparticles, answer to the multidrug resistance. Mater Today Chem 19:100391. https://doi.org/10.1016/j.mtchem.2020.100391

Shittu KO, Bankole MT, Abdulkareem AS, Abubakre OK, Ubaka AU (2017) Application of gold nanoparticles for improved drug efficiency. Adv Nat Sci Nanosci Nanotechnol 8:035014. https://doi.org/10.1088/2043-6254/aa7716

Shubhashree KR, Reddy R, Gangula AK, Nagananda GS, Badiya PK, Ramamurthy SS, Aramwit P, Reddy N (2022) Green synthesis of copper nanoparticles using aqueous extracts from Hyptis suaveolens (L.). Mater Chem Phys 280:125795. https://doi.org/10.1016/j.matchemphys.2022.125795

Singh R, Smitha MS, Singh SP (2014) The role of nanotechnology in combating multidrug resistant bacteria. J Nanosci Nanotechnol 14:4745–4756. https://doi.org/10.1166/jnn.2014.9527

Singh R, Shedbalkar UU, Wadhwani SA, Chopade BA (2015) Bacteriagenic silver nanoparticles: synthesis, mechanism, and applications. Appl Microbiol Biotechnol 99:4579–4593. https://doi.org/10.1007/s00253-015-6622-1

Singh H, Du J, Singh P, Yi TH (2018) Ecofriendly synthesis of silver and gold nanoparticles by Euphrasia officinalis leaf extract and its biomedical applications. Artif Cells, Nanomed Biotechnol 46:1163–1170. https://doi.org/10.1080/21691401.2017.1362417

Singh J, Vishwakarma K, Ramawat N, Rai P, Singh VK, Mishra RK, Kumar V, Tripathi DK, Sharma S (2019) Nanomaterials and microbes' interactions: a contemporary overview. 3 Biotech 9:68. https://doi.org/10.1007/s13205-019-1576-0

Singh J, Kukkar P, Sammi H, Rawat M, Singh G, Kukkar D (2019a) Enhanced catalytic reduction of 4-nitrophenol and congo red dye by silver nanoparticles prepared from Azadirachta indica leaf extract under direct sunlight exposure. Part Sci Technol 37:430–439. https://doi.org/10.1080/02726351.2017.1390512

Singh J, Kumar V, Kim KH, Rawat M (2019b) Biogenic synthesis of copper oxide nanoparticles using plant extract and its prodigious potential for photocatalytic degradation of dyes. Environ Res 177:108569. https://doi.org/10.1016/j.envres.2019.108569

Sinha R, Karan R, Sinha A, Khare SK (2011) Interaction and nanotoxic effect of ZnO and Ag nanoparticles on mesophilic and halophilic bacterial cells. Bioresour Technol 102:1516–1520. https://doi.org/10.1016/j.biortech.2010.07.117

Sirelkhatim A, Mahmud S, Seeni A, NHM K, Ann LC, SKM B, Hasan H, Mohamad D (2015) Review on zinc oxide nanoparticles: antibacterial activity and toxicity mechanism. Nanomicro Lett 7:219–242. https://doi.org/10.1007/s40820-015-0040-x

Smerkova K, Dolezelikova K, Bozdechova L, Heger Z, Zurek L, Adam V (2020) Nanomaterials with active targeting as advanced antimicrobials. Wiley Interdiscip Rev Nanomed Nanobiotechnol 12:e1636. https://doi.org/10.1002/wnan.1636

Sondi I, Salopek-Sondi B (2004) Silver nanoparticles as antimicrobial agent: a case study on E. coli as a model for gram-negative bacteria. J Colloid Interface Sci 275:177–182. https://doi.org/10.1016/j.jcis.2004.02.012

Song CX, Labhasetwar V, Murphy H, Qu X, Humphrey WR, Shebuski RJ, Levy RJ (1997) Formulation and characterization of biodegradable nanoparticles for intravascular local drug delivery. J Control Release 43:197–212. https://doi.org/10.1016/S0168-3659(96)01484-8

Sultan I, Rahman S, Jan AT, Siddiqui MT, Mondal AH, Haq QMR (2018) Antibiotics, resistome and resistance mechanisms: a bacterial perspective. Front Microbiol 9:2066. https://doi.org/10.3389/fmicb.2018.02066

Sunny NE, Mathew SS, Chandel N, Saravanan P, Rajeshkannan R, Rajasimman M, Vasseghian Y, Rajamohan N, Kumar SV (2022) Green synthesis of titanium dioxide nanoparticles using plant biomass and their applications- a review. Chemosphere 300:134612. https://doi.org/10.1016/j.chemosphere.2022.134612

Sur DH, Mukhopadhyay M (2019) Role of zinc oxide nanoparticles for effluent treatment using Pseudomonas putida and Pseudomonas aureofaciens. Bioprocess Biosyst Eng 42:187–198. https://doi.org/10.1007/s00449-018-2024-y

Surendra TV, Roopan SM (2016) Photocatalytic and antibacterial properties of phytosynthesized CeO2 NPs using Moringa oleifera peel extract. J Photochem Photobiol B Biol 161:122–128. https://doi.org/10.1016/j.jphotobiol.2016.05.019
Surendra TV, Roopan SM, Arasu MV, Al-Dhabi NA, Rayalu GM (2016) RSM optimized Moringa oleifera peel extract for green synthesis of M. oleifera capped palladium nanoparticles with antibacterial and hemolytic property. J Photochem Photobiol B Biol 162:550–557. https://doi.org/10.1016/j.jphotobiol.2016.07.032
Sutradhar P, Saha M, Maiti D (2014) Microwave synthesis of copper oxide nanoparticles using tea leaf and coffee powder extracts and its antibacterial activity. J Nanostruct Chem 4:86. https://doi.org/10.1007/s40097-014-0086-1
Tak YK, Pal S, Naoghare PK et al (2015) Shape-dependent skin penetration of silver nanoparticles: does it really matter. Sci Rep 5:16908. https://doi.org/10.1038/srep16908
Tatarchuk T, Peter A, Al-Najar B, Vijaya J, Bououdina M (2018) Photocatalysis: activity of nanomaterials. Nanotechnol Environ Sci 1:209–292. https://doi.org/10.1002/9783527808854.ch8
Ting SW, Lee CH, Liu JW (2018) Risk factors and outcomes for the acquisition of carbapenem-resistant gram-negative bacillus bacteremia: a retrospective propensity-matched case control study. J Microbiol Immunol Infect 51:621–628. https://doi.org/10.1016/j.jmii.2016.08.022
Tippayawat P, Phromviyo N, Boueroy P, Chompoosor A (2016) Green synthesis of silver nanoparticles in aloe vera plant extract prepared by a hydrothermal method and their synergistic antibacterial activity. PeerJ 4:2589. https://doi.org/10.7717/peerj.2589
Tiwari PM, Vig K, Dennis VA, Singh SR (2011) Functionalized gold nanoparticles and their biomedical applications. Nano 1:31–63. https://doi.org/10.3390/nano1010031
Tolaymat TM, El Badawy AM, Genaidy A, Scheckel KG, Luxton TP, Suidan M (2010) An evidence-based environmental perspective of manufactured silver nanoparticle in syntheses and applications: a systematic review and critical appraisal of peer-reviewed scientific papers. Sci Total Environ 408:999–1006. https://doi.org/10.1016/j.scitotenv.2009.11.003
Trifan A, Luca SV, Greige-Gerges H, Miron A, Gille E, Aprotosoaie AC (2020) Recent advances in tackling microbial multidrug resistance with essential oils: combinatorial and nano-based strategies. Crit Rev Microbiol 46:338–357. https://doi.org/10.1080/1040841X.2020.1782339
Tsao LH, Hsin CY, Liu HY, Chuang HC, Chen LY, Lee YJ (2018) Risk factors for healthcare-associated infection caused by carbapenem-resistant Pseudomonas aeruginosa. J Microbiol Immunol Infect 51:359–366. https://doi.org/10.1016/j.jmii.2017.08.015
Valodkar M, Modi S, Pal A, Thakore S (2011) Synthesis and anti-bacterial activity of Cu, Ag and Cu–Ag alloy nanoparticles: a green approach. Mater Res Bull 46:e384–e389. https://doi.org/10.1016/j.materresbull.2010.12.001
Velmurugan P, Hong S-C, Aravinthan A et al (2017) Comparison of the physical characteristics of green-synthesized and commercial silver nanoparticles: evaluation of antimicrobial and cytotoxic effects. Arab J Sci Eng 42:201–208. https://doi.org/10.1007/s13369-016-2254-8
Ventosa A, Nieto JJ, Oren A (1998) Biology of moderately halophilic aerobic bacteria. Microbiol Mol Biol Rev 62:504–544. https://doi.org/10.1128/mmbr.62.2.504-544.1998
Verma A, Mehata MS (2016) Controllable synthesis of silver nanoparticles using neem leaves and their antimicrobial activity. J Radiat Res Appl Sci 9:109–115. https://doi.org/10.1016/j.jrras.2015.11.001
Vijayan R, Joseph S, Mathew B (2018) Indigofera tinctoria leaf extract mediated green synthesis of silver and gold nanoparticles and assessment of their anticancer, antimicrobial, antioxidant and catalytic properties. Artif Cells, Nanomed Biotechnol 46:861–871. https://doi.org/10.1080/21691401.2017.1345930
Wang ZL (2004) Zinc oxide nanostructures: growth, properties and applications. J Phys Condens Matter 16:R829. https://doi.org/10.1088/0953-8984/16/25/R01
Watnick P, Kolter R (2000) Biofilm, city of microbes. J Bacteriol 182:2675–2679. https://doi.org/10.1128/JB.182.10.2675-2679.2000

Westmeier D, Hahlbrock A, Reinhardt C, Fröhlich-Nowoisky J, Wessler S, Vallet C, Pöschl U, Knauer SK, Stauber RH (2018) Nanomaterial-microbe cross-talk: physicochemical principles and (patho)biological consequences. Chem Soc Rev 47:5312–5337. https://doi.org/10.1039/c6cs00691d

Wigginton NS, Haus KL, Hochella MF Jr (2007) Aquatic environmental nanoparticles. J Environ Monit 9:1306–1316. https://doi.org/10.1039/B712709J

Xu C, Sun S (2013) New forms of superparamagnetic nanoparticles for biomedical applications. Adv Drug Deliv Rev 65:732–743. https://doi.org/10.1016/j.addr.2012.10.008

Xu H, Zeiger BW, Suslick KS (2013) Sonochemical synthesis of nanomaterials. Chem Soc Rev 42: 2555–2567. https://doi.org/10.1039/C2CS35282F

Yabu H, Higuchi T, Ijiro K, Shimomura M (2005) Spontaneous formation of polymer nanoparticles by good-solvent evaporation as a nonequilibrium process. Chaos 15:047505. https://doi.org/10.1063/1.2137621

Yun H, Kim JD, Choi HC, Lee CW (2013) Antibacterial activity of CNT-Ag and GO-Ag nanocomposites against gram-negative and gram-positive bacteria. Bull Korean Chem Soc 34:3261–3264. https://doi.org/10.5012/bkcs.2013.34.11.3261

Zhang L, Ding Y, Povey M, York D (2008) ZnO nanofluids-a potential antibacterial agent. Prog Nat Sci 18:939–944. https://doi.org/10.1016/j.pnsc.2008.01.026

Zhang S, Mu Y, Zhang JZH, Xu W (2013) Effect of self-assembly of fullerene nanoparticles on lipid membrane. PLoS One 8:77436. https://doi.org/10.1371/journal.pone.0077436

Zhang XF, Liu ZG, Shen W, Gurunathan S (2016) Silver nanoparticles: synthesis, characterization, properties, applications, and therapeutic approaches. Int J Mol Sci 17:1534

Zhao Y, Jiang X (2013) Multiple strategies to activate gold nanoparticles as antibiotics. Nanoscale 5:8340–8350. https://doi.org/10.1039/c3nr01990j

Zhao X, Zhao F, Wang J, Zhong N (2017) Biofilm formation and control strategies of foodborne pathogens: food safety perspectives. RSC Adv 7:36670–36683. https://doi.org/10.1039/c7ra02497e

Zhou Y, Kong Y, Kundu S et al (2012) Antibacterial activities of gold and silver nanoparticles against Escherichia coli and bacillus Calmette-Guérin. J Nanobiotechnol 1:1. https://doi.org/10.1186/1477-3155-10-19

Ziarati A, Safaei-Ghomi J, Rohani S (2013) Sonochemically synthesis of pyrazolones using reusable catalyst CuI nanoparticles that was prepared by sonication. Ultrason Sonochem 20: 1069–1075. https://doi.org/10.1016/j.ultsonch.2013.01.005

Zinjarde S (2012) Bio-inspired nanomaterials and their applications as antimicrobial agents. Chron Young Sci 3:74. https://doi.org/10.4103/2229-5186.94314

MIX
Papier aus verantwortungsvollen Quellen
Paper from responsible sources
FSC® C105338

If you have any concerns about our products,
you can contact us on
ProductSafety@springernature.com

In case Publisher is established outside the EU,
the EU authorized representative is:
Springer Nature Customer Service Center GmbH
Europaplatz 3, 69115 Heidelberg, Germany

Printed by Libri Plureos GmbH
in Hamburg, Germany